MATERIALS AND MANUFACTURING TECHNOLOGY
J. Paulo Davim, Series Editor

METAL CUTTING: RESEARCH ADVANCES

MATERIAL AND MANUFACTURING TECHNOLOGY

J. Paulo Davim, Series Editor

Tribology Research Advances
J. Paulo Davim (Editor)
2009. ISBN: 978-1-60692-885-1

Drilling of Composite Materials
J. Paulo Davim (Editor)
2009. ISBN: 978-1-60741-163-5
2009. ISBN: 978-1-60876-584-3 (E-book)

Artificial Intelligence in Manufacturing Research
J. Paulo Davim (Editor)
2010. ISBN: 978-1-60876-214-9

Metal Cutting: Research Advances
J. Paulo Davim (Editor)
2010. ISBN: 978-1-60876-207-1

Tribology of Composite Materials
J. Paulo Davim (Editor)
2010. ISBN: 978-1-61668-319-1

Micro and Nanomanufacturing Research
J. Paulo Davim (Editor)
2010. ISBN: 978-1-61668-488-4

MATERIALS AND MANUFACTURING TECHNOLOGY

J. Paulo Davim, Series Editor

METAL CUTTING: RESEARCH ADVANCES

J. PAULO DAVIM

EDITOR

Nova Science Publishers, Inc.

New York

For permission to use material from this book please contact us:
Telephone 631-231-7269; Fax 631-231-8175
Web Site: http://www.novapublishers.com

LIBRARY OF CONGRESS CATALOGING-IN-PUBLICATION DATA

Metal cutting : research advances / editor, J. Paulo Davim.
p. cm.
Includes index.
ISBN 978-1-60876-207-1 (hardcover)
1. Machining. 2. Metal-cutting. I. Davim, J. Paulo.
TJ1185.M465 2009
671.5'3--dc22

2009037468

Published by Nova Science Publishers, Inc. † New York

CONTENTS

PREFACE

The science and technology of metal cutting is of great interest for several important industries such as automotive, aeronautics, aerospace, alternative energy, moulds and dies, biomedical, and others. Metal cutting constitutes manufacturing processes in which parts are shaped by the removal of unwanted material. Interest in this topic has increased over the last twenty years, along with rapid advances in materials science, automation and control, and computer technology. Recently, micro- and nanocutting also have grown in importance as manufacture of microcomponents has become increasingly important for the development of new products for modern industry.

This book presents research and reviews studies on metal cutting. The first chapter provides information on system outlook of research and application aspects in metal cutting. Chapter 2 focuses on prediction of heat partition in metal cutting – a state-of-the-art technique. Chapter 3 discusses experiments in mechanical and thermal cutting processes for new behaviour law. Chapter 4 focuses on finite element modelling of machining aluminium alloy and experimental validation. Chapter 5 deals with diamond-coated cutting tools for machining applications. Chapter 6 reviews generation and modelling of the surface roughness in machining using geometrically defined cutting tools. Chapter 7 considers chatter modelling in drilling and microdrilling. Finally, chapter 8 focuses on assisted machining processes.

The research presented here may be used in conjunction with final undergraduate engineering courses (for example, mechanical, manufacturing, and materials engineering, etc) or for postgraduate level metal cutting study. This book may also serve as a reference for academic or manufacturing researchers, mechanical, manufacturing, and industrial engineers, and for other industries related to machining and machinability of materials.

Editor acknowledges his gratitude to Nova Publishers for the opportunity to present this material and for their professional support. Finally, the editor would like to thank all the chapter authors for making their research available for this work.

Aveiro, Portugal
J. Paulo Davim
May 2009

In: Metal Cutting: Research Advances
Editor: J. Paulo Davim pp. 1-22
ISBN: 978-1-60876-207-1

Chapter 1

METAL CUTTING: SYSTEM OUTLOOK OF RESEARCH AND APPLICATION ASPECTS

Viktor P. Astakhov*

Michigan State University; Department of Mechanical Engineering
2453 Engineering Building; East Lansing, MI 48824-1226, USA

ABSTRACT

Discrepancies between the results obtained in modeling and laboratory testing and those from field applications present a serious problem in metal cutting research and development. This chapter explains these discrepancies as resulting primarily from non-system considerations made at the research and applications stages. It is also suggested that the development of an applied similarity theory and corresponding similarity number can significantly simplify metal cutting research, while making research results more widely applicable, i.e., applicable for many practical machining systems. This article also discusses the physical implications and proper use of basic similarity numbers in metal cutting, with an emphasis on experimental studies. Physical implications and beneficial application of the following similarity numbers are discussed: the chip compression ratio, and Péclet and Poletica criteria. The use of these similarity numbers indicates a significant reduction in the extent of the experimental program. The use of similarity numbers to assess machinability is also discussed.

1. INTRODUCTION

Reduction of direct manufacturing costs associated with machining operations is a never-ending challenge for manufacturing plants. The increased use of special alloys with advanced properties, the increasing number of quality requirements for machined parts, the reduction of cycle time, the increased use of high-speed and near-dry machining—all of these factors add additional considerations to cost reduction issues. To address these issues, leading tool and machine manufacturers have developed a number of new products—new tool materials and

* Corresponding author, e-mail: astakhov@msu.edu

coating, new cutting insert and tool designs, new tool holders, powerful precision machines, part fixtures, advanced controllers, which provide a wide spectrum of information on cutting processes and other aspects of machining. and increase efficiency of machining operations in industry through increasing working speeds, feed rates, tool life, and reliability. For example, modern grades of high speed steels combined with advanced coatings allow high cutting speeds that previously were considered suitable only for carbide tools. Today, it is common to use dry hobbing with high speed steel hobs at 80 m/min. Modern grades of polycrystalline diamond (PCD) tool material allow milling, drilling, and reaming of high-silicon aluminum alloys at speeds of 1000-4000 m/min. Modern grades of carbide tools combined with advanced coatings allow machining of alloyed steels at speeds of 300-600m/min.

The challenges of machining impose enormous pressure on metal cutting research and manufacturing because new manufacturing facilities, for example in automotive industry, eliminate issues associated with underpowered machines, weak spindles, poor part fixtures, poorly maintained coolants supplied by flooding, and other ordinary problems. In other words, the common excuses cannot be used to explain poor or inefficient processes and performance. A new level of university-based industrial research is needed to address the emerging challenges in the field to develop new ideas, advanced experimental techniques, and computer-based technology , and to provide helpful assistance in designing new cutting tools and machining processes to meet ever-growing requirements of quality, productivity, and efficiency in machining operations.

The further text discusses some common issues that have never been discussed openly in the literature in the field. In the author's opinion, these issues limit the implementation of the results obtained in many metal machining researches. Unless these issues are well understood by all parties involved in research on metal cutting and application of the results, there will be always a noticeable gap between the metal cutting research (theory and its conclusion) and many practical machining operations.

There are great discrepancies between the results obtained in laboratory testing and field application results. Manufacturing companies spend a great deal of money, valuable manufacturing and engineering time to adjust even proven tools, coolants, cutting operations to their particular manufacturing conditions. All this accomplished by trial-and-error methods which often fail causing much frustration. In the author's opinion, the prime reason for that is the difference in the system properties of the test and a given application machining system. Therefore, researches and application specialist should learn how to minimize the influence of the properties of a particular machining system on the research results in studies, on one hand, and adjust these results to any practical machining system, on the other hand. In the author's opinion, this can be accomplished if the theory of similarity is applied to metal cutting thus the corresponding similarity number are worked out to assess similarity between testing and application conditions.

2. The Importance of Machining System Coherence in Research and Application Practices

Although the following section discusses a system approach to drilling, its main ideas are applicable to any cutting tool and tooling. In the author's opinion, such an approach should be used in metal cutting research, proper reporting of research results, machining system design, retrofitting, components and tool selection, i.e. in any aspect of research and application.

Systemic analysis and application of metal cutting research is based on the theory of engineering systems. To obtain and apply experimental results for particular machining systems in a meaningful way, one must consider the complete machining system, which includes everything related to the operation. This approach—the systems engineering approach, requires the definition and analysis of the machining system and an understanding of the coherency of its components. An example of the drilling machining system (common in the automotive and mold-making industries) is show in figure 1.

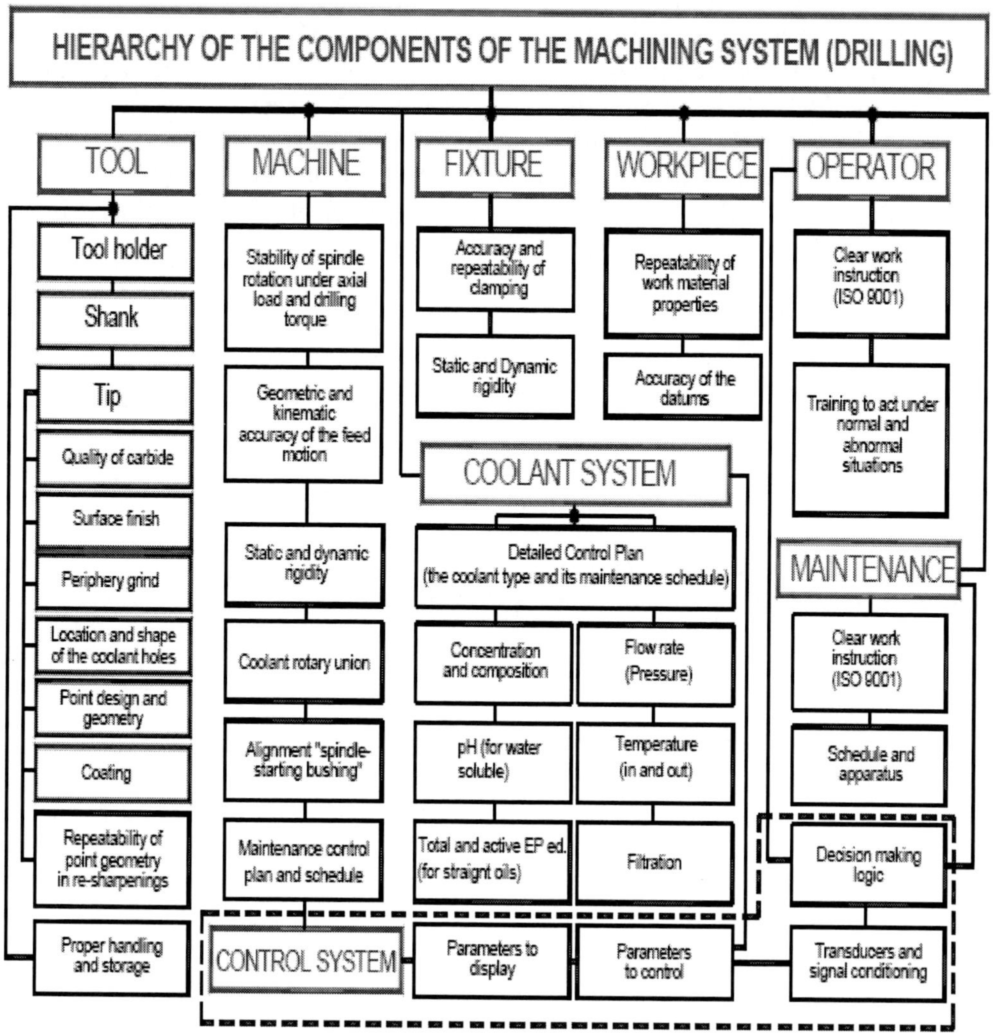

Figure 1. Structure of the gun-drilling machining system.

Systems engineering theory dictates that it is improper to consider any component of machining operations separately, thereby ignoring system properties. Nevertheless, this practice, the so-called "component approach," is a common in today's manufacturing environment, where several different manufacturers produce the various components of the machining system; but no one seems to be responsible for system coherence. Tool failure (i.e., poor tool life, poor quality of machined part, tool breakage, etc.) is a direct result of such an approach, as cutting tools are normally the weakest link in the machining system. In other words, lack of coherence in the overall machining system may lead to tool failure; as well as incorrect conclusions in experimental research if the tests are carried out using a non-engineered machining system. Thus the component approach can easily ruin the implementation of the results of the best research.

Figure 1 illustrates the structure, of a gun-drilling system. An understanding of the performance of this system, and of the root cause of many drilling-related problems, requires an understanding of the following drilling system components (see figure 1): the tool, the drilling machine and its control unit, coolant parameters and coolant delivery unit, fixtures, and accessories; the workpiece, the operator, and the system maintenance.

In understanding the drilling process, it is important to note that the same drill used in different drilling machines will show a wide range of outcomes, from tool breakage to excellent performance. The same drill used on the same machine exhibits different results for different work materials; the same drill used on the same machine for drilling the same work material performs differently depending upon the particular brand of coolant used for the operation, the coolant flow rate, filtration, and temperature. Furthermore, the performance of the same drill used on the same machine for drilling the same work material using the same coolant parameters would depend largely on the extent of the operator's experience and training. The latter is particularly true if the control system provides relevant information to the operator. Thus, each system component can affect system performance dramatically. As a result, system coherency—the condition when all system components work as a 'team' to achieve the ultimate system's objective—is of great importance.

A superior tool designed using promising research results may easily fail at its implementation due to the improper performance of one of the various components of the machining system. For example, a certain manufacturer of gun drills for the automotive industry was blamed for gun drill breakage occurring at the tip-shank brazed joint. Five years of a university- based research into the matter resulted in a significant improvement in the strength of this joint. When the joint's strength was sufficient, breakage of the gun drill carbide tip began to occur. An analysis of the root cause of this new problem showed that insufficient coolant flow rate to the drilling zone caused drill breakage and not the strength of the fortified brazed joint. Because the root cause was not properly determined initially, increasing strength in the brazed joint only shifted the breakage point to the carbide tip as a new weakest link. Another example of inaccurate experimental data or improper application of experimental results concerns the coolant—in terms of its method of delivery, flow rate (pressure), concentration, pH, viscosity, and purity. When coolant is used improperly in testing, the test results are inadequate; when it is used improperly in a machining system implemented on the basis of results of the study, the actual tool performance does not correspond to that obtained in the study. Unfortunately, this type of scenario often undermines the benefits of metal cutting research.

Although research professionals in the field can suggest the best way to carry out a particular machining operation by implementing research results obtained in laboratory conditions, the test method may not be suitable for a given application, which the researcher may not fully understand. For example, if advanced sidewall design tires specially designed for use at the Indianapolis 500 with the Speedway's official "Wing and Wheel" logo in full color are installed on an old, broken-down car, the performance of the car will be even worse than that with its old tires. While the high-performance tires are probably the best that the tire manufacturers can offer (not to mention their cost), they are not suitable for an old car. The same is true of the results of metal cutting studies of newly developed cutting tools and systems: The best and most expensive tool will not perform well if the overall machining system does not support its performance. Therefore, a systems approach in research on metal cutting is essential both at the verification stage of developed models and in the practical implementation of the results.

3. Dealing with System Issues

Metal cutting is embedded into a complex system of interacting physical, chemical, biological, and other phenomena over a wide variety of spatial and temporal scales. Hypothetically, applying the basic laws of the corresponding science areas, the entire range of machining processes can described in differential equations. For each particular case of machining, certain unique conditions should be defined as geometrical configurations of the system components, their physical constants, and initial and boundary conditions. Because nearly all processes taking place in the machining system are interrelated, the corresponding differential equations and unique conditions should be considered as a mathematical model of the machining system. Unfortunately, the present stage of development in metal cutting studies does not allow construction of such a model with any reasonable degree of accuracy. Even if such a model were developed, it would require substantial experimental confirmation of its validity. As a result, the experiment remains the essential instrument of knowledge in metal cutting studies.

Although experimental studies may also entail high cost, questionable accuracy, and substantial commitment of time and resources, their major problem is the particularity of experimental result in metal cutting; that is to say, the experimental results are valid only under exactly the same conditions as when it was obtained and even a small departure from test conditions can interfere with the conclusions.

The so-called “similarity theory” offers a better way to obtain a sound mathematical model of the complicated processes taking place in a complex technical system [1]. Today, the similarity theory is used largely in the area of thermodynamic, fluid flow,and the like. This theory combines various types of knowledge and information about a complex process under study. Its basic principle involves the separation of a group of similar phenomena from a larger class of phenomena by a least common denominator. In such a context, a similarity may be geometrical, physical,or of another type.

The similarity theory is of considerable importance in modeling various phenomena, where amodeling replaces the study of the natural phenomenon with the study of an analogous phenomenon in a model—that is, a replication of the phenomenon in a controlled setting, with well-defined parameters of the components and under special laboratory conditions, where it is relatively easier to measure the system parameters with necessary accuracy). Modeling is often based on the analysis of physically similar phenomena. The physical similarity may be considered as a generalization of geometrical similarity. Geometrical similarity assumes that the two geometric figures are similar if the ratios of all corresponding dimensions are identical. If the similarity ratio —the scale—is known, then simple multiplication of the dimensions of one geometric figure by the scale factor yields the dimensions of the other—its similar, geometric figure.

The numerical characteristics of two different but similar phenomena can be considered as the numerical characteristics of the same phenomenon expressed in two different systems of units. All the dimensionless characteristics (dimensionless combinations of dimensional quantities) of a set of similar phenomena have the same numerical value. Of great importance is that the converse is also correct: that is, if all dimensionless characteristics of two phenomena are identical then the phenomena are physically similar even though they may take place in considerable different physical systems. The latter property is significant for metal cutting, where a great variety of ~~physical~~ systems such as turning, milling, broaching, drilling,and others, exist. Because the cutting model and theory are commonly developed using one model of chip formation, and the experimental results are obtained using one machine and one process (turning, drilling, etc.), a key question here is how to apply the obtained theoretical and experimental results to each given particular system. The proper way is through application of the similarity theory.

The first theorem of similarity enables the transformation of differential equations that describe physical phenomena and their presentation as functions of criteria passing over analytical solution. The second theorem, which solves the problem, deals with presentation of the solution of the differential equations as functions of similarity criteria called numbers. The third theorem of similarity formulates the necessary and sufficient conditions determining the similarity of phenomena in two systems.

Unfortunately, the similarity theory is not yet developed for use in metal cutting studies. However, some useful similarity criteria (numbers) were developed that can be used in the modeling of metal cutting process and they are discussed below[2, 3].

3.1. Chip Compression Ratio

3.1.1. Definition

The chip compression ratio (hereafter, CCR) assures similarity of deformation processes in metal cutting. Because the energy spent on plastic deformation of wide variety of ductile work material is within the range of 70-80% [4], the chip compression ratio is probably the most important similarity criterion in metal cutting. Astakhov shows that the CCR can be used to assess the work of plastic deformation in metal cutting [5], calculate the cutting force [4], verify the validity of a FEM [6], and for other purposes.

The chip compression ratio (ζ) is determined as the ratio of the length of cut (L_l) to the corresponding length of the chip (L_c), or as the ratio of the chip thickness (t_2) to the uncut chip thickness (t_1), or as the ratio of the cutting speed (v) to the chip velocity (v_1) i.e.

$$\zeta = \frac{L_l}{L_c} = \frac{t_2}{t_1} = \frac{v}{v_1} \tag{1}$$

3.1.2. Significance

Astakhov also shows that [5], for a given work material, the CCR assures the similarity of the deformation process in various machining systems regardless of various other process and regime parameters. Indeed, CCR is probably the major determining criterion of similarity in metal cutting. This is because the CCR directly defines the elementary work spent over plastic deformation (A_u) of a unit volume of the work material as

$$A_u = 2K\left(\ln \zeta\right)^{n+1} \tag{2}$$

where K is the strength coefficient (N/m^2) and n is the hardening exponent of the work material. Figure 2 shows that CCR is greatest in the machining of copper and lowest in the machining of steel. This fact confused earlier researchers as it well known that the cutting energy and force are much greater in the machining of steel. Equation (2) resolves this long-standing contradiction. Figure 2 shows that the elementary work is the greatest for steel. In other words, the energy per unit volume spent in the machining of steel is the greatest, which results in a much higher amount of heat generated and in more profound tool wear. This conclusion is supported in multiple ways known from the everyday practice of machining.

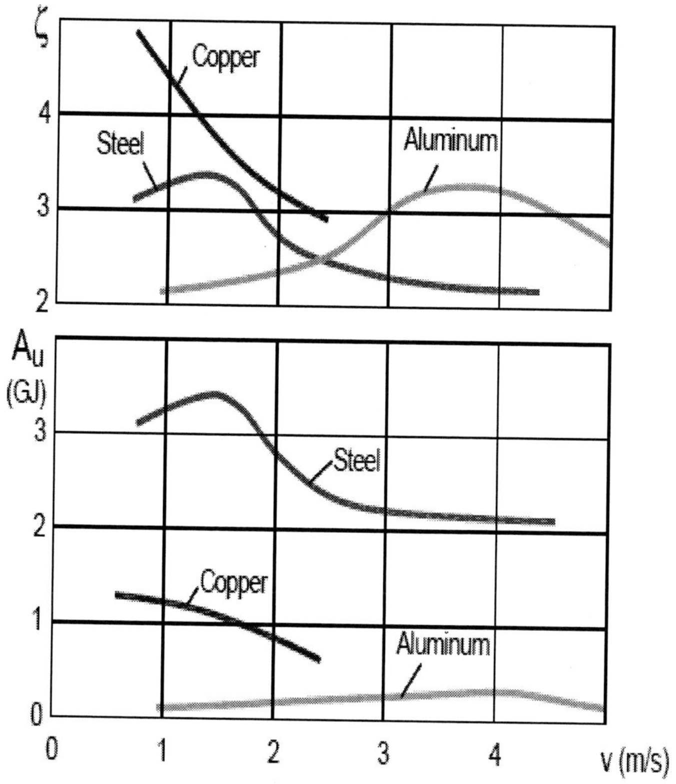

Figure 2. Influence of cutting speed on CCR and the work done in plastic deformation for: (1) AISI steel E52100, (2) copper, and (3) aluminum 1050–0.

Therefore, a necessary condition for the similarity of two deformation processes in metal cutting is the same plastic deformation value—A_u.

This result is of great significance to experimental studies in metal cutting because it correlates in a simple and physically grounded manner the work on plastic deformation done in cutting with a measurable, post-process characteristic of the cutting process such as CCR.

Knowing CCR, one can not only assure the similarity of the deformation process but also calculate the power spent on the plastic deformation of the layer being removed and the power spent due to friction at the tool-chip interface. These power requirements are the two major contributors to the total power required by the cutting system [4].

The power spent on the plastic deformation of the layer being removed, P_{pd}, can be calculated from the chip compression ratio and parameters of the deformation curve of the work material as follows [5, 7]:

$$P_{pd} = \frac{K\left(1.15\ln\zeta\right)^{n+1}}{n+1} vA_w \tag{3}$$

Where A_w is the uncut chip cross-sectional area (m^2)

$$A_w = d_w f \tag{4}$$

where d_w is the depth of cut (m) and f is the cutting feed per revolution (m/rev).

CCR represents the true strain in plastic deformation and can be used to calculate the elementary work spent over plastic deformation of a unit volume of the work material. If the elementary work is known, the total work done by the external force applied to the tool can then be calculated. As a result, CCR can be used as the prime parameter for optimization of the metal cutting process. Considered together with the work of plastic deformation, CCR reveals the energy spent in cutting. Moreover, CCR is a post-process parameter, and thus a number of simple though forgotten ways exist to measure this parameter accurately in metal cutting. The simple practical methods of experimental determination of the chip compression ratio ζ are discussed above [7].

3.2. Péclet number

3.2.1. Definition

The metal cutting system includes moving heat sources, that is, the tool moves relative to the workpiece and the chip moves relative to the tool. As a result, similarity numbers used in thermodynamics to deal with moving heat sources [8-10] should be utilized also for thermodynamic analysis of metal cutting. Unfortunately, this is not the case in traditional studies on thermal aspects of metal cutting.

To avoid typical misrepresentation of the experimental data on CCR, the CCR should be determined as a function of the Péclet criterion. Such a representation allows accounting for the combined influence of the cutting regime and physical properties of the work material. Moreover, knowing the actual Péclet number, one can assess the influence of the thermal energy generated in machining due to both plastic deformation and friction on the mechanical properties of the work materials ahead of the cutting edge so that a proper work materials model can be used in modeling (analytical, numerical, or physical modeling) the tribological interfaces where traditional steady-state thermodynamic analysis of heat transfer commonly is used today as, for example, in [11-13]. The Péclet criterion, often referred to as the Péclet number (Pe), is widely used in the thermal analysis of technical systems subjected to moving heat sources [8-10, 14]. It is a dimensionless number expressing the ratio of advection to thermal diffusion and is expressed as

$$Pe = \frac{\mathbf{UL}}{w_w} \tag{6}$$

where U is the velocity scale, L is the horizontal length scale, w_w is the thermal diffusivity.

In metal cutting, the Péclet criterion is represented in terms of machining process parameters as follows [3, 5]

$$Pe = \frac{vt_1}{w_w} \tag{7}$$

where v is the velocity of a moving heat source (the cutting speed) (m/s), w_w is the thermal diffusivity of the work material (m^2/s),

$$w_w = \frac{k_w}{(c_p\rho)_w} \tag{8}$$

where k_w is the thermoconductivity of the work material, ($J/(m\cdot s\cdot {}^oC)$), $(c_p \cdot \rho)_w$ is the volume specific heat of work material, ($J/(m^3\cdot {}^oC)$).

The Péclet number is a similarity number, which characterizes the relative influence of the cutting regime (vt_1) with respect to the thermal properties of the workpiece material (w_w). If Pe>10, then the heat source (the cutting tool) moves over the workpiece faster than the velocity of thermal wave propagation in the work material in such as way that the thermal energy generated in cutting due to the plastic deformation of the work material and due to friction at the tool-chip interface does not affect the work material ahead of the tool [15]. If Pe<10, then the thermal energy generated due to the plastic deformation and friction makes a strong contribution to the process of plastic deformation during cutting, as it affects the mechanical properties of the work material.

3.2.2. Practical Use in Modeling

As an example, consider machining of AISI 1040 steel under the following typical machining conditions: operation – turning; Tool – MTJNR-1616H-09 (ISO 5608:1995) with a carbide insert; cutting speed v=3m/s (180m/min); cutting feed f=0.25mm/rev. The thermal diffusivity of the work material is $6.67\cdot 10^{-6}$ m^2/s. For the J-style tool holder, the tool cutting edge angle is $\kappa_r=93^o$. Thus, the uncut chip thickness calculates as [7] $t_1=f\cdot\cos(\kappa_r-90^0)=0.25\cdot\cos(93^o-90^o)$ = $0.24965\approx0.25$mm. Thus, the Péclet criterion calculates as $Pe=(3\cdot0.25\cdot10^{-3})/6.67\cdot10^{-6}=112$. As Pe>>10, one must conclude that the thermal energy generated in cutting due to plastic deformation of the work material and due to friction at the tool-chip interface does not affect the properties of the work material ahead of the tool. In other words, the cutting process in this case is a cold-working process. Thus, there is no need to use temperature-dependant work material models (for example, the Johnson-Cook model) in the numerical modeling of the process.

For comparison, consider another typical operation in the automotive industry—a helical spline broaching where the standard machining regime is as follows: cutting (puller) speed v=0.1m/s (6m/min); cutting feed (the rise per tooth) f=0.03mm, helical angle $\omega=15^o$. Thermal diffusivity of the work material is $6.67\cdot10^{-6}$ m^2/s. The tool cutting edge angle is $\kappa_r=\omega=15^o$, thus the uncut chip thickness calculates as [7] $t_1=f\cdot\cos(\kappa_r)=0.03\cdot\cos(15^o)\approx0.029$mm. Thus, the Péclet criterion calculates as $Pe=(0.1\cdot0.0.029\cdot10^{-3})/6.67\cdot10^{-6}=0.44$. As Pe<<10, one must conclude that the thermal energy generated in cutting due to the plastic deformation of the work material and due to friction at the tool-chip interface affects the work material ahead of the tool. Thus, temperature-dependant work material models should be used in the numerical modeling of the considered broaching process.

3.2.3. Practical Use in Testing

Figure 3 shows the influence of the cutting speed on CCR for different feeds in six complete tests. Figure 4 shows what happens if the Péclet criterion is used as the independent variable. Figure 5 presents another example of the experimental data obtained in the machining of tool steel H13. Such a representation allows reduction of the number of cutting tests needed to study the amount of plastic deformation occurring in the metal cutting process. Moreover, it helps in revealing the mutual influence of the cutting regime, tool geometry and physical properties of the work material on this plastic deformation. For example, it is clearly seen that the amount of plastic deformation in cutting for a work material having low thermoconductivity is greater than that in cutting a work material having higher thermoconductivity if the other cutting parameters remain the same. It can be stated also that the experimental results for the first time indicate the correlation between the thermoconductivity of the work material and the amount of plastic deformation that occurs during its machining. In other words, the results quantify one of the most important qualitative characteristics of the work material as "machinability." Using the work of plastic deformation which is a sole property of the work material involved in the cutting process, one can determine machinability of various other work materials quantitatively.

However, by analyzing the results, one also may argue that the influence of the tool geometry is not accounted for. To address this concern, the following should be explained. The tool rake angle is the only tool geometry parameter that might affect the work of plastic deformation. The influence of the tool rake angle on CCR is shown in figure 6. As seen, CCR does not depend on the tool rake angle if the Péclet criterion is used as an independent variable in tests.

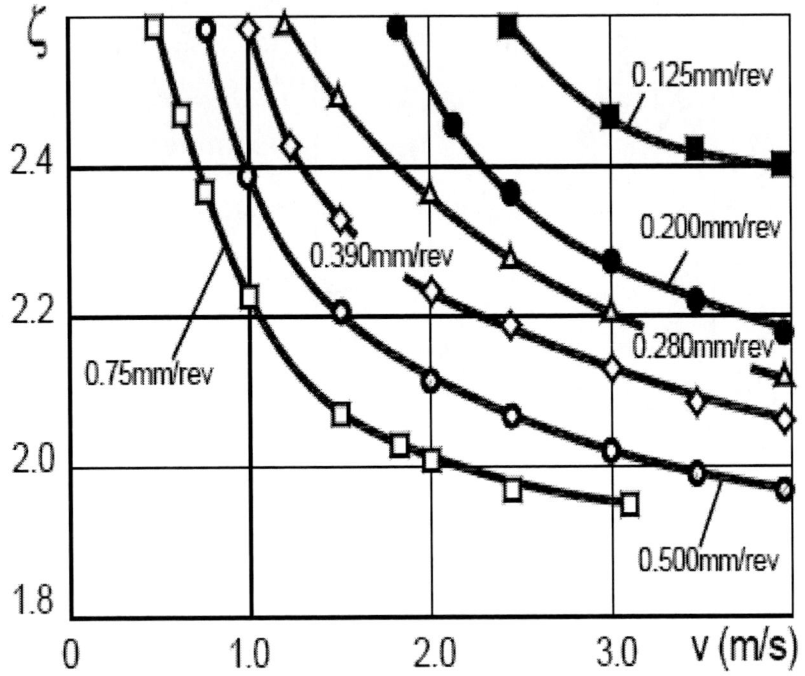

Figure 3. Influence of the cutting speed on CCR for different speeds, for: Work material – steel AISI 1030, tool material – carbide P20, rake angle $\gamma_n = 10°$, cutting edge angle $\kappa_r = 60°$, and depth of cut d_w = 2mm.

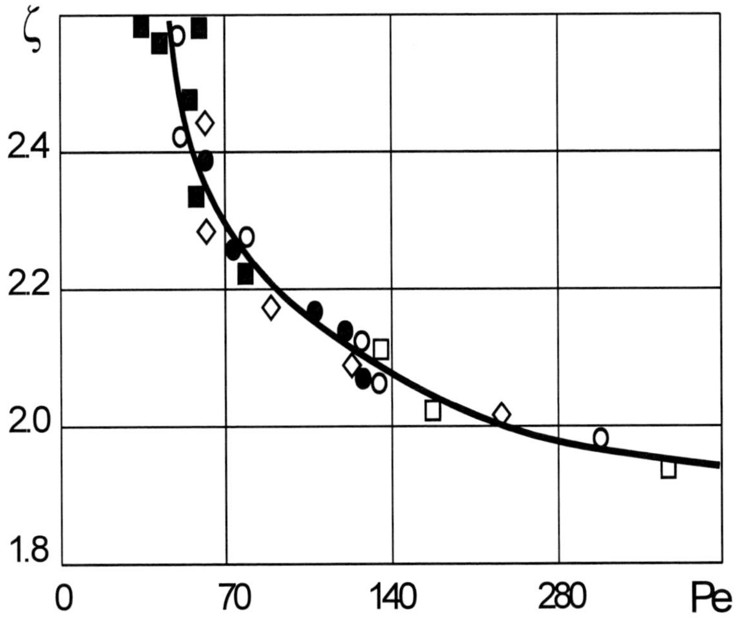

Figure 4. Generalization of the experimental data using the Péclet number

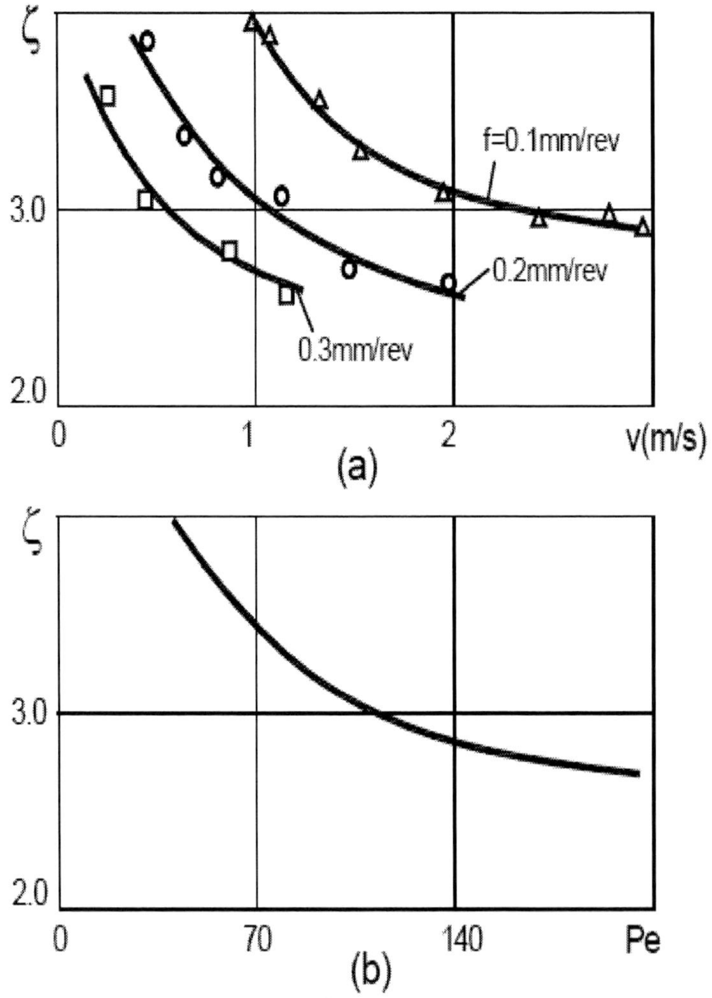

Figure 5. CCR vs. (a) the cutting speed for different feeds and (b) Pe criterion. Work material – tool steel H13, tool material – carbide M10, rake angle $\gamma_n = -10^{\circ}$, cutting edge angle $\kappa_r = 60^{\circ}$, depth of cut $d_w = 2$mm.

3.3 Poletica number

3.3.1 Definition

In metal cutting, the tool–chip contact length known as the length of the tool–chip interface determines major tribological conditions at this interface as temperatures, stresses, tool wear, etc. Moreover, all the energy required by the cutting system for chip removal passes through this interface. Therefore, it is of great interest to find out a way to asses this length.

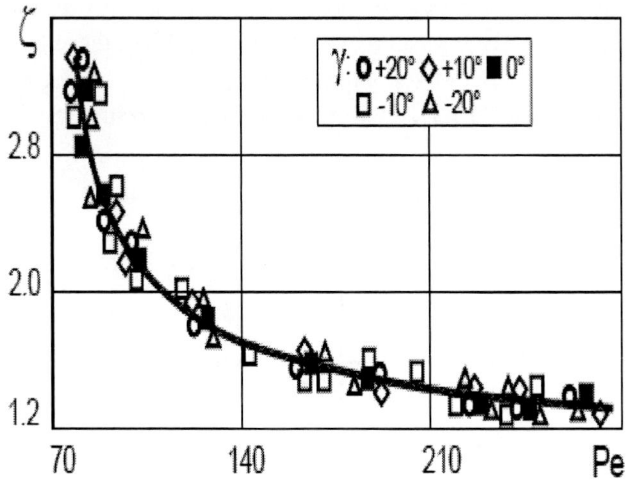

Figure 6. CCR vs. Pe criterion for different rake angles. Work material – steel AISI 1045, tool material – carbide P20, cutting edge angle $\kappa_r = 60°$, depth of cut $d_w = 2$mm.

To deal with the problem, the Poletica criterion (Po-criterion) is introduced [16] as the ratio of the contact length, l_c to the uncut chip thickness, t_1

$$Po = \frac{l_c}{t_1} \tag{9}$$

3.3.2. Practical Use in Modeling and Testing

The Po criterion defines the similarity of the tribological processes at the tool chip interface. Astakhov has shown[7] that, among the major factors that affect the tool-chip contact length—namely, the uncut chip thickness, cutting speed, and rake angle, only the uncut chip thickness affects this length directly. Figure 7 illustrates experimental data indicating that the Po-criterion strongly depends on CCR and weakly depends on the rake angle. . As shown, the influence of rake angle is within the normal experimental scatter. Moreover, the Po-criterion is shown to be invariant to changes in the mechanical and physical properties of the work material. Figure 8 indicates that the hardness of the work material does not affect the dependence of Po-criterion on CCR. It should be noted that beryllium copper is an excellent test material for this particular test because its mechanical properties can be changed significantly by heat treatment, while its phase composition and microstructural parameters remain practically unchanged.

Figure 9 presents the results of cutting tests using various work and tool materials having a wide range of physical and mechanical properties. In this figure, the normalized chip compression ratio, ζ_t, is used instead of ζ. Thus, the CCR ratio is calculated as

$$\zeta_t = \frac{\zeta}{d_{w1}/d_w} \tag{10}$$

where $d_{w1}$is the width of chip. The calculation is done in this way to include the result of cutting tests with copper and Armco iron may be included. When these two materials are cut, the chip width changes considerably compared to the width of cut (the depth of cut), whereasfor many other work materials, the ratio $d_{w1}/d_w \approx 1$, if measured properly.

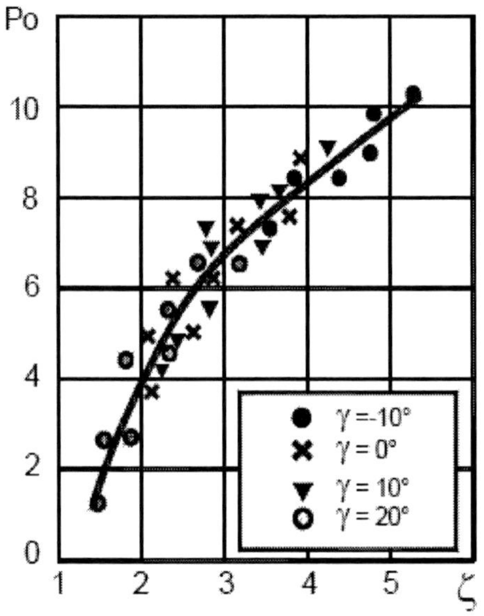

Figure 7. Influence of chip compression ratio on Po-criterion in machining steel AISI E9310, tool material P20 (79%WC, 15%TiC, 6%Co), cutting feed f = 0.07 – 0.43mm/rev, and cutting edge angle κ_r = 70°.

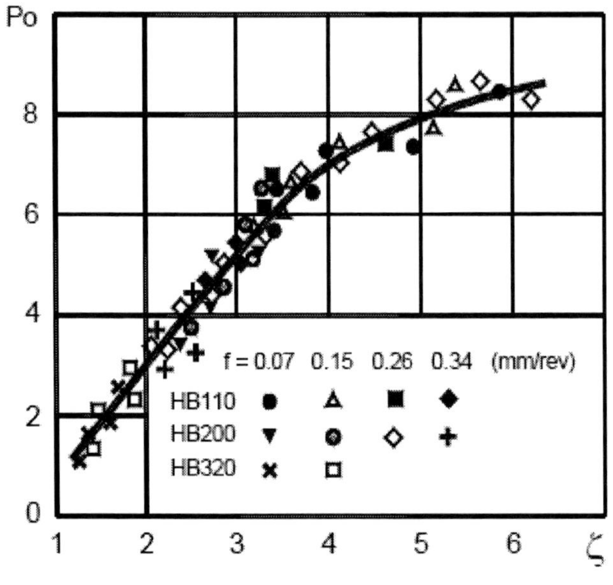

Figure 8. Influence of CCR on Po-criterion in machining beryllium copper UNSC17000 of varying hardness. Tool material – M30 (92%WC, 8%Co).

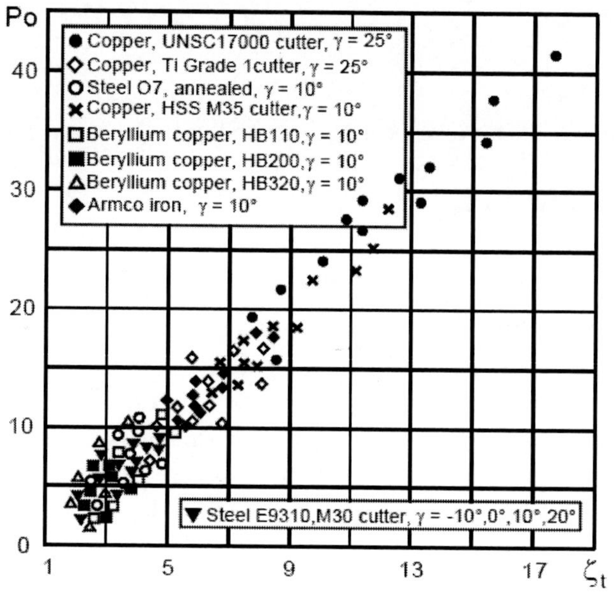

Figure 9. Influence of chip compression ratio on Po-criterion in machining various work materials using different tool materials and tool rake angles.

3.4. Some Other Similarity Numbers: A, Silin, D, E Criterion

Other useful similarity numbers in metal cutting were introduced by Silin [2, 3] for use in metal cutting research. One of the most important of these is the A-criterion (A). As it was first derived and studied by Silin, it is referred to as Silin criterion. It calculates as

$$A = \frac{t_1 b_{1T} c_p \theta_c}{F_p} \tag{11}$$

and characterizes the part of the thermal energy (heat) absorbed by the chip relative to the whole amount of heat generated in the deformation zone. In the equation above, $t_1$and b_{1T} are the uncut chip thickness and the true chip width, respectively; m, c_p is the volumetric heat capacity of the work material, $J/(m^3\ {}^oK)$; θ_c is the cutting temperature, ^{o}C; F_p is the power component of the force, *N*.

The D-criterion which calculates as

$$D = \frac{t_t}{b_{1T}} \tag{12}$$

and characterizes the uncut chip cross-section.

The E-criterion, or relative sharpness of the cutting edge, calculates as

$$E = \frac{\rho_1}{t_1} \tag{13}$$

and characterizes the influence of the cutting edge radius ρ_1 (*m*) with respect to the uncut chip thickness t_1 (m).

The F-criterion calculates as

$$F = \frac{k_t}{k_w}\beta_n \varepsilon_{tn} \tag{14}$$

and characterizes the influence of the tool geometry with respect to the thermal conductivity of tool and work materials. In Eq. (14), k_t and k_w are thermal conductivity of tool and work materials, *J/(m s °C)*, respectively; β_n is the normal tool wedge angle; ε_{tn} is the acute angle in the reference plane between the major (side) and minor cutting edges.

3.5. Machinability Studies Using Similarity Numbers

In the literature on metal cutting and in various dictionaries, the machinability of a material is defined as the ease with which it can be machined. It is often pointed out, however, that machinability depends on the physical properties of the material, as well as on the cutting conditions. Therefore, it is not clear if machinability is an inherent property of the material or if it is a property of the material in the particular machining system of components. This issue is the first and foremost question that should be resolved for proper definition of machinability and its assessment.

Mills and Redford [17] published the only existing book on machinability. As pointed out by Mills and Redford [17], machinability considered simply as a property of a work material, has no generally accepted parameter for its measurement. As a result, machinability tends to remain a term that means "all things to all people." However, the authors suggest that this term should be understood to be at least some indication of the way in which a material wears away a cutting tool when it is being machined.

In the author's opinion, Mills and Redford [17] built a logical trap for themselves for the following reasons:

- Considering machinability as a property of the work material, the authors were forced to find a specific characteristic of the work material responsible for tool wear. Moreover, this characteristic should be measurable.
- Having selected a measure of the way in which the work materials wears away the cutting tool, the authors assume that this measure is a property of the work material; so all known tool materials, ranging from high carbon tool steel to PCD, should be subjected to the same work material characteristic, not to mention a great range of cutting conditions, resulting in great variety of contact pressures and temperatures at tool-chip and tool-workpiece interfaces.

- The authors shift their attention to the place of wear (flank wear, crater wear, etc) instead of to the physics of wear. They admit, however, that the definite lack of a fundamental understanding of the process of tool wear makes it impossible to combine basic properties of the work and tool materials, as well as the cutting conditions, to arrive at a measure of machinability.
- Having realized that the experimental determination of tool wear is highly uncertain, the authors conclude that known tests and experimental data machinability are valid only for the test conditions. As a great variety of the machining conditions exist in terms of tool materials and coatings, work materials and their state, coolant parameters and grades, machining regimes, machine tool properties, part design, and so on, the machinability index becomes next to meaningless. The authors point out that even if the machinability test attempts to compare machinability of two different work materials for a given set of cutting conditions, there is no guarantee that when cutting conditions change the machinability ranking will remain the same.

Two works are often referred to as machinability standards, International standard ISO 3685— "Tool-Life Testing with Single-Point Turning Tools," and its ANSI/ASME analog, "Tool Life Testing With Single-Point Turning Tools" (B94.55M-1985). However, these works can hardly be considered as such, because their standards represent an obsolete methodology for tool life testing wherein one parameter is changed at a time [7].

Standard ASTM E618-07, "Standard Test Method for Evaluating Machining Performance of Ferrous Metals Using an Automatic Screw/Bar Machine," was written to fill a requirement for determining the machinability of ferrous metals using automatic screw/bar machines. The work is based on producing pairs of standard design in automatic screw/bar machines to uniform levels of quality with respect to surface roughness and size variation. The standards intend to simulate mass production conditions in a controlled environment using a single- or multi-spindle automatic screw machine. The machinability of a work material is judged by the number of standard parts that can be produced based on one form tool insert change in eight hours. The tool geometry and the tool materials (HSS M2 for the form tools and M7 for the drills) are fixed by the standard. The machining sequence is specified by the standard and is based on normal industrial practices. Recommendations are also given for the calibration of a screw machine to be used for testing. The standard admits that testing according to this standard is neither simple nor inexpensive. Substantial quantities of test material are required, varying from a few hundred of kilograms to a few thousand kilograms. Significant time also is required to compare different work materials, including finding the proper regimes, the machining process itself, and part quality evaluation. This standard is not in wide use even in the automotive industry not because it requires extensive and thus expensive testing, but primarily because of lack of meaningful results.

All these issues considered, it is this author's opinion that machinability is a system property and must be addressed as such. However, a given combination of the work and tool materials has probably the strongest impact on this property. Therefore, the most objective way to assess machinability at the most general level, leaving apart all other properties of the machining system, is to determine the optimal cutting temperature for the given combination of the tool and work materials. The concept, physics, and determination of the optimal cutting temperature were discussed by Astakhov [7].

It is understood that the optimal cutting temperature can be achieved using a machine-system-specific cutting speed known as the optimal cutting speed. Therefore, the optimal cutting temperature is an independent system parameter, while the optimal cutting speed depends on many particularities of a given machining system.

In experimental studies of machinability, when a specific tool (tool material, tool holder etc) and workpiece (dimensions and work material) are selected for testing, it is often sufficient at the first stage of the study to consider the following relationship:

$$Pe = n_1 A^{m_1} \quad \text{or} \quad \frac{vt_1}{w_w} = n_1 \left(\frac{t_1 b_{1T} c_p \theta_c}{F_p} \right)^{m_1} \tag{15}$$

where n_1 and m_1 are constants to be determines experimentally.

Equation (15) can be rewritten for the optimum cutting speed v_o (the speed that corresponds to the optimal cutting temperature θ_o) as

$$v_o = \frac{n_1 w_w}{t_1} \left(\frac{t_1 b_{1T} c_p \theta_o}{F_p} \right)^{m_1} \tag{16}$$

To determine constants n_1 and m_1, the power components of the force, F_p, and the cutting temperature, θ_c, are measured simultaneously. If the test results are plotted on a double logarithmic A versus Pe diagram (the same module along both axes) as shown in figure 10, then n_1 = Pe when A = 1 and $m_1 = \tan \delta_1$.

For data shown in figure 10(a), Eq. (16) becomes

$$v_o = \frac{9.1 w_w}{t_1} \left(\frac{t_1 b_{1T} c_p \theta_o}{F_p} \right)^{2.3} \tag{17}$$

and for data shown in figure 10(b), Eq. (16) becomes

$$v_o = \frac{8.35 w_w}{t_1} \left(\frac{t_1 b_{1T} c_p \theta_o}{F_p} \right)^{2.0} \tag{18}$$

The foregoing analysis leads to a new approach to determining machinability using the following procedure. Before starting the test, it should be ascertained that machine tool (lathe) and workpiece fulfill all the requirements of ANSI/ASME B94.55M - 1985 standard. Between five and seven different cutting feeds should be selected for study. The depth of cut should be kept the same for all tests. The number of tests should correspond to the number of the selected cutting feeds. In each test, the cutting speed is varied and the cutting force and cutting temperature are measured. These readings should be plotted on the cutting force and cutting temperature (ordinates) versus cutting speed (abscissa) diagram as shown in figure 11. Such diagrams should show at least fifteen experimental points for each curve so that the cutting speed at which the cutting force assumes the minimum stabilized value can be assessed with sufficient accuracy. Under no circumstances should the cutting force and/or cutting temperature be determined by extrapolating corresponding curves.

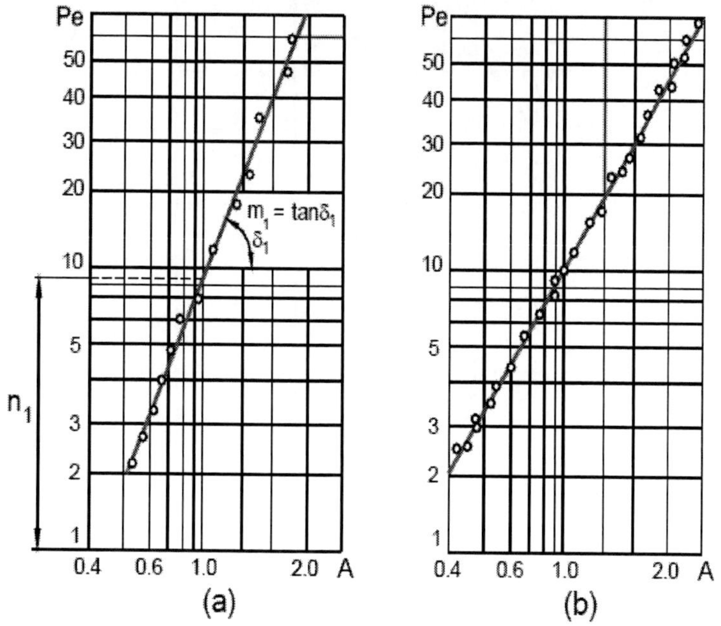

Figure 10. Experimental determination of the constants of Eq. (15): (a) work material - stainless steel AISI 303, tool material: carbide P01 (66%WC30%TiC4%Co), tool geometry: γ_n = 12 deg, α_n = 10 deg, κ_r = 45 deg, κ_{r1} = 25 deg, r_n = 1 mm), similarity numbers: F = 1.48, D = 0.0126-0.1500, E = 0.06 - 0.76; (b) workpiece material - steel AISI 1045, tool material carbide P10 (79%WC 15%TiC6%Co), tool geometry: γ_n = 12 deg, α_n = 12 deg, κ_r = 45 deg, κ_{r1} = 15 deg, r_n = 1 mm, similarity numbers: F = 2.31, D = 0.022-0.122, E = 0.018 - 0.105.

The optimum cutting speed is defined for each feed using points on each curve F_p (v), where F_p assumes the minimum stabilized value corresponding to the optimum cutting temperature θ_o. To set the test properly (calibration stage), it is recommended to determine the optimum cutting temperature θ_o using the point where F_p assumes the minimum stabilized value for cutting feeds f = 0.2 - 0.4 mm/rev. Plotting the coordinated values (?) (v_{o1} ,t_{1-1}), (v_{o2} ,t_{1-2}), etc, obtained from figure 11 on a double logarithmic with the true uncut chip thickness versus cutting speed (same module along both axes), one can obtain a t_1 - v curve, as show in figure 12. The true uncut ship thickness should be calculated as such [7]. The t_1 - v curve may be considered linear within a certain range of the uncut chip thickness. The equation for this linear portion of the curve is written

$$v_o = n_2 t_1^{m_2} \tag{19}$$

In which constants n_2 and m_2 are determined from figure 12 to be n_2=0.034 and m_2 = 0.81.

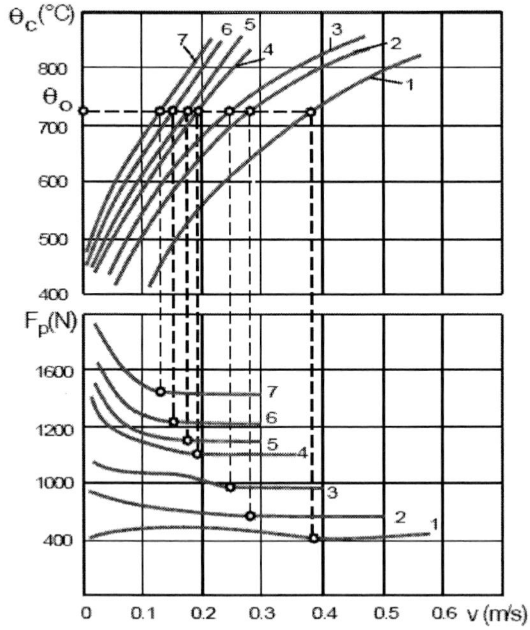

Figure 11. Experimental determination of OCS by the minimum stabilized cutting force. Workpiece material: nickel-based high alloy (0.08%C1%Cr56%Ni1%Co1%Al), tool material: carbide M30 (92%WC8%Co), tool geometry: γ_n = 12 deg, α_n = 12 deg, κ_r = 45 deg, κ_{r1} = 45 deg, r_n = 1 mm, cutting regime: t_1 = 1 mm, f, mm/rev, =1- 0.074, 2- 0.11, 3- 0.15, 4- 0.25, 5- 0.30, 6- 0.34, 7- 0.39.

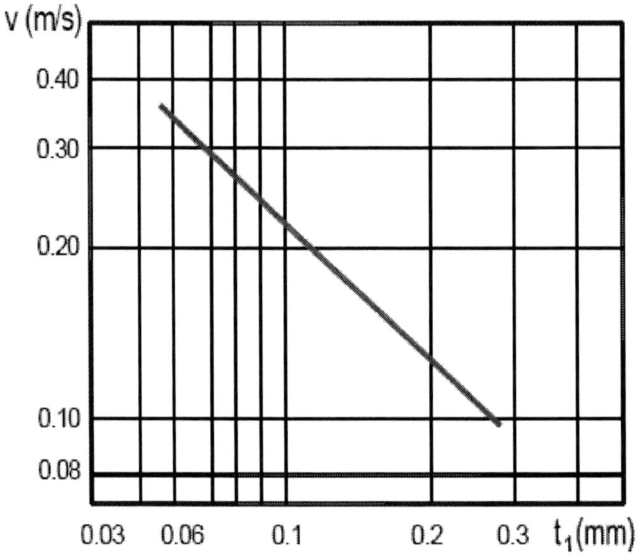

Figure 12. The optimum cutting speed vs. true uncut chip thickness.

This method does not require expensive and time-consuming tool-life testing to define machinability as required by *ANSI/ASME B94.55M - 1985* standard.

On the contrary, analysis of a large body of experimental data shows that the simplified relationship between similarity numbers defined by Eq. (15) provides satisfactory results over a wide range of the cutting speed variation. However, it is valid only for a comparison of the cutting tools of the same geometry when the cutting feed $f \geq 0.08$ mm/rev and the depth of cut $t_1 \geq 1$ mm. To obtain results more general than Eq. (15), requires determination of experimental relationships between process parameters in metal cutting to assess machinability, a more sophisticated relationship between the similarity numbers as

$$A = \frac{n_3 Pe^{m_4}}{F^{m_5} D^{m_6} E^{m_7}} \tag{20}$$

where constants n_3, $m_4 - m_7$ are to be determined experimentally using a suitable design of experiment techniques [7].

REFERENCES

[1] Gukhman, A.A., Introduction to the Theory of Similarity. 1965, New York: Academic press.

[2] Silin, S.S., Similarity Methods in Metal Cutting (in Russian). *Mashinostroenie,* Moscow. 1979.

[3] Astakhov, V.P., Metal Cutting Mechanics. 1998/1999, Boca Raton, USA: CRC Press.

[4] Astakhov, V.P., Xiao, X., A methodology for practical cutting force evaluation based on the energy spent in the cutting system. *Machining Science and Technology,* 2008. 12: p. 325 - 347.

[5] Astakhov, V.P., Shvets, S., The assessment of plastic deformation in metal cutting. *Journal of Materials Processing Technology,* 2004. 146: p. 193-202.

[6] Astakhov, V.P., Outeiro, J. C., Metal cutting mechanics, finite element modelling, in *Machining Fundamentals and Recent Advances,* J.P. Davim, Editor. 2008, Springer: London. p. 1-27.

[7] Astakhov, V.P., Tribology of Metal Cutting. Triblogy and Interface Engineering Series, No. 52, ed. B.J. Briscoe. 2006, London: Elsevier.

[8] Incopera, F.P., De Witt, D.P.,, Fundamentals of Heat and Mass Transfer. 5th ed. 2001, New York: John Wiley and Sons.

[9] Manca, O., Morrone, D., Nardini, S., Thermal analysis of solids at high peclet numbers subjected to moving heat sources. *ASME Journal of Heat Transfer,* 1999. 121-1: p. 182-186.

[10] Muzychka, Y.S., Yovanovich, M. M., Thermal resistance models for non-circular moving heat sources on a half space. *Journal of Heat Transfer,* 2001. 123(4): p. 624-632.

[11] Childs, T.H.C., Maekawa, K., Obikawa, T., Yamane, Y., *Metal Machining. Theory and Application.* 2000, London: Arnold.

[12] Boothroyd, G., Knight, W.A., Fundamentals of Machining and Machine Tools, Third ed. 2006, Boca Raton: CRCPress.

[13] Shaw, M.C., Metal Cutting Principles. 2nd Edition. 2004, Oxford: Oxford University Press.

[14] Ostafiev, V.A., Noshchenko, A.N., Numerical analysis of three-dimensional heat exchange in oblique cutting. *Annals of the CIRP,* 1985. 34(1): p. 137-140.

[15] Astakhov, V.P., Shvets, S.V., A novel approach to operating force evaluation in high strain rate metal-deforming technological processes. *Journal of Materials Processing Technology,* 2001. 117(1-2): p. 226-237.

[16] Poletica, M.F., Contact Loads on Cutting Tool Interface Surfaces (em Russo). *Mashinostroenie,* Moscow, 1969.

[17] Mills, B., Redford, A.H., Machinability of Engineering Materials. 1983, London: Applied Science Publishers.

In: Metal Cutting: Research Advances
Editor: J. Paulo Davim pp. 23-75

ISBN: 978-1-60876-207-1

Chapter 2

PREDICTION OF HEAT PARTITION IN METAL CUTTING – A STATE-OF-THE-ART REVIEW OF CONVENTIONAL TO HIGH-SPEED MACHINING

F. Akbar**, P. T. Mativenga and M. A. Sheikh
School of Mechanical, Aerospace and Civil Engineering, The University of Manchester,
Sackville Street Building, Manchester M60 1QD, UK

ABSTRACT

Temperature and heat partition along the tool rake face are critical parameters that influence the metal cutting process and thus determine machining quality and tool stability, especially in high-speed cutting. Therefore, determining and reducing the temperature along the tool rake face is of particular importance in machining. The problem has been approached in numerous ways, including experimental, analytical, and numerical analysis. Although considerable research exists on the thermal problem in metal cutting, there is little agreement on the basic principles, which makes it extremely difficult to predict heat partition in a precise manner. This chapter first presents a comprehensive review of previous research on heat generation and heat partition in orthogonal machining at speeds ranging from conventional to high-speed machining (HSM). The next section evaluates experimental methods and various analytical approaches for determining the fraction of heat flowing into the tool from the secondary deformation zone. The final section discusses the Finite Element Method (FEM), which may be applied in conjunction with the experimental data to predict the amount of heat entering into the tool. This application has been demonstrated for conventional to HSM of AISI/SAE 4140 high strength alloy steel with uncoated, as well as TiN and TiAlN-coated tools. It is concluded that existing models are quantitatively inadequate for predicting heat partition into the cutting tool, particularly in high-speed machining. Moreover, it is proposed that important thermal properties of the tool and the workpiece, cutting process variables, and contact phenomena should all be considered in predicting heat partition.

* Corresponding author: E-mail: faraz.akbar@postgrad.manchester.ac.uk

Nomenclature

A_a	area shape factor
a_p	depth of cut (mm)
C_p	specific heat (J/kg/K)
D	damage parameter
D_W	diameter of the workpiece (m or mm as defined)
F_c	cutting force (N)
F_{fr}	frictional force at the rake face (N)
F_f	feed force (N)
F_N	normal force on tool face (N)
F_R	resultant tool force (N)
f	feed rate (mm/rev)
h	convective heat transfer coefficient (W/m^2/C)
l_{ch}	tool-chip contact length (mm)
l_{sl}	sliding length of the tool-chip contact zone (mm)
l_{st}	sticking length of the tool-chip contact zone (mm)
N	spindle speed (rpm)
N_T	dimensionless velocity diffusivity factor
N_p	dimensionless parameter
Pe_T	peclet numbers for the tool
Pe_W	peclet numbers for workpiece
q	frictional heat flux (W/mm^2)
R_{ch}	fraction of heat conducted into the chip
R_T	fraction of heat conducted into the tool
R_t	non-dimensional thermal number
t	uncut chip thickness (mm)
t_c	actual chip thickness (mm)
V_c	cutting velocity (m/min or m/sec as defined)
V_{ch}	chip velocity (m/s)

Greek Symbols

α_W	thermal diffusivity of the workpiece material (mm^2/s)
α_T	thermal diffusivity of the tool material (mm^2/s)
γ	rake angle (deg.)
λ_W	thermal conductivity of the workpiece material (W/m/K)
λ_T	thermal conductivity of the tool material (W/m/K)
λ_h	chip compression ratio
μ	mean coefficient of sliding friction
θ_o	initial tool and workpiece temperature (^{o}C)
$\Delta\theta_t$	mean temperature rise at tool-chip interface (^{o}C)
ϕ	shear angle (deg.)
ρ	density (kg/m^3)

τ_{sh} shear stress (N/mm^2)

Subscripts

ch chip
T tool
W workpiece

1. INTRODUCTION

Metal cutting, or machining, is one of the most important methods of removing unwanted material in the production of mechanical components. Of all the processes used to shape metals, it is in machining that the conditions of operation are most varied. Almost all metals and their alloys—hard or soft, cast or wrought, ductile or brittle, with high or low melting point, are machined. The final shapes of most mechanical parts are obtained by machining operations [1].

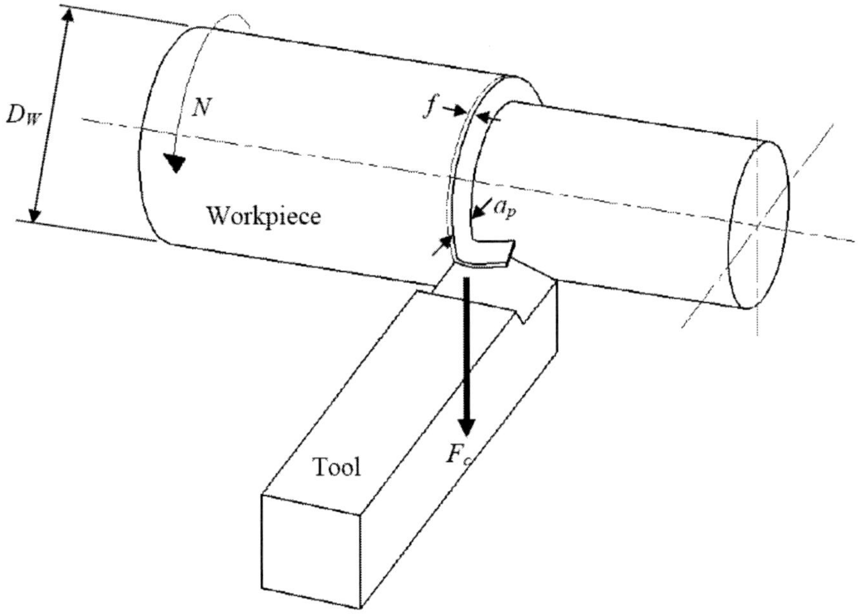

Figure 1. The turning process.

Turning is the most widespread metal cutting process used in manufacturing industry. It is also the most commonly employed operation in experimental work on metal cutting. The machine tool on which turning is accomplished is a lathe, where the workpiece is held in a chuck and rotated. The turning tool is held rigidly in the tool post and moved at a constant rate along the axis of a bar, cutting away a layer of metal to form a cylinder or a surface of a more complex profile. Figure 1 illustrates a turning operation, where in one revolution of the bar, the tool moves an axial distance f (the feed distance) to reduce the bar radius by an amount a_p (the depth of cut). Figure 1 also shows the cutting force F_c acting on the tool, the diameter D_W of the workpiece being cut, and the angular spindle speed N at which the bar rotates.

2. High-Speed Machining (HSM)

High-speed machining (HSM) has gained importance over the last few years following technological developments enabling its application. As a result of advances in machine tools and cutting tool technology, HSM has become a cost-effective manufacturing process for producing parts with high precision and surface quality. Until recently, high-speed machining was applied in machining aluminum alloys for manufacturing complicated parts used in the aircraft industry. This technology was successfully applied with significant improvements in machine tools, spindles and controllers [2]. Recently, with the advance of cutting tool technologies, HSM also has been employed for machining alloy steels for making dies/molds used in the production of a wide range of automotive components, as well as plastic molding parts [3].

High-speed machining (HSM) can be defined as machining at considerably higher cutting speeds and feed rates compared to those in traditional methods. HSM processes enable high material removal rates, low cutting forces, reduction in lead times, and improved part precision [4, 5]. The distinction between conventional and high-speed machining is based on the workpiece material being machined, type of cutting operation, and the cutting tool used [6]. Figure 2 shows the generally accepted cutting speeds in high-speed machining for various materials [6]. For instance, a cutting speed of 500 m/min is considered high-speed machining for cutting alloy steel, whereas this speed is considered conventional in cutting aluminum.

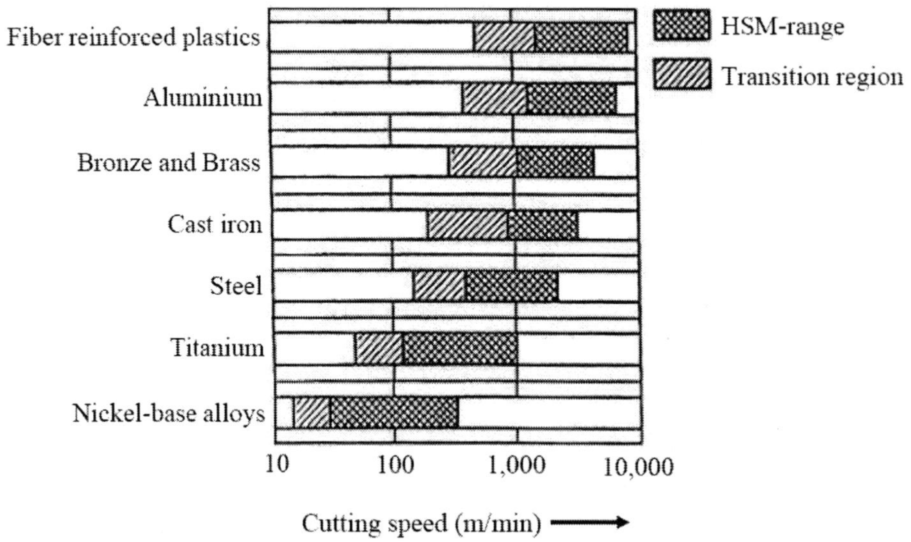

Figure 2. High-speed cutting range in machining of various materials [6].

Many advantages of HSM have been cited. The most common advantages reported are high metal removal rates, low cutting forces, minimal workpiece distortion, the ability to machine thin-walled sections, use of simple fixtures, little or no damage to workpiece surface integrity, reduction in cutting tool variety, burr-free components, and easier chip disposal [5-8]. Figure 3 shows some characteristics of the high-speed machining process [9]. Figure 3 also shows that with increasing cutting speed the cutting forces can be reduced and better surfaces can be generated, but that the tool life is reduced.

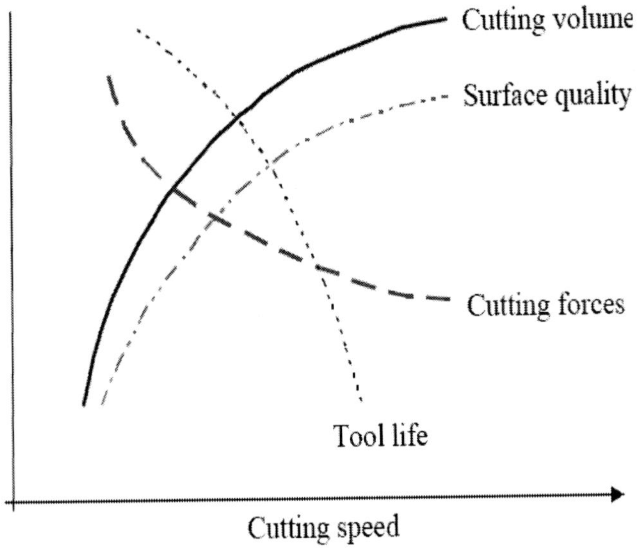

Figure 3. Characteristics of high-speed machining [9].

Recently, Denkena et al. [10] developed a mathematical model to predict high-speed cutting ranges based on cutting energy, tensile strength, and thermal effects. The model is based on the theory that cutting energy can be approximated as the sum of deformation energy and friction energy at the rake face. A power factor (k) was used to determine the relationship between the mechanical and thermal effects. The authors claimed that the model results, presented here in figure 4, were in good agreement with the experimental results.

Schulz [9] summarizes the advantages of high-speed machining given here in table 1. A detrimental effect of high-speed machining may be that the rate of tool wear increases; however, Schulz [9] reports that by optimizing the cutting parameters, contact conditions, the workpiece material, the cutting tool, and the machining strategy, the tool life can be extended. Hence, parallel to the increase in high-speed machining applications, there is also an increase in research on the development of new cutting tool materials, improved design of cutting tool inserts, and improvement of cutting process conditions. The common disadvantages of high-speed machining are also identified, and include high tool wear, the need for special and expensive machine tools with advanced spindles and controllers; fixturing, balancing the tool holder, and last but most importantly, the need for advanced cutting tool materials and coatings [11].

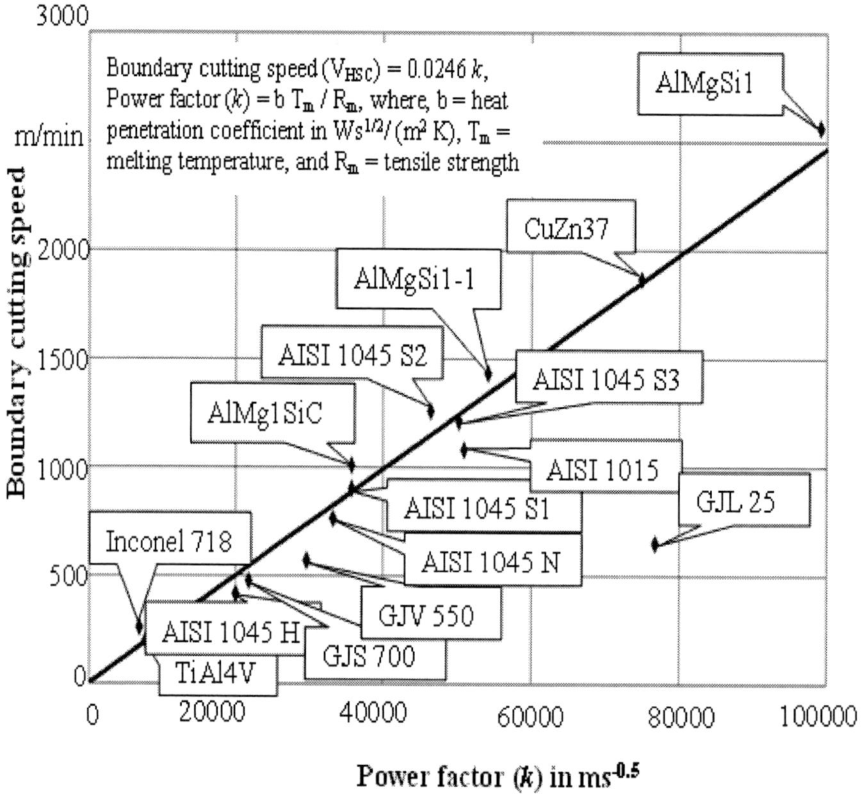

Figure 4. Influence of the power factor k on the boundary cutting speed (V_{HSC}) [10].

Table 1. Advantages and applications of high-speed machining [9]

Technological advantage	Application field	Application examples
Large cutting volume per machining time	Light metal alloys	Aircraft and aerospace production
	steel and cast iron	Tool and die mould manufacturing
High surface quality	Precision machining	Optical industry, fine mechanical parts
	Special workpiece	Spiral compressor
Low cutting forces	Processing thin-walled workpieces	Aircraft and aerospace products, automotive industry, house-hold equipment
High frequencies of excitation	Non-vibration machining of difficult-to-machine workpieces	Precision mechanics and optical industry
Heat transported by the chips	Non-warping machining	Precision mechanics
	Cold workpieces	Magnesium alloys

3. Tribological Tool Coatings

HSM requires special cutting tool materials engineered with extreme hot hardness, improved fracture toughness, as well as resistance to abrasion, thermal shock, adhesion, and diffusion. The use of advanced tool materials with sufficient heat isolation between the chip and the tool has resulted in the development of dry machining for typical cutting operations. Currently metal cutting surfaces have coatings of different structures containing layers of titanium carbide (TiC), titanium nitride (TiN), titanium carbonitride (TiCN), aluminum oxide (Al_2O_3) and titanium aluminum nitride (TiAlN) [12, 13]. Over the past decades, TiN coatings have gained increasing importance for cutting tools, dies, and many mechanical parts to increase their lifetime and performance. This increase is due to their attractive properties, such as high hardness, good wear resistance, and chemical stability [14, 15]. Recently, Zieberta and Ulrich [16] reviewed hard, multilayer coatings containing TiN and/or ZrN coatings. More recently, TiAlN coatings have been developed as hard coatings with excellent properties for high temperature use [17, 18].

Rapid improvements in coating deposition techniques and the development of more advanced coating materials have led to their wider application in many industries including automotive, aerospace, cutting tool, and precision manufacturing. The development of thin, hard and soft coatings deposited on high-speed steel, sintered carbides cermets, and ceramics has been hailed as a major innovation in cutting tool technology, resulting in significant improvements in cutting process performance [19-21]. Figure 5 shows the possible influence through contact conditions, interfacial effects, volume effects, and surface integrity, as well as the structural effects by which tool coating improves the wear resistance of the tool [22]. In this chapter, special attention is focused on the contact zone (friction, heat generation, and heat flow) in order to establish the desired properties of coating materials for specific machining tasks.

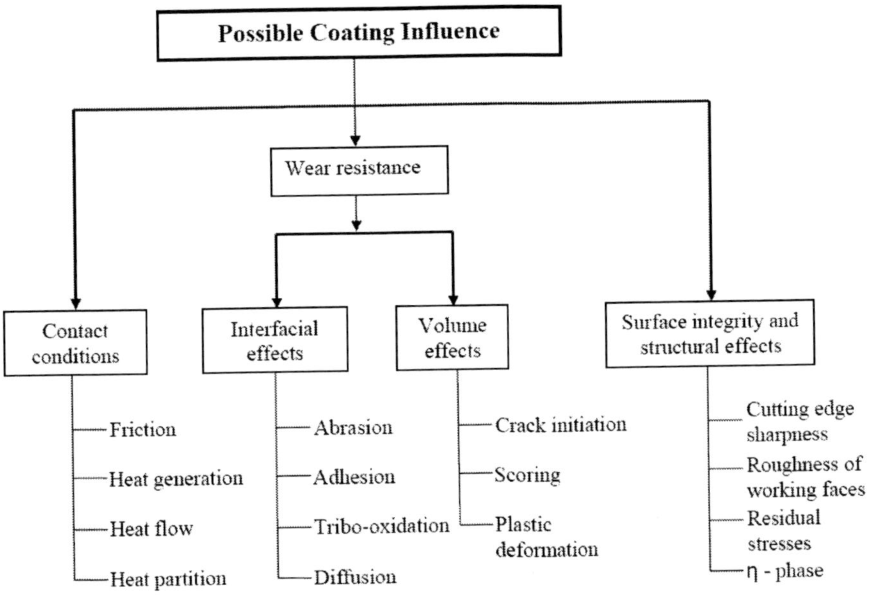

Figure 5. The effects of coatings in cutting process performance [22].

Generally, altering the properties of the tool material by depositing hard or hard/soft protective coatings substantially slows down some wear processes, reduces friction, prolongs tool life, improves hardness to combat the wearing action, and improves hot hardness to withstand the heat involved. As a result, more than 70 per cent of all cemented carbide inserts currently feature coatings applied by Physical Vapour Deposition (PVD), Chemical Vapour Deposition (CVD), or a combination of these deposition techniques [15, 23]. The commercial coatings most widely used in metal cutting manufacturing are TiC, Al_2O_3 (CVD), TiAlN (PVD), TiN, TiCN (CVD and PVD) [13, 23].

In general, cutting tool manufacturers use four major groups of hard coating components. These are as follows: (1) a group of titanium-based materials (TiN, TiC and TiCN often supplemented by Al or Cr [TiAlN, CrCN]), ceramic coatings like Al_2O_3, and (2) two special groups of super-hard coatings and solid lubricant coatings (MoS_2, MoST or pure graphite WC/C) [12, 13]. Such coatings are said to reduce the forces and power in machining due to lower friction at the tool-chip interface, which in turn reduces required cutting energy, contact temperatures, and tool wear. In addition, the increased resistance of hard coatings to abrasion and diffusion results in significant improvements in tool-life [19]. Yen et al. [24] report a rank matrix for different coatings in terms of their material properties such as hardness, hot hardness, chemical stability, and oxidation resistance, as shown in table 2.

Table 2. Rank matrix for different coatings in terms of major material properties [24]

	Chemical stability	Oxidation resistance	Hardness at 20 °C	Hot hardness
Best	Al_2O_3	Al_2O_3	TiC	Al_2O_3
	TiAlN	TiAlN	TiCN	TiAlN
	TiN	TiN	Al_2O_3	TiN
	TiCN	TiCN	TiAlN	TiCN
Poor	TiC	TiC	TiN	TiC

The Al_2O_3 coating has the best chemical resistance and hot hardness of all the hard coatings, but has poor layer adhesion and thermal shock resistance due to its high brittleness. For this reason, the TiAlN coating is used as a substitute for Al_2O_3 in cutting applications. Another unique property of the Al_2O_3 coating is its decreasing thermal conductivity with increasing temperature, which makes it suitable for high-speed (high-temperature) applications. Balaji et al. [25] report that TiC aids in reducing cratering, whereas TiCN aids in reducing cratering and also provides higher hardness. TiN aids in lowering interfacial friction by acting as a diffusion barrier.

Major improvements in the performance of metal cutting operations can be achieved only by fully understanding the physical potential provided by cutting tool coatings. Attention needs to be focused especially on the role of the tribological and thermal properties of the workpiece and the tool materials in achieving favourable tool-chip contact conditions. To this end, a more realistic model for practical machining applications would include temperature-dependent thermal properties as one of the process variables.

Most prior research [26-28] has, for given combinations of coating and workpiece materials, recommended rake face configurations in agreement with efficient chip control, coatings influence on chip-formation and the tribological interaction at the chip-tool contact area. From a tribological point of view, the most important factor is the influence of thermal properties of the tribo-pair components and the coating structure on the behaviour of the tool-chip contact. One of the difficulties in modelling coated tools is obtaining accurate and reliable readings for thermal properties, most importantly thermal conductivity of the coatings. It is not unusual to observe a wide range of data in the available literature. Figure 6 summarises the temperature-dependant thermal conductivity values of coating materials, collected from various literature sources [22, 24, 25, 29, 30]. Thermal conductivity values for tungsten carbide (WC) and all titanium-based coating materials specified in figure 6 increase steadily with temperature, but for Al_2O_3 based ceramics these values decrease rapidly. For example, at 1000 °C, thermal conductivity values are 26 and 5 W/m/K for TiN and Al_2O_3, respectively.

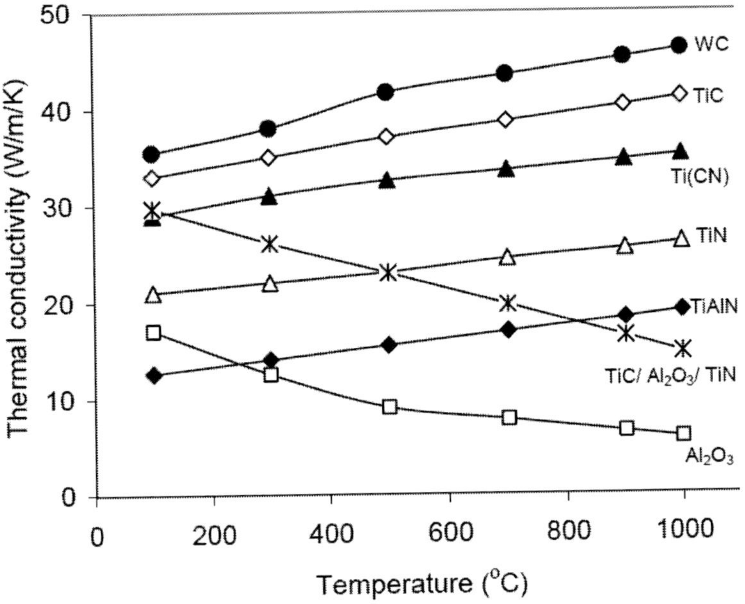

Figure 6. Thermal conductivity variation with temperature [22, 24, 25, 29, 30].

According to Grzesik [17], the maximum working temperature for cutting tools can be increased by depositing TiAlN based coatings because this hard layer maintains the hot hardness up to 800 °C. In contrast, the cutting speed for titanium nitride (TiN) coating, with a substantially lower hardness, should be reduced because its maximum working temperature is 600 °C. These values of maximum working temperature are established based on observations that TiN oxidises intensively at a temperature higher than 600 °C, whereas TiAlN shows a high oxidation resistance up to 800 °C. In addition [31], thermal effects caused by coatings are an important influence on the tribological behaviour between the cutting tool and the chip. The amount of thermal energy transferred into the chip body depends on the thermal conductivity of the coating film deposited on the cutting tool insert. When the thermal conductivity of the cutting tool is low, the temperature in the chip is high and consequently the fraction of heat entering the cutting tool is low. Furthermore, a direct correlation exists between the tool-chip contact length, the properties of the cutting tool, and the performance of the cutting process. For this reason, cutting tool materials with low thermal conductivity produce more curled chips because of a thermal bi-metallic effect.

4. Heat Generation in Metal Cutting Process

Metal cutting is a coupled thermo-mechanical process. During this process, heat generation occurs as a result of plastic deformation and friction along the tool-chip and the tool-workpiece interfaces. A schematic of the heat sources in orthogonal metal cutting is shown in figure 7. It can be safely assumed that almost all the energy required for cutting is converted to heat in the cutting domain [32]. This is because there is friction to be overcome. In a simple manner, the total work carried out can be divided into three quantities. Firstly, heat is generated in the primary deformation zone due to plastic work done at the shear plane. Secondly, heat is generated in the secondary deformation zone due to work done in deforming the chip and in overcoming the sliding friction at the tool-chip interface zone. Finally, the heat generated in the tertiary deformation zone, i.e,. the tool-workpiece interface, is due to the work done to overcome friction, which occurs at the rubbing contact between the tool flank face and the newly machined surface of the workpiece. Heat generation in this zone is a function of tool flank wear. The temperature rise in the cutting tool is mainly due to the secondary heat source, but the primary heat source also contributes towards the temperature rise of the cutting tool and indirectly affects the temperature distribution on the tool rake face. The fraction of heat that flows into the cutting tool, which is stationary relative to the heat source, is based on the determination of heat partition coefficient (R_T), which defines the percentage of heat entering the cutting tool (figure 7). The fraction $R_{ch} = 1 - R_T$ defines the heat energy going into the moving chip.

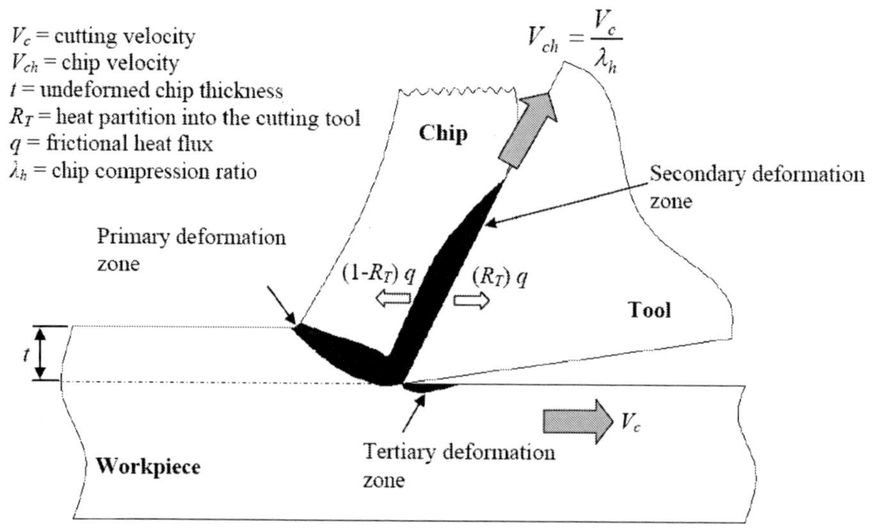

Figure 7. Illustration of heat generation zones in orthogonal metal cutting.

During the cutting process, part of the heat generated at the shear plane flows by convection into the chip and then through the interface zone into the cutting tool. Therefore, the heat generated at the shear zone affects the temperature distributions of both the tool and the chip sides of the tool-chip interface, and the temperature rise on the tool rake face is due to the combined effect of the heat generated in the primary and secondary zones. The work done will depend on the material being cut. Ductility, hardness, work hardening and thermal properties all have definite effects on the tool forces, as well as affecting the temperature. These properties, in addition to the tool material characteristics, cutting conditions, contact phenomena, and geometry of the process, will define the power and therefore the heat generation.

As has been shown [33-35], the heat generated at the tool-work interface is of major importance in relation to tool performance, and is particularly critical in limiting the rates of metal removal. Along the greater part of the contact between the chip and the tool, there is a seizure zone where these two materials are in close contact. There is no relative movement between the chip and the tool at the surface of the tool and, depending on the cutting speed, a flow zone in the chip with high strains and strain rates or a built-up edge (for low cutting speed) is established. Contact between the workpiece material and the flank face of the tool will occur even for a tool without wear and may contribute to an increase in the workpiece temperature. At conventional cutting speeds, the largest part of heat is dissipated in the chip and only a low percentage of heat flows towards the tool, as well demonstrated [36].

The heat generated at the tool-work interface will is lost to the tool holder, which can affect the dimensional accuracy of the workpiece. Depending on the temperatures achieved and the material of the tool holder [37], the accuracy of the operation is affected. The increase in the temperature of the workpiece is due to the work done in the primary deformation zone and the tertiary deformation zone.

It follows from the above discussion that two aspects must be carefully taken into account with regard to heat flow analysis: namely the evaluation of the heat partition between the tool and the chip, and the tool-chip contact phenomena.

4.1. Evaluation of Heat Generation in Metal Cutting Processes

As mentioned above, nearly all the work that takes place in the process of metal cutting is converted into heat. The total heat generated, Q, in a cutting operation is the product of the cutting force, F_c and cutting speed V_c, given as equation (1).

$$Q = F_c V_c \tag{1}$$

Q is the sum of the heat generated on the shear plane Q_{sh}, and in moving the chip over the tool Q_f.

$$Q = Q_{sh} + Q_f \tag{2}$$

The amount of heat generated due to work done in the secondary deformation zone along the rake face is given by the following equations (3) or (4).

$$Q_f = F_{fr} V_{ch} \tag{3}$$

or

$$Q_f = \frac{F_{fr} V_c}{\lambda_h} \tag{4}$$

Where V_{ch} is the chip velocity, V_c is the cutting velocity, F_{fr} is the frictional force at the rake face, and F_{fr} is calculated from equation (5), as shown in figure 8.

$$F_{fr} = F_c \sin\gamma + F_f \cos\gamma \tag{5}$$

where, λ_h is the chip compression ratio, F_f is the feed force and γ is the rake angle.

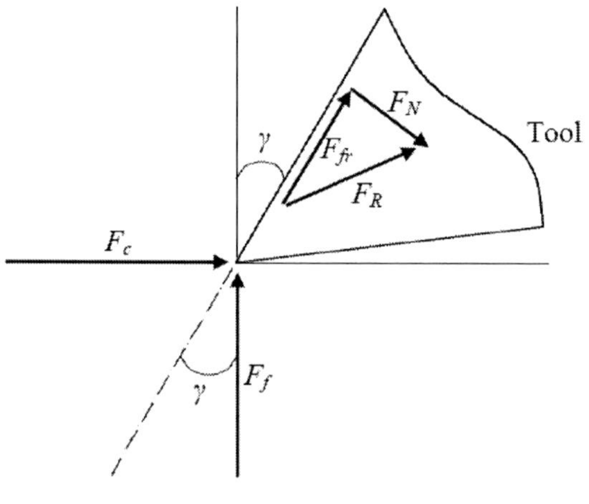

Figure 8. Forces acting on the considered system in orthogonal cutting (F_c = cutting force, F_f = feed force, F_{fr} = frictional force, F_N = normal force on tool face, F_R = resultant tool force and γ = rake angle).

4.2. Estimation of Non-Uniform Heat Flux Along the Tool-Chip Interface

In machining, contact phenomena can be characterised as a sticking zone and a sliding zone. The distinction between the two is highly dependent on the tool and workpiece materials, as well as cutting conditions. In the sticking zone, the shear stress (τ_{sh}) can be assumed to be constant along the entire sticking (seizure) contact length, up to $x = l_{st}$, where x is the distance from the cutting tool edge. In the sliding zone, this shear stress reduces to zero at the point at which the chip departs from the tool rake face. According to Wright et al. [38], the shear stress in the sticking region, τ_{sh} can be computed from equation (6).

$$\tau_{sh} = \frac{F_{fr}}{a_p(l_{st} + 0.5 l_{sl})} \tag{6}$$

where l_{st} is the sticking length of the tool-chip contact zone, l_{sl} is the sliding length of the tool-chip contact zone, a_p is the depth of cut, and F_{fr} is the frictional force that can be calculated from equation (5). The heat flux $q(x)$ along the sticking zone is defined as $\tau_{sh}V_{ch}$ ($x \leq l_{st}$), which then linearly decreases to zero along the sliding contact zone ($l_{st} < x \leq l_{ch}$), as shown in figure 9. This is the heat generation model used in this study for thermal modelling of the tool-chip interface. It follows the works of Tay et al. [39] and Wright et al. [38], who also treat the heat distribution in the sliding region as linear. Recently, Huang and Liang [40] used a similar methodology to evaluate the non-uniform heat intensity, but adopted sticking and sliding length values from other studies. An important input variable used in evaluating the heat intensity at the tool-chip interface is the chip velocity (V_{ch}), which is computed using equation (7).

$$V_{ch} = \frac{V_c}{\lambda_h} \tag{7}$$

where V_c is the cutting speed and λ_h is the chip compression ratio, which is defined as:

$$\lambda_h = \frac{t_c}{t} \tag{8}$$

where t_c is the actual chip thickness and t is the undeformed chip thickness.

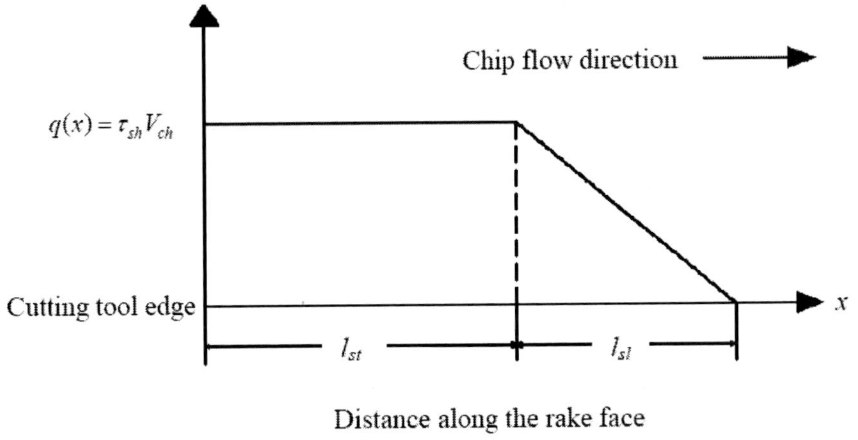

Figure 9. Non-uniform heat intensity model along the rake face of the cutting tool according to Wright et al. [38] and Tay et al. [39].

5. A Review of Previously Reported Values of Heat Partition between the Tool and the Chip

Many dynamically loaded engineering components and tribological applications involve the relative sliding of two bodies with thermal energy generated at the contact region due to either plastic deformation or friction, e.g., heat generated in machining due to plastic deformation in the primary shear zone or the sliding of the chip past the tool face resulting in frictional heat at the interface. The thermal aspects around the contact region can be studied by examining the heat transfer between a stationary body and a moving body (relative to the heat source). In metal cutting operations, temperature elevates at the tool-chip interface and the heat transferred from this zone is crucial in determining key process issues such as tool wear, tool life, and surface integrity. These influences are predominant in dry machining at high speeds, which becomes an economical and ecological imperative. While heat generation depends mainly on process parameters and the machinability rate of the work material, the thermophysical properties of the cutting tool material determine the distribution of temperature fields and heat dissipation [1].

The heat that flows into the cutting tool during dry turning is an important factor which influences tool wear mechanisms, tool performance, and quality of the machined part. Blok's principle [41] has been widely used in the analytical investigation of temperatures generated in metal cutting. It models two bodies, one stationary and the other moving with a relative velocity. This principle can be employed to solve the problem of heat partition at the tool-chip interface by matching the temperature between the chip and the tool. Chao and Trigger [42] used Blok's partition principle but were not successful in achieving an interface temperature rise on the chip in agreement with that on the tool. Komanduri and Hou [43] furthered the functional analysis approach based on the idea of Chao and Trigger [42]. Huang and Liang [40] used a non-uniform heat intensity along the tool-chip interface in their model and determined heat partition by using the length of sticking and sliding zones from other studies. Liu and Chou [44, 45] used uniform heat flux along the tool-chip interface in their model and determined heat partition without considering sticking and sliding phenomena at the tool rake face. Yvonnet et al. [46] used the simple inverse procedure to determine the convection heat transfer coefficient and the average heat flux flowing into the region of the tool closer to the tool tip. These authors [46] also used a piecewise constant function to calculate the approximate heat flux distribution. Karpat and Ozel [47, 48] determined heat partition based on a non-uniform heat flux and calculated the actual lengths of sticking and sliding zones empirically for a carbide tool.

Francis [49] developed an analytical solution for the steady-state temperature distribution within a sliding contact. This author considered a circular disk heat source with an elliptic heat intensity distribution at the interface, and approximated the temperature field in the circular contact area by half the harmonic mean of the two temperature fields, which were the temperature of the moving and stationary bodies. However, the author also assumed that the true value of the heat partition will be the one that best satisfies the condition that the temperature fields at the two surfaces are identical over the contact area.

Bos and Moes [50] investigated heat partition for circular and elliptic heat sources and developed a numerical approach to solve the steady-state heat partition and the associated temperatures for arbitrarily shaped contacts by matching the surface temperature of two contacting solids at points in the contact area. Komanduri et al. [51] considered heat partition between the stationary and the moving bodies in sliding contact using the integral transform method. Gecim and Winer [52] investigated steady-state temperatures in a rotating cylinder at high-speeds (large Peclet numbers) subjected to surface heating and convective cooling.

Reported numerical values for heat partition in machining are summarised in table 3. According to Loewen and Shaw [53], heat transmitted into the tool varies between 40% and 20% when machining SAE B1113 free cutting steel with a K2S carbide tool at cutting speeds of 30 to 182 m/min. Takeuchi et al. [54] have reported that 10% to 30% of the total heat generated enters the tool. They applied this assumption in machining of carbon steel (C = 0.55%) using P15 carbide tooling at a cutting speed of 100 m/min. According to Wright et al. [38], 10% to 20% of heat generated at the rake face flows into the tool when machining annealed low carbon iron with M34 high-speed steel at cutting speeds ranging from 10-175 m/min. According to Ghani et al. [55], heat transmitted into the tool for conventional cutting speed is 22% and for high-speed machining region it is around 14% when machining H13 tool steel with a Cubic Boron Nitride (CBN) cutting tool. The authors used uniform heat flux along the tool-chip interface in their model without Considering sticking and sliding phenomena at the tool rake face.

Trigger and Chao [56] predicted that the fraction of heat going into the tool varies from 5.5% to 32% when machining NE 9445 steel with cemented carbide cutting tools at cutting speeds of 30.48 to 182.88 m/min. Stephenson et al. [57] report that about 17% to 24% of the heat generated enters the tool when machining gray cast iron and 2024 aluminium. According to Trent [58], the proportion of heat entering the tool is 3%. Wang and Liu [59] use Blok's heat partition principle to calculate the heat partition into the cutting tool. They estimate that the fraction of heat going into the tool vary from 15% to 30% when machining AISI 52100 steel with CBN cutting tool at a cutting speed of 100 m/min. They conclude that the fraction of heat that goes into the chip increases when flank wear becomes larger.

Casto et al. [60] report that between 56% and 24% of heat is transmitted into the tool when machining AISI 1040 with a P10 sintered carbide insert at cutting speeds of 99 to 240 m/min. Abukhshim et al. [61] evaluate the heat partition coefficient for the tool in machining of AISI/SAE 4140 steel using P20 carbide tools to vary between 50% and 15% in conventional machining at cutting speeds of 200 to 600 m/min. Grzesik and Nieslony [62] estimate that the fraction of heat going into the tool varies from 35% to 20% when using multilayer coated tools, but changes from 50% to 40% for uncoated carbide tools in the range of cutting speeds of 50 to 210 m/min for the workpiece material AISI 1045.

According to Komanduri and Hou, the heat partition fraction is 10% into the tool and 90% into the chip [43]. Recently, Cotterell and Byrne [63] use Shaw's analytical solution to calculate heat partition into the cutting tool and chip when machining of titanium alloy Ti-6Al-4V with WC insets (ISO Grade S15) for cutting speeds from 4 to 120 m/min. The maximum value of heat partition into the chip was found to be 20% and the lowest value was negative when cutting at speeds below 12.5 m/min. They conclude that this represents a situation where heat generated at the chip in the shear zone flows into the tool and results in a temperature drop at the tool-chip interface.

Table 3. Reported values of heat partition into the cutting tool (R_T) for specified cutting velocities, workpiece, and cutting tool combinations

Authors and reference	Workpiece material	Cutting tool material	Cutting velocity (m/min)	Heat partition into the cutting tool (R_T)
Loewen and Shaw [53]	SAE B1113 Steel	K2S carbide	30 to 182	40% down to 20% (Calculated analytically)
Takeuchi et al. [54]	Carbon steel (C = 0.55%)	P15 carbide	100	10% to 30% (Assumed)
Wright et al. [38]	low carbon iron	M34 high-speed steel	10-175	10% to 20%
Casto et al. [60]	AISI 1040	Sintered carbide type P10	99 to 240	56% down to 24%
Grzesik and Nieslony [62]	AISI 1045	Multilayer coating	50 to 210	35% down to 20% (Calculated analytically)
Grzesik and Nieslony [62]	AISI 1045	P20 uncoated carbide	50 to 210	50% down to 40% (Calculated analytically)
Shih [64]	AISI 1020	High-speed steel	35	50% (Assumed)
Lei et al. [65]	AISI 1020	tungsten carbide	240	50% (Assumed)
Xie et al. [66]	AISI 1045	Uncoated carbide WC-Co	300	50% (Assumed)
Arrazola et al. [67]	AISI 4140	C. C. grade P10	300	50% (Assumed)
Arrazola et al. [68]	AISI 4140	P10	200-600	50% (Assumed)
Abukhshim et al. [61]	AISI 4140	Uncoated carbide	200 to 600	46% down to 15%
Reginaldo et al. [69]	AISI 4340	Polycrystalline cubic boron nitride (PCBN) substrate with TiAlN coating	150	50% (Assumed)
Shet and Deng [70]	AISI 4340	-	152	50% (Assumed)
Shet et al. [71]	AISI 4340	-	152	50% (Assumed)
Shi et al. [72]	AISI 4340	-	152	50% (Assumed)
Shet and Deng [73]	AISI 4340	-	152	50% (Assumed)
Shi and Liu [74]	AISI 52100	PCBN	182	50% (Assumed)
Shi and Liu [75]	AISI 52100	SPG-422 (PCBN)	107	50% (Assumed)

Kar and Melkote [76]	AI5083-H116	Single crystal and Polycrystalline diamond tools	10 and 200	50% (Assumed)
Shi and Liu [77]	HY-100	TPG 322 inserts	107	50% (Assumed)
Li et al. [78]	Oil hardening O1 steel	Flat and two types of cratered tools	244	50% (Assumed)
Mabrouki and Rigal [79]	AISI 4340	Tungsten carbide	100	65% (Calculated analytically)
Ng et al. [80]	AISI H13	PCBN	75, 150 and 200	48% and 59% (Calculated analytically)
Ghani et al. [83]	AISI H13	CBN cutting tool	144.26 and 288.52	22 to 14%
Zong et al. [81]	OFHC copper	diamond tool	6, 30, 60 and 120	72% (Calculated analytically)
Mohamed et al. [82]	AISI 316L	Kennametal K313	125	60% (Calculated analytically)
Authors and reference	Workpiece material	Cutting tool material	Cutting velocity (m/min)	Heat partition into the cutting tool (R_T)
Trigger and Chao [56]	NE 9445 steel	cemented carbide	30.48 and 182.88	5.5% to 32%
Wang and Liu [59]	AISI 52100 steel	Cubic Boron Nitride (CBN)	100	15% to 30%

A better understanding of heat partitioning between the tool and the chip is required in order to produce more realistic Finite Element (FE) models of machining processes, with the objective of using FE models to optimise the cutting process for longer tool life and better surface integrity. From table 3, it is noted that for FE models, most researchers assume the value of heat partition into the cutting tool and chip as 50% [64-78]. Mabrouki and Rigal [79] calculate heat partition into the cutting tool as 65% and use this value in their simulations with ABAQUS software to model the turning of AISI 4340 steel. Ng et al. [80] use heat partition into the cutting tool as 48% and 59% and use these values in their simulations with the code FORGE2 to model the turning of hardened AISI H13, in which continuous chips were obtained. Zong et al. [81] calculated heat partition into the diamond tool as 72% and use this value in their simulations with the updated lagrangian formulation to simulate the turning of OFHC copper. Mohamed et al. [82] calculated heat partition values analytically between the tool and the chip to be 60% and 40% respectively and use these values in their simulations with ABAQUS to model the turning of AISI 316L steel.

A critical review of the available literature reveals that whilst FEM codes are widely utilised nowadays in metal forming companies, there is still a lack of relevant knowledge regarding heat partition into the cutting tool at the tool-chip interface and heat transfer conditions, which seriously limits their application to the design of cutting processes.

From a review of literature, it is clear that two aspects must be carefully taken into account for chip formation and temperature distribution—namely, the evaluation of the interface heat partition coefficient between the chip and the tool, and temperature dependant thermal properties of the tool and the workpiece. Accurate heat partition prediction is challenging due to the complexity of the contact phenomena in the cutting process. Table 3 indicates that most researchers [64-78] assume heat partition values in their analysis for machining applications, but these values differ considerably from experimental values. Based on these considerations, it is possible to postulate that a consistent model that is able to describe the temperature evolution in a tool-chip system, must consider the following:

1. An accurate procedure for the determination of the heat partition coefficient between the chip and the tool along the tool-chip interface.
2. An accurate dataset of mechanical and temperature-dependant thermal properties of the tool and the workpiece materials.
3. An accurate quantitative determination of sticking and sliding lengths on the tool rake face.

From these considerations, it may be concluded that an effective and consistent numerical procedure for the coupled thermo-mechanical analysis is possible with proper input parameters.

6. Existing Analytical Models for The Prediction of Heat Partition

Prediction of the fraction of heat that flows into the chip for a body with a moving heat source is based on a determination of the heat partition coefficient (R_{ch}), which defines the percentage of the heat entering the moving chip. The fraction $R_T = 1 - R_{ch}$ defines the heat energy going into the tool, which is stationary relative to the heat source. The analytical methods used to estimate the heat partition depend on the frictional heat flux distribution, real tool-chip contact area, the velocity of the moving heat source, and the thermo-physical properties of the tool and workpiece materials. In the literature, seven different methods are reported for the analytical calculation of heat partition coefficient. The following sections review the heat partition models and give the expressions for directly computing the heat flux into the cutting tool, which are briefly described below:

6.1. Loewen and Shaw Heat Partition Model

Loewen & Shaw's model [53] incorporates two modifications to the classical solutions of Jaeger's [84] moving band (for the chip) and stationary rectangular (for the tool) heat sources for application to metal cutting. It takes into account appropriate boundaries (besides the tool-chip contact interface) and considers non-uniform distribution of the heat partition fraction along the tool-chip interface for the purpose of matching the temperature distribution both on the chip side and on the tool side. For metal cutting, the frictional interface heat source relative to the chip is considered as a moving band heat source. Finally, according to Loewen & Shaw, the fraction of heat going into the chip R_{ch} can be calculated according to equation (9).

$$R_{L(ch)} = \frac{\dfrac{q\,l_{ch} A_a}{\lambda_T} - \Delta\theta_S + \theta_o}{\dfrac{q\,l_{ch} A_a}{\lambda_T} + \dfrac{0.377 q\,l_{ch}}{\lambda_W \sqrt{N_p}}} \tag{9}$$

where, λ_W and λ_T are the thermal conductivities of workpiece and tool materials respectively, q is the frictional heat flux, N_p is a non-dimensional parameter, θ_o is the initial temperature of the tool, and A_a is the area shape factor which describes the thermal and geometric features of the interface. The non-dimensional parameter N_p [53] is estimated by equation (10).

$$N_p = \frac{V_{ch} \dfrac{l_{ch}}{2}}{2\alpha_W} \tag{10}$$

and the area shape factor A_a is given by equation (11) [53].

$$A_a = \frac{2}{\pi}\left[\sinh^{-1}\left(\frac{m}{l}\right) + \left(\frac{m}{l}\right)\sinh^{-1}\left(\frac{l}{m}\right) + \frac{1}{3}\left(\frac{m}{l}\right)^2 + \frac{1}{3}\left(\frac{l}{m}\right) - \frac{1}{3}\left\{\left(\frac{l}{m}\right) + \left(\frac{m}{l}\right)\right\}\left\{1 + \left(\frac{m}{l}\right)^2\right\}^{0.5}\right] \tag{11}$$

where, $m/l = a_p/l_{ch}$ [53], $m = a_p$ is the depth of cut and l_{ch} is the tool-chip contact length. The average shear plane temperature rise ($\Delta\theta_s$) due to plastic deformation in the primary deformation zone is calculated based on the theory of similarity, elaborated for metal cutting by Silin [85]. In this approach, first the maximum temperature rise in the shear plane, θ_{smax}, and then the average temperature occurring at the shear plane are calculated. Grzesik and Neilsony [86] recently acknowledged that the theory of similarity facilitates the calculation of this temperature more accurately than the Oxley [87] model. The maximum temperature rise in the shear plane can computed by equation (12) [85].

$$\theta_{s\max} = \frac{\tau_{sh}}{C_p \rho \tan\phi} erf \sqrt{\frac{R_t \tan\phi}{4}} \tag{12}$$

where, τ_{sh} is the shear stress, ϕ is the shear angle, and erf is the error function. The non-dimensional thermal number R_t is calculated using equation (13) [88, 89].

$$R_t = \frac{\rho C_p V_c t}{\lambda_W} \tag{13}$$

where, t is the undeformed chip thickness. The average temperature rise at the shear plane is then calculated by the approximate expression in equations (14a) to (14c) [85].

$$\Delta\theta_s = 0.685\,(R_t \tan\phi)^{0.07}\,\theta_{s\max} \quad (\text{for } R_t \tan\phi \le 5) \tag{14a}$$

$$\Delta\theta_s = 0.620\,(R_t \tan\phi)^{0.13}\,\theta_{s\max} \quad (\text{for } 5 < R_t \tan\phi \le 20) \tag{14b}$$

$$\Delta\theta_s = 0.820\,(R_t \tan\phi)^{0.04}\,\theta_{s\max} \quad (\text{for } R_t \tan\phi > 20) \tag{14c}$$

6.2. Reznikov Heat Partition Model

Reznikov [90] proposed a model for the amount of heat flowing into the chip using the thermal conductivity and thermal diffusivity values of the tool and workpiece materials according to equation (15)

$$R_{R(ch)} = \frac{1}{1 + [(3\lambda_T / 2\lambda_W)\sqrt{(\alpha_W / \alpha_T)}]} \tag{15}$$

where, λ_W and λ_T are the thermal conductivity values of the workpiece and the tool materials respectively, and $\alpha_W = (\lambda / \rho C_p)_W$ and $\alpha_T = (\lambda / \rho C_p)_T$ are the corresponding thermal diffusivity values.

6.3. Gecim and Winer Heat Partition Model

Gecim and Winer [91] studied transient temperatures in the vicinity of an asperity contact for a small stationary circular heat source. The heat source was assumed to be acting on the surface of a much larger body which could be considered as semi-infinite. The authors used successive integral transforms to obtain solutions for the partial differential equation of heat conduction. The study was intended to represent the thermal behaviour of a single asperity in an apparent area of contact. The authors derived the heat partition coefficient for the tool by equating the average temperatures of the moving and stationary circular heat sources, given as:

$$R_{GW(T)} = \frac{\lambda_T}{\lambda_T + 0.807\, b_W \sqrt{r_o V_{ch}}} \tag{16}$$

where b_W is the heat transmission coefficient, which can be calculated from equation (17) as:

$$b_W = \sqrt{(C_P \rho \lambda)_W} \tag{17}$$

λ_T is the thermal conductivity of the tool, r_o is the heat source radius. In this article, r_o is assumed to be equal to the contact length for simplicity because, as Gecim and Winer [91] point out, square and circular heat sources having the same characteristic dimensions give similar results.

6.4. Shaw Heat Partition Model

Shaw [92] proposed that the slider is not a perfect insulator and part of '*q*', i.e. ($R_{ch}\, q$), will flow into the chip and ($1-R_{ch}$) q will flow into the tool, so that the percentage of heat flowing into the chip is given by equation (18) as:

$$R_{SH\,(ch)} = \frac{1}{1 + [0.754\,(\lambda_T / \lambda_W)/(A_a \sqrt{N_T})]} \tag{18}$$

where, λ_W and λ_T are the thermal conductivity values of the workpiece and the tool materials, respectively, N_T is a dimensionless velocity diffusivity factor and A_a is the area shape factor, which describes the thermal and geometrical features of the interface, as evaluated from equation (11). The dimensionless velocity diffusivity factor N_T is given by equation (19) [92].

$$N_T = \frac{V_{ch}\, l_{ch}}{2\alpha_W} \tag{19}$$

Thus, from equation (18), '$R_{SH(ch)}$' is seen to be a function of three non-dimensional quantities: a geometric factor (A_a), a ratio of thermal conductivities (λ_T/λ_W), and a velocity-diffusivity factor (N_T).

6.5. Berliner and Krainov Heat Partition Model

Berliner and Krainov [93] developed a model for the amount of heat flowing into the cutting tool using thermal properties of the tool and workpiece materials, chip velocity, and contact length, given by equation (20) as:

$$R_{BK(T)} = 1 - \frac{1}{1 + 0.45 \dfrac{\lambda_T}{\lambda_W} \sqrt{\dfrac{\pi \, \alpha_W}{V_{ch} \, l_{ch}}}} \tag{20}$$

where, λ_W and λ_T are the thermal conductivity values of the workpiece and the tool materials respectively, α_w is the thermal diffusivity of the workpiece material, and l_{ch} is the tool-chip contact length. All thermal properties are temperature dependent. Recently, Li and Shih [94] used this model with a tungsten carbide cutting tool in their transient finite element analysis.

6.6. Tian and Kennedy Heat Partition Model

Tian and Kennedy [95] investigated the maximum and average flash temperatures in sliding contacts for a wide range of Peclet numbers. Using Blok's postulate [41] of heat partition between two contacting bodies, they analysed the interface temperature rise by matching the maximum surface temperature at the real area of contact of two contacting bodies and assumed a uniform square heat source sliding over a semi-infinite body. Tian and Kennedy developed the heat partition model given in (21) as:

$$R_{TK(ch)} = \frac{1}{1 + \dfrac{\lambda_T}{\lambda_W} \sqrt{\dfrac{1 + Pe_T}{1 + Pe_W}}} \tag{21}$$

where, λ_T and λ_W are the thermal conductivity values of the tool and the workpiece materials respectively and Pe_T and Pe_W are the Peclet numbers for the tool and the workpiece, respectively. In fact, the Peclet number is the same as the non-dimensional parameter in Loewen and Shaw's model. These Peclet numbers for the tool and the workpiece are computed from equation (22) [95] as:

$$Pe_T = \frac{V_c \frac{l_{ch}}{2}}{2\,\alpha_T} \text{ and } Pe_W = \frac{V_{ch} \frac{l_{ch}}{2}}{2\,\alpha_W} \tag{22}$$

6.7. Kato and Fujii Heat Partition Model

Kato and Fujii [96] determined the heat partition coefficient for conventional surface grinding based on the values of the heat transmission coefficients (triple product of C_p, λ and ρ) for the workpiece and the grinding wheel materials. Their proposed expression was originally defined for conventional surface grinding, of which a modified version was used. Grzesik and Nieslony [62] used a similar approach in metal cutting with uncoated and multilayer coated tools. By analogy, the heat partition coefficient for given thermal properties of workpiece (W) and tool (T) materials, is given by equation (23) as:

$$R_{KF(ch)} = \frac{1}{1+\sqrt{(C_p\rho\lambda)_T / (C_p\rho\lambda)_W}} \tag{23}$$

where, C_p is the specific heat, λ is the thermal conductivity, and ρ is the density of the tool or workpiece material.

By further manipulating the heat partition models mentioned above, the fraction of heat going into the tool, R_T, can be directly computed as $R_T = 1 - R_{ch}$. These models are assessed below for the prediction of heat partition in orthogonal turning using cemented carbide tools.

7. Some Comments on the Existing Heat Partition Models and Their Limitations

1. Some of the heat partition models described are based purely on the thermal properties of the tool and the workpiece materials (for example, Reznikov [90] and Kato and Fujii [96] heat partition models). While this methodology has been used for conventional machining, its relevance to HSM has not been validated.
2. These models do not account for the influence of the cutting speed and frictional heat flux on heat partition, with the exception of the Loewen and Shaw model [53]. Indeed, most of the studies on steel were performed at cutting speeds of less than 100 m/min with some studies at 300 m/min, all of which are in the conventional cutting range where the effect of speed may be neglected.
3. The tool-chip contact length (contact phenomenon), in Reznikov [90] and Kato and Fujii [96] heat partition models, was found to have no influence on heat partition. However, other factors such as tool nose radius, wear, and cutting speed also affect the contact length. Thus, heat partition may be critical in HSM.

4. The coating thickness in all heat partition models was found to have no influence on heat partition. However, this factor can significantly affect heat partition into the cutting tool.
5. The main concern about Block's procedures [41] is the unrealistic assumption of a uniform distribution of heat flux over the contact region, especially when considering the existence of sticking and sliding zones.
6. Most of these models assume steady-state conditions when partitioning the heat generated in the primary and secondary zones. This means that the amount of heat flowing into the process components is constant during the process. This leads to an underestimation of heat partition into the cutting tool, especially in the transient regime.
7. There is a lack of consensus in the methodology for calculating heat partition and the effect of the remote primary heat source on the rake face temperature field. In HSM, it could be argued that the low interaction time diminishes the significance of the primary heat source on rake face temperatures.
8. In general, there is still a lack of an accurate heat partition model that uses the experimental data.

Some factors discussed above may lead to the overestimation, and some may lead to the underestimation, of heat partition and temperature distribution. The estimated heat partition and temperature profiles may be higher or lower than the measurements, depending on how strong these factors are. A combination of these factors determines the final error. While these models are useful in the qualitative assessments of the thermal properties of the tool and workpiece materials and of the cutting variables on heat partition into the cutting tool, they do not yield quantitative agreements. In fact, for the models reviewed here, previous research suggests that heat partition is independent of the cutting speed and contact phenomena. This is the prime reason further experimental investigation is required to elucidate the effects of cutting speeds and contact phenomena (with coated tools) on heat partition into the cutting tool. It is obvious that the issue of heat partition into the cutting tool needs a more scientific study, in particular, for HSM.

8. Finite Element Modelling

A commercial finite element software ABAQUS/Standard (6.6-1) [97] has been used to solve the transient heat transfer problem and simulate heat transfer into the cutting tool insert. The element used to model the whole cutting assembly was a four-node tetrahedral heat transfer element DC3D4. The tool was modelled as a perfectly rigid body, as tool materials have significantly high elastic moduli. The analysis was based on the estimated but realistic uniform and non-uniform heat flux where appropriate, applied over the experimentally evaluated tool-chip contact area on the rake face of the cutting insert.

The heat transfer taking place in the tool-chip-work system under transient heat conduction in three dimensions is governed by the partial differential equation (24) as

$$\frac{\partial^2\theta}{\partial x^2}+\frac{\partial^2\theta}{\partial y^2}+\frac{\partial^2\theta}{\partial z^2}+\frac{q(x,y,z,t)}{\lambda_T}=\frac{1}{\alpha_T}\frac{\partial\theta}{\partial t} \tag{24}$$

where α_T and λ_T are thermal diffusivity and thermal conductivity values for the tool respectively, and θ is the temperature. During the cutting process, some heat penetrates into the shim seat, the clamping parts, and the holder; and the computational domain cannot be considered as the cutting insert only. Therefore, models were developed for uncoated cemented carbide, TiN (5 μm) and TiAlN (5 μm) inserts, which consisted of the shim seat, the tool holder, and the clamp screw. Small details of the clamping screw parts were neglected in the models due to modelling limitations of the analysis package. Figure 10 shows the finite element mesh with refinement of the mesh near the cutting edge.

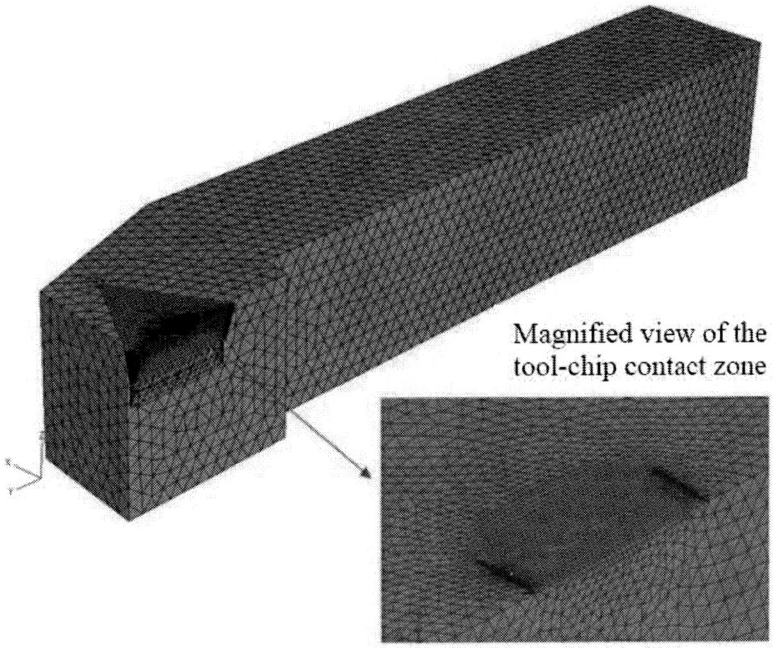

Figure 10. Finite element mesh for the assembly tool, shim, and tool holder in ABAQUS.

The heat generation module available in ABAQUS/Standard was employed for heat partition between the tool and the chip. Heat partition into the cutting tool ($R_{T\ (FEM\text{-}EXP)}$) was obtained using the hybrid FEM-experimental approach as described by the authors in [**4**]. FEM was used as a tool in an inverse procedure aimed to identify the percentage of heat flowing into the cutting tool.

The algorithm was based on the following steps:

1. Compute for each experimental test the non-uniform heat intensity value at the tool-chip interface.
2. Apply 100% of the heat flux in the FE model (uniform or non-uniform depending on contact phenomenon) as thermal load on the tool-chip contact area.

3. Reduce the applied heat flux until the measured and simulated temperatures are matched at three specified points. This value is taken as heat partition into the cutting tool.

The following boundary conditions were applied to the model for thermal analyses:

- The cutting insert, shim seat, and tool holder interaction surfaces were assumed to be smooth and held together with a rigid clamping system and were assumed to be in perfect contact.
- Initially the whole model was set at room temperature as 25 °C.
- For non-interactive surfaces of the cutting insert, shim seat, and workpiece, heat losses due to heat convection were calculated using a convective heat transfer coefficient $h = 20$ W/m^2/K [98].

8.1. Thermal Properties of the Workpiece, and Uncoated and Coated Tool Materials

Variations in thermal properties of the cutting tool and the workpiece materials with temperature significantly affect the modelling of machining processes. It is clear that a more realistic model should be developed for metal cutting applications in which the temperature dependency of fundamental thermal properties such as thermal conductivity, specific heat, and thermal diffusivity, is taken into account. Variations in thermal properties with temperature are likely to have a significant effect on the tool-chip interface temperature and ultimately on heat partition into the cutting tool. The thermal properties of the materials used in the FE model are given in table 4. The thermal diffusivity of the cutting tool and workpiece materials can be calculated from the other variables (ρ, λ and C_p) given in table 4.

9. Experimental Tests

9.1. Experimental Set-Up

9.1.1. Machining Conditions

A series of cutting tests were carried out on a Dean Smith & Grace Lathe machine under dry cutting conditions. A Kistler cutting force dynamometer model 9263 was used to measure the cutting forces. The experimental setup is shown in figure 11. The cutting tests were performed at eight different cutting speeds of 100, 197, 314, 395, 565, 628, 785, and 880 m/min. These cutting speeds were set by the available RPM on the conventional lathe. Experimental conditions were similar for all cutting tool materials. A new cutting edge was used for every cutting speed. The feed rate and depth of cut were kept constant at 0.1 mm/rev and 2.5 mm, respectively. The process parameters were selected as appropriate to cutting tests in the HSM range. The used tools were analysed using Scanning Electron Microscope (SEM) equipped with Energy-Dispersive X-ray Analysis (EDXA) to quantify the sticking and sliding zones.

Table 4. Summary of temperature-dependent mechanical and thermal properties of workpiece, cutting inserts, and tool holder materials [4, 24, 29, 30, 61, 69, 99-103]

	Value at the following temperatures (°C)									
		100	200	300	400	500	600	700	800	900
Properties of workpiece (AISI 4140)										
Young's modulus (GPa)	210									
Poisson's ratio	0.29									
Density (kg/m^3)	7850									
Thermal conductivity (W/m/K)		42.67	41.83	40.58	38.91	36.40	33.89	30.96	26.36	27.19
Specific heat (J/kg/K)		473.10	489.80	506.60	527.50	548.41	577.70	611.27	690.82	678.26
Properties of uncoated cemented carbide cutting tool										
Young's modulus (GPa)	534									
Poisson's ratio	0.22									
Density (kg/m^3)	11900									
Thermal conductivity (W/m/K)		40.15	44.35	48.55	52.75	56.95	61.15	65.35	69.55	73.75
Specific heat (J/kg/K)		346.01	358.01	370.01	382.01	394.01	406.01	418.01	430.01	442.01
Properties of TiN-coated cutting tool										
Young's modulus (GPa)	250									
Poisson's ratio	0.25									
Density (kg/m^3)	5420									
Thermal conductivity (W/m/K)		21	21.47	22	22.52	23.0	23.72	24.38	25.01	25.5
Specific heat (J/kg/K)		702.60	752.70	783.40	801.16	818.9	833.46	846.39	856	857.60
Properties of TiAlN-coated cutting tool										
Young's modulus (GPa)	370									
Poisson's ratio	0.22									
Density (kg/m^3)	1892									
Thermal conductivity (W/m/K)		12.61	13.31	14.01	14.71	15.41	16.11	16.81	17.51	18.21
Specific heat (J/kg/K)		639.89	692.82	727.28	751.50	769.46	783.30	794.29	803.24	810.67
Properties of tool holder (AISI 1045)										
Young's modulus (GPa)	207									
Poisson's ratio	0.3									
Density (kg/m^3)	7844									
Thermal conductivity (W/m/K)		50.70	48.20	45.30	41.90	38.10	33.90	30.10	24.70	25.75
Specific heat (J/kg/K)		470.40	520.80	571.20	621.60	672.00	722.40	772.80	823.20	873.60

Figure 11. Setup for cutting experiments.

9.1.2. Cutting Tool Materials

Tungsten-based uncoated cemented carbide, TiN and TiAlN-coated tools were used in this investigation. The tool was coated on tungsten-based with TiN (5 μm) and TiAlN (5 μm) applied by the Physical Vapour Deposition (PVD) process. The coatings' micro-hardness was measured experimentally on a Buehler 5114 micro hardness testing machine with 300 g load for TiN (5 μm) and TiAlN (5 μm) coated tools. The results are shown in table 5. All coated tools had geometries designated as TCMW 16T304 (ISO specification) [Triangular shape insert with 0.4 mm nose radius, 3.97 mm thickness and 7° clearance angle]. Figures 12 (a) and (b) show the SEM images of cross-sectional views of the TiN and TiAlN-coated tools, respectively. In these images, TiN and TiAlN coatings and tungsten-carbide-based substrate (WC) materials show colour and brightness peculiar to the materials. Coating thickness was measured from cross-sectional views of the tools using SEM.

Table 5. Coatings and their mechanical properties

Coating	Coating thickness (μm)	Hardness (HV)
Uncoated	-	2074
TiN	5	1409
TiAlN	5	1800

9.1.3. Workpiece Material

A pre-bored workpiece of AISI/SAE 4140 high-tensile alloy steel (EN19 in British Standard) with an external diameter of 200 mm, length 200 mm, and tube thickness 2.5 mm was used in the cutting tests. The hardiness of the AISI/SAE 4140 steel was 23.9 HRC. AISI/SAE 4140 steel was chosen as it has many diverse applications in the manufacture of automobile components such as crankshafts, piston rods, steering components, axles, gears, and high-resistance screws. The chemical composition of AISI/SAE 4140 steel was checked by SEM and is presented in Table 6.

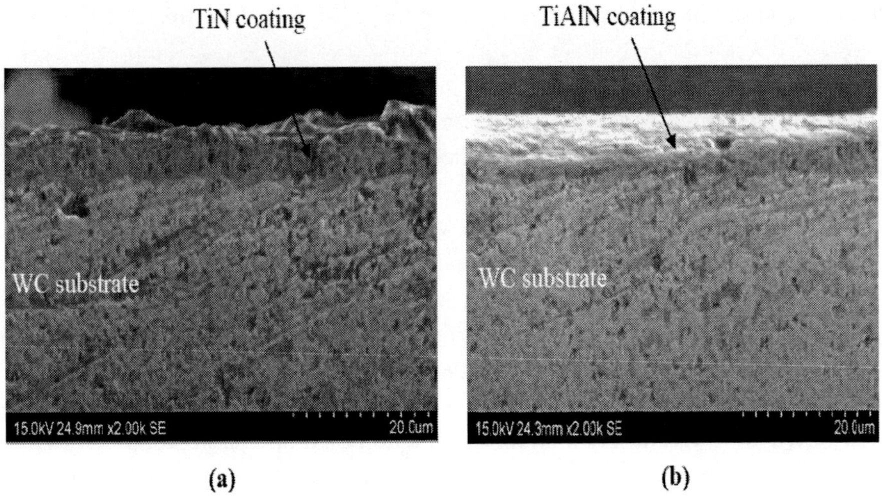

Figure 12. The SEM images of typical cross-sectional views of coating samples for (a) TiN and (b) TiAlN.

Table 6. Chemical Composition of AISI/SAE 4140 by weight %

AISI/SAE	C	Cr	Mn	Mo	P	Si	S	Fe
4140	0.43	1.1	0.95	0.25	0.035	0.25	0.04	Balance

9.2. Temperature Measurements

The temperatures generated in machining are important in that they can influence the properties of the workpiece material being machined and also play an important part in determining the effective cutting life of the tool. In metal cutting processes, it is desirable and often necessary to have knowledge of the amount of heat generated and consequent temperature rise, as well as its distribution in the tool, chip, and workpiece. For example, the maximum temperature on the tool rake face of a cutting tool will determine its life. Optimal cutting conditions, especially the cutting speed, depend on this as well as on the characteristics of the cutting tool material in rotation to the workpiece material. Similarly, metallurgical alterations of the machined surface and residual stresses in the finished part depend on the maximum temperature, the temperature gradient, and the rate of cooling of the part. The development of new tool materials as well as the advancement of machining technology will depend to a large extent on the knowledge and limitations of the cutting temperatures on the tool material as they influence the life and performance of the tool.

Various experimental techniques have been developed and utilised to study the cutting temperatures during turning operations. The most common experimental techniques for temperature measurement in metal cutting can be classified as: direct conduction, indirect radiation, and metallographic. These techniques were reviewed by Da Silva et al. [32], Herchang et al. [104] and more recently by Komanduri et al. [105], O'Sullivan [106], and Sutter et al. [107]. Generally, these techniques include tool-work thermocouples, radiation pyrometers, metallographic techniques, and the use of powders that melt at a defined temperature.

Boothroyd [108] in an early work used a photographic technique to study the temperature distribution. Wanigarathne et al. [109] used an infrared imaging system to measure transient temperature at different points (Spots) on the cutting tool during orthogonal machining of AISI 1045 steel with coated grooved tools. An inverse method, together with infrared temperature measurements, was applied by Kwon et al. [110] to calculate the temperatures at the tool-chip interface. Davies et al. [111] used the finite element method in modelling the cutting tool temperature field and correlating it with the experimental infrared measurements of the temperature fields at the tool-chip interface during orthogonal machining of AISI 1045 steel. M'Saoubi et al. [112] used an infrared thermo-video camera to measure temperatures in the deformation zone ahead of the cutting tool tip. M'Saoubi and Chandrasekaran [113] successfully used an infrared (IR) imaging technique based on charge-coupled device (CCD) sensors to determine temperature distributions in orthogonal turning of SS2541 (~SAE 4337) steel with uncoated (grade P20 S6) and TiN-coated tools. The authors concluded that TiN-PVD coatings decreased the maximum rake temperatures by 30-35 ^{o}C as observed using the IR-CCD based imaging technique. The size of the heat source was reduced as a smaller tool-chip plastic contact length was observed when machining with the coated tool, resulting in less heat propagation into the cutting tool. Davies et al. [114] obtained infrared thermal images from machining experiments with tungsten carbide tools and compared the temperature distributions with those obtained from a finite element analysis. Recently, Davies et al. [115] reviewed several temperature measurement methods and described how they can be applied to temperature monitoring during material removal. The authors detailed the sources of uncertainty in the measurement/evaluation of temperature using these techniques. Filice et al. [116] obtained infrared thermal images from turning experiments with uncoated carbide tools and compared the temperature distributions with those obtained numerically. Most recently, Faraz et al. [4] successfully used an infrared (IR) imaging technique to measure temperatures at multiple points (spots) on the cutting tool rake face during orthogonal turning of AISI/SAE 4140 steel with uncoated and TiN coated tools. Thus, it appears from the available literature that infrared thermal imaging is a useful technique for obtaining temperature distributions in cutting.

In more recent work, temperatures are measured using an infrared thermal imager (FLIR thermal CAM SC3000), a long wave, and a self-cooling analysis system with a cool-down time of < 6 min. The imager has a temperature range of -20 to +2000 ^{o}C, with an accuracy of ±2% or ±2 ^{o}C for measurement ranges above 150 ^{o}C. This camera can capture and store thermal images and data at high rates of up to 750 Hz PAL/900 Hz NTSC with a ThermaCAM ResearcherTM HS package. Complete technical specifications of the thermal imaging system are given in table 7. This package allows extensive analysis of highly dynamic objects and events typically found in metal machining research applications. The infrared thermal imager was mounted near the machine turret and placed directly over the tool rake face during cutting tests. Cutting tool temperatures were measured at three different points (Spots P_1-P_3) on the tool rake face, as shown in figure 13. These three spots were appropriately selected as prominent observable locations on the rake face of the cutting tool. The high-resolution thermal imaging system was utilised, making it possible to generate thermal maps of the cutting tool insert rake face at distances away from the chip flow zone for the points (P_1-P_3) in order to observe the thermal images clearly and avoid the effect of chip obstruction during machining, as shown in figure 13.

Table 7. Thermo-graphic system technical specifications [4]

IR detector	Quantum Well Infrared Photodetector (QWIP)
Spectral range	8 to 9 μm
Image frequency	50/60 Hz non-interlaced (standard), up to 750/900 Hz (optional and with Researcher HS option)
Thermal sensitivity	20 mK at 30 °C
Temperature range	-20 to +2000 °C
Accuracy	±1% or ±1 °C (for measurement ranges up to +150 °C) ±2% or ±2 °C (for measurement ranges above +150 °C)
Spatial resolution	1.1 mrad
Pixel per image	320 ξ 240
Zoom function	4 ξ
File format	14-bit radiometric IR digital image (IMG) 8-bit standard bitmap (BMP)

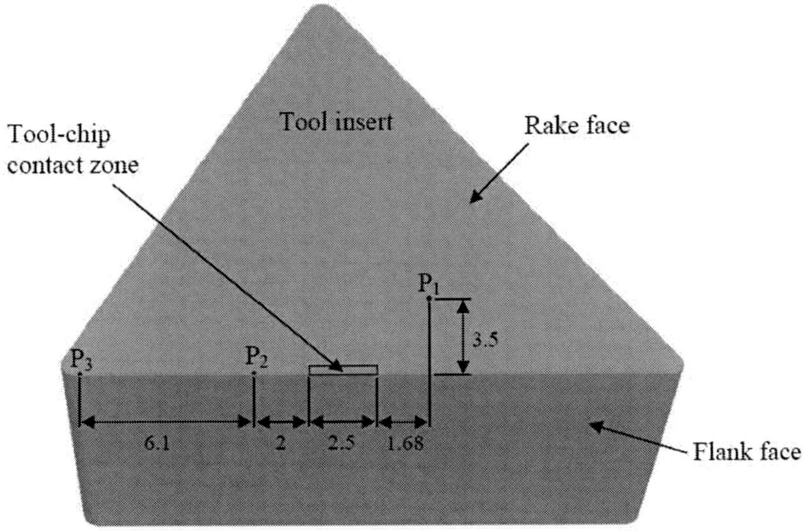

Figure 13. Multiple-points temperature measurement locations (all dimensions are in mm).

The ThermaCAM SC3000 was set up, and the location of the analysis area was chosen to be on the cutting tool insert. The stored images were recalled and analysed by using the accompanied analysis software. When placed on the image the spotmeter gave the temperature value at the required points (P_1-P_3).

The real temperature of an object depends strongly on the emissivity of the material, which is of particular concern when a thermal imaging camera is used. An accurate calibration of the thermo-graphic system was carried out to find the emissivity values of the cutting tool materials (uncoated, TiN-, and TiAlN-coated tools). Samples were heated to temperatures ranging from 100 °C up to 900 °C in an oven. This range possibly encompasses the entire temperature range expected during the actual machining. A thermal imaging camera was used to read the inserts' temperature, and the emissivity was adjusted until the temperature reading of the thermal imager matched a thermocouple reading. Fifteen emissivity values were obtained at each temperature level. A mean of these values was then used for the camera setting. According to M'Saoubi et al. [117], the calculation of the error is less than 10 per cent for an emissivity of 0.2, lower than 7.5 per cent for an emissivity higher than 0.3, and lower than 5 per cent for an emissivity value higher than 0.5. The emissivity results obtained in the present study for the uncoated, and TiN- and TiAlN-coated tools produced typical variations of emissivity with temperature, as shown in figure 14. The average emissivity of uncoated, TiN-, and TiAlN-coated inserts was determined to be 0.48, 0.21, and 0.45, respectively at 700 °C. A small variation in tool emissivity was observed in the temperature range of 450 - 900 °C given by 0.48 ± 0.01020 for uncoated, 0.21 ± 0.00546 for TiN-coated, and 0.45 ± 0.00648 for TiAlN-coated tools.

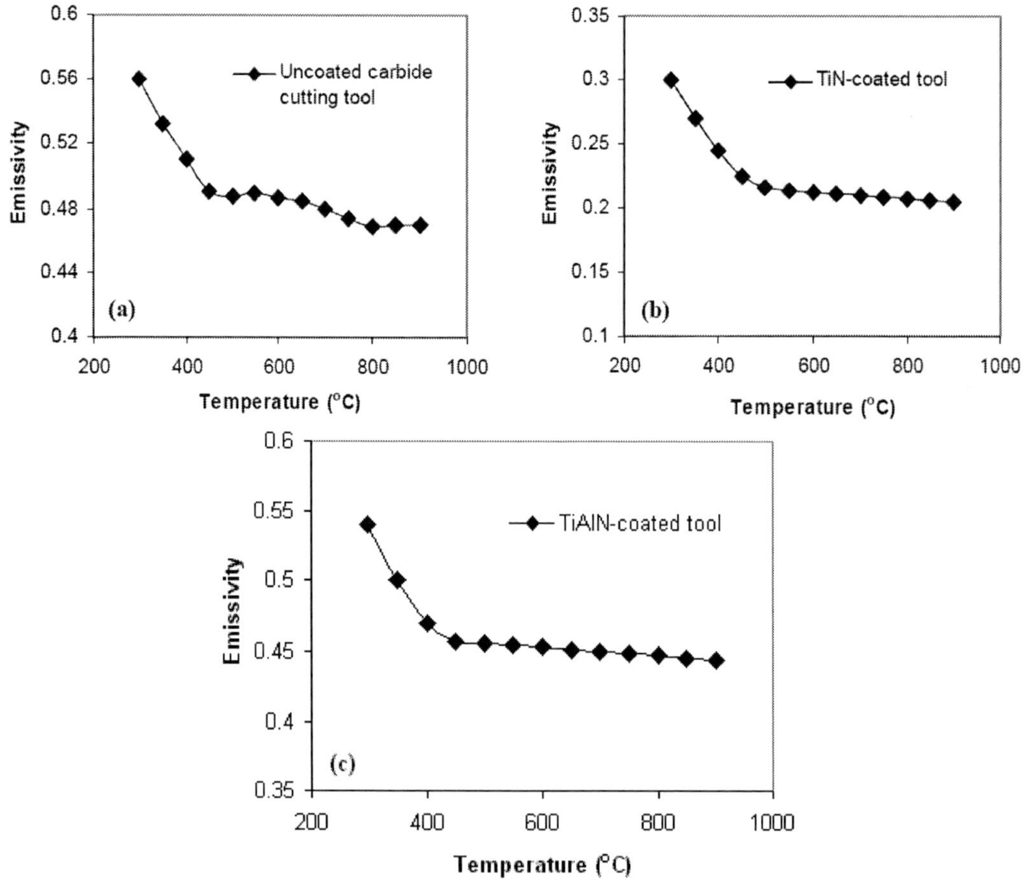

Figure 14. Variation of emissivity with temperature for (a) uncoated carbide, (b) TiN-coated, and (c) TiAlN-coated tools.

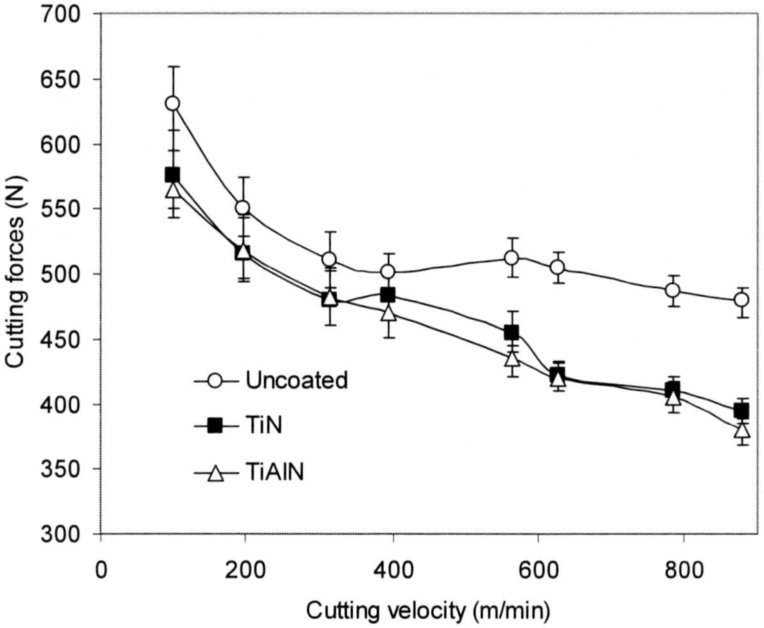

Figure 15. Variation of the cutting force with the cutting speed.

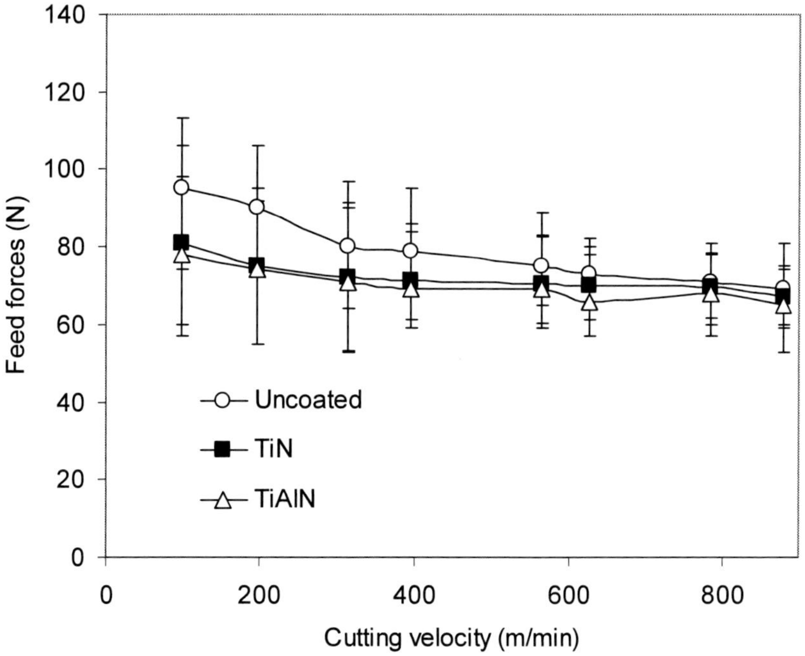

Figure 16. Variation of the feed force with the cutting speed.

10. Results and Discussions

10.1. Cutting Forces

Figures 15 and 16 show the variation of cutting and feed forces with cutting speed. For uncoated tools, the cutting force decreases when the cutting speed is increased from 100 to 395 m/min, marginally increases with the cutting speed up to 565 m/min, and subsequently decreases in the high-speed cutting region. For TiN-coated tools, the cutting force decreases when the cutting speed is increased from 100 to 314 m/min, marginally increases with the cutting speed up to 395 m/min, and subsequently decreases in the high-speed cutting region. On the other hand, for TiAlN-coated tools, cutting forces gradually decrease when the cutting speed is increased from 100 to 880 m/min. In general, the experimentally measured cutting and feed force values for the TiN-coated tools are higher than for the TiAlN-coated tools over the whole range of cutting speeds. The feed force also decreases gradually when the cutting speed is increased in all cutting tests for all the cutting tools.

10.2. Tool-Chip Contact Area

In this study, the tool-chip contact area of the worn uncoated, TiN-coated, and TiAlN-coated inserts was obtained at different cutting speeds from a Scanning Electron Microscope (SEM) equipped with image-processing software. Sample optical images of the worn inserts are shown in figure 17. Figure 18 presents the calculated tool-chip contact area obtained from the SEM images of the worn inserts at different cutting speeds for uncoated, and TiN- and TiAlN-coated tools. It is observed that the contact area is affected by the cutting speed. For uncoated tools, there is a decrease in the contact area when the cutting speed is increased from 100 to 395 m/min. Beyond 395 m/min, the contact area increases with the cutting speed. For TiN-coated tools, there is a consistent decrease in the contact area when the cutting speed is increased from 100 to 880 m/min. On the other hand, with TiAlN-coated tools, the contact area decreases with increasing cutting speed in both conventional and HSM regions. However, it should be noted that TiAlN-coated inserts present larger numerical values for all cutting speeds, compared to TiN-coated inserts. This could be due to the thermal bimetallic effect. These results clearly show that TiN coatings mitigate the expansion of contact area as compared to the uncoated and TiAlN-coated tools.

10.3. Tool-Chip Contact Phenomena

The nature of the interface formed between cutting tool and workpiece material is of fundamental importance to the efficiency of the metal removal process. It is well known [58, 88, 118] that frictional conditions at the tool-chip interface have a significant influence on the phenomena of tool wear, built-up-edge (BUE), and surface finish. The action of cutting creates exceptional tribological conditions at the interface, such that plastic contact (seizure) between tool and work material involving atomic bonding over most of the apparent contact is a normal condition in industrial machining operations [58, 119]. Moreover, the region of the contact is intensively heated by the energy generated in plastic deformation and friction, and new surfaces with very high levels of cleanliness and chemical activity are created. From all these considerations it is evident that seizure (adhesion) at the tool-chip interface could not be avoided.

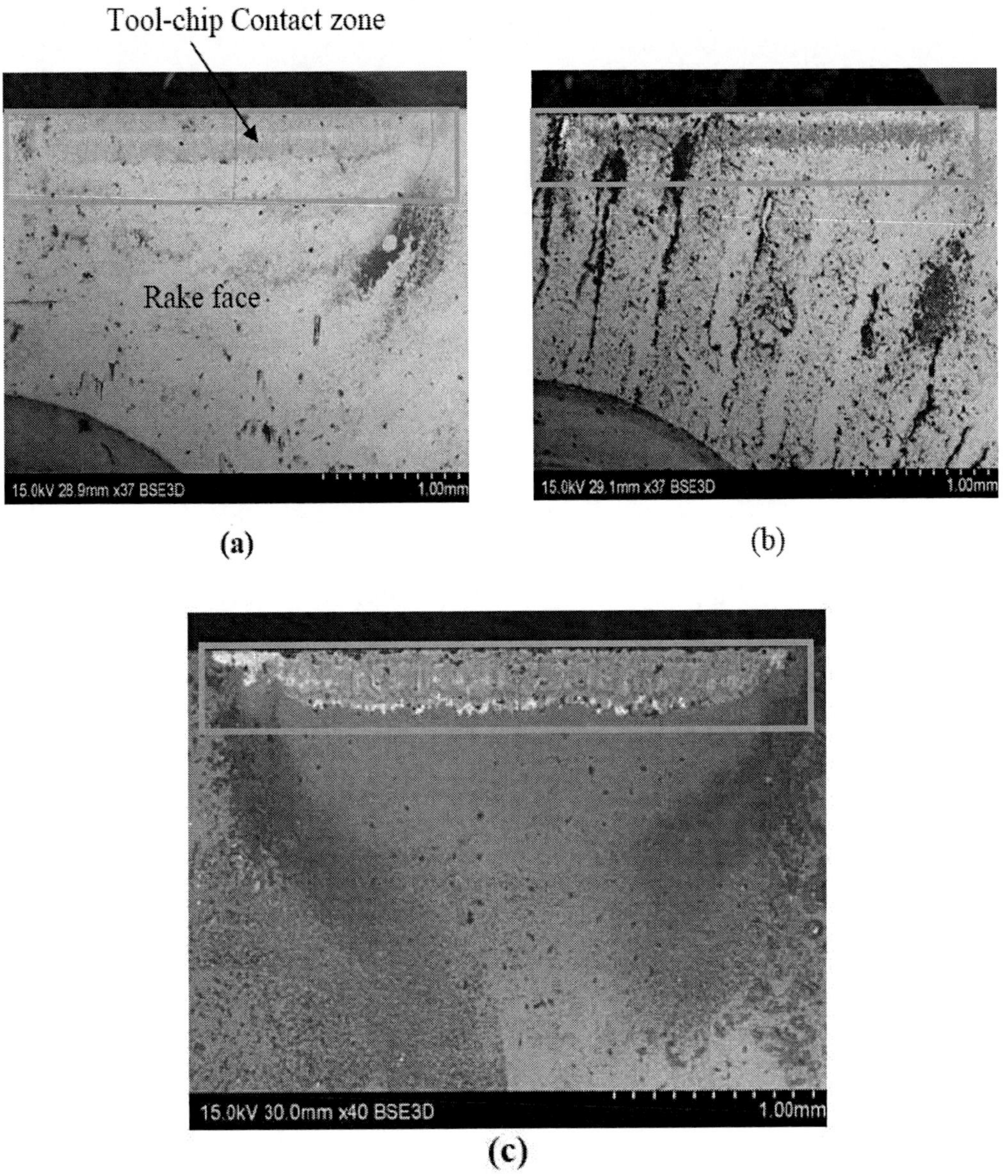

Figure 17. Scanning Electron Microscopic images of the tool-chip contact area at a cutting speed of 880 m/min for (a) uncoated (b) TiN-coated, and (c) TiAlN-coated tools.

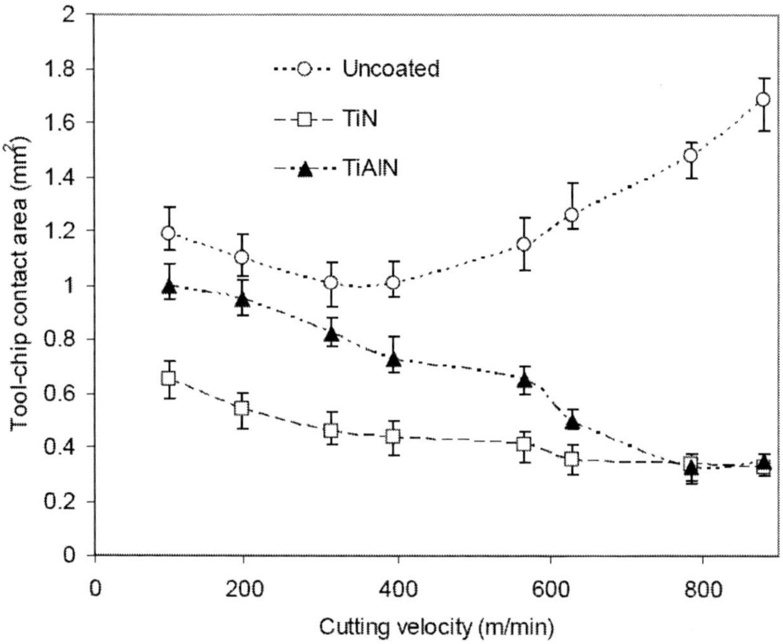

Figure 18. Variation of the tool-chip contact area with the cutting speed.

10.3.1. Determination of Sticking and Sliding Zones for Uncoated, and TiN- and TiAlN-Coated Tools

As mentioned earlier, at any instant during the cutting process, a part of the interface may be in seizure, while the rest of the contact undergoes interfacial sliding. These interfacial conditions are highly dependent on the cutting conditions and properties of the tool and the workpiece materials [33, 120]. In this article, Scanning Electron Microscopy (SEM) and Energy-Dispersive X-ray Analysis (EDXA) are used to distinguish between the sticking and the sliding regions from an examination of the tool-chip contact region for uncoated, and TiN- and TiAlN-coated tools. The worn inserts were examined by SEM using a backscattered imaging mode for the range of cutting speeds. The contact length was scanned for evidence of iron (Fe) transfer from the chip. The evaluation of this Fe transfer intensity was taken on a cross-sectional plane of the contact area in the direction of chip flow.

Figure 19 shows the EDXA of Fe concentration maps of the rake face of the cutting inserts for the uncoated tools for cutting speeds ranging form 100 to 880 m/min. All the micrographs show a low concentration of iron about the cutting edge, and this is attributed to the effect of the tool edge radius. Figure 19 shows that at 100 m/min the tool rake face is not dominated by Fe transfer from the chip; hence, providing evidence that seizure does not occur at this cutting speed. As the cutting speed is increased to 565 m/min, the nature of the contact changes and the tool rake face is dominated by a significant amount of Fe transfer, thereby suggesting that the tribological phenomenon of seizure is taking place. For higher cutting speed of 880 m/min, there are high peaks of Fe appear near the edge of the rake face within the contact length. Iron (Fe) peaks support a dominant sticking condition. For uncoated carbide cutting tools, it is justifiable to conclude from Figure 19 that no significant sticking or seizure occurred for 100 m/min, with 85% sticking and 15% sliding at 565 m/min, and 100% sticking at 880 m/min.

For TiN-coated tools (figure 20), at low cutting speed 100 m/min, no significant concentration of Fe is present (less than 10%), indicating the absence of material transfer or sticking and thus supporting a dominant sliding condition. When the cutting speed increases to 565 m/min, still no dominant Fe concentration is present, again supporting a prevalent sliding condition. At the highest cutting speed of 880 m/min, about 32% of the contact area shows significant Fe transfer, and hence a 32% sticking (seizure) and 68% sliding is established for this case.

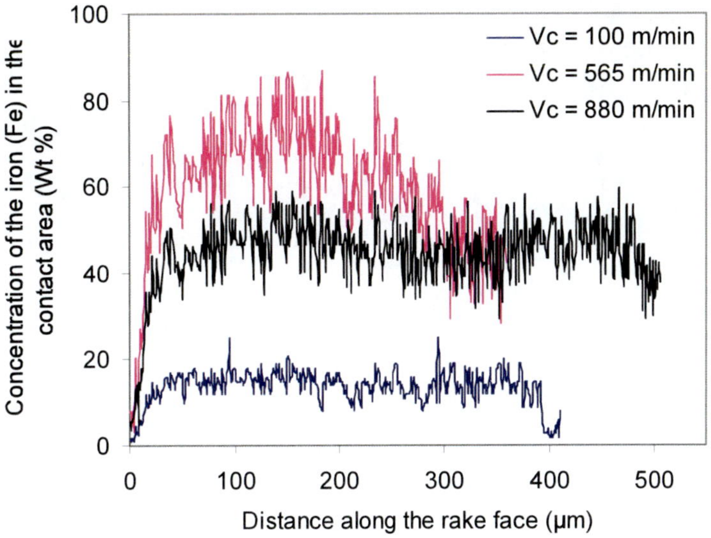

Figure 19. SEM-EDXA Fe density maps along the rake face of uncoated carbide cutting inserts.

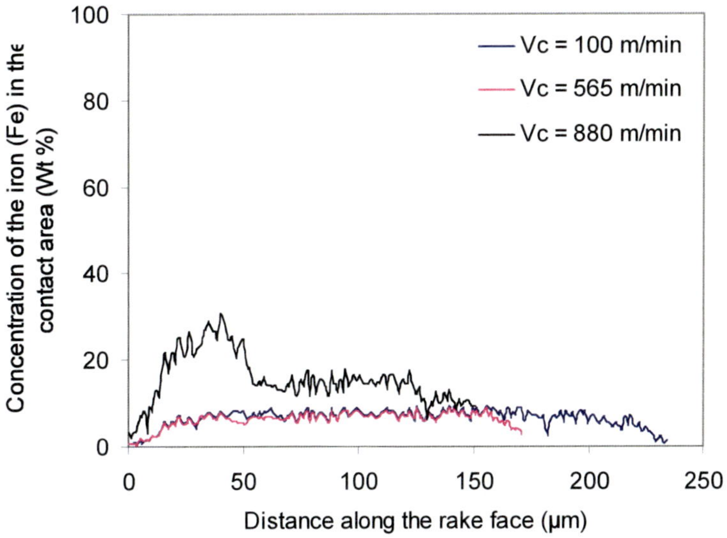

Figure 20. SEM-EDXA Fe density maps along the rake face of TiN-coated cutting inserts.

For TiAlN-coated tools (figure 21), at low cutting speed 100 m/min, no significant concentration of Fe is present (less than 10%), indicating the absence of material transfer or sticking, and thus supporting a dominant sliding condition. When the cutting speed increases to 565 m/min, still no dominant Fe concentration is present, again supporting a prevalent sliding condition. At the highest cutting speed of 880 m/min, about 48% of the contact area shows a significant Fe transfer, and thus a 48% sticking (seizure) and 52% sliding is established for this case. It is worth noting that TiN-coated tools reduce the sticking of the work material as compared to the TiAlN-coated tools in the high-speed machining region.

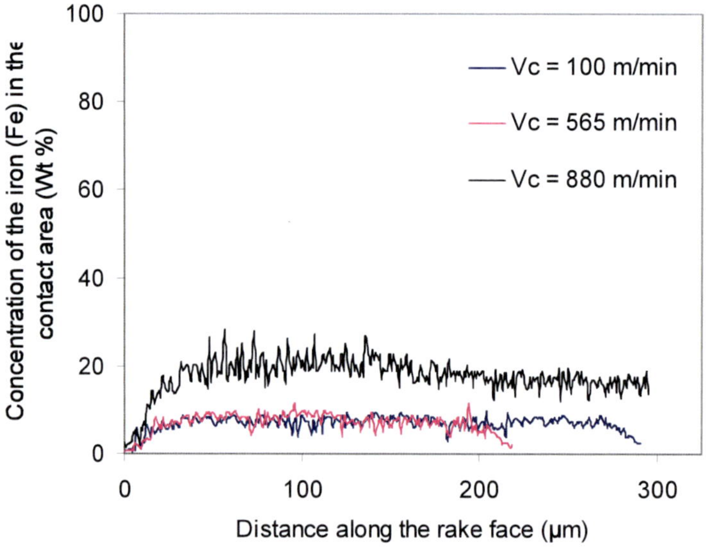

Figure 21. SEM-EDXA Fe density maps along the rake face of TiAlN-coated cutting inserts.

10.4. Heat Partition into the Cutting Tool

Figure 22 shows typical FE simulation results for temperature distribution at the cutting insert for a non-uniform heat input at a cutting speed of 565 m/min for uncoated, and TiN- and TiAlN-coated tools. These distributions clearly show the presence of high temperature gradients near the cutting edge. Very high temperatures are obtained, as expected, due to the high strength of the machined workpiece alloy and high depth of cut, which was set at 2.5 mm.

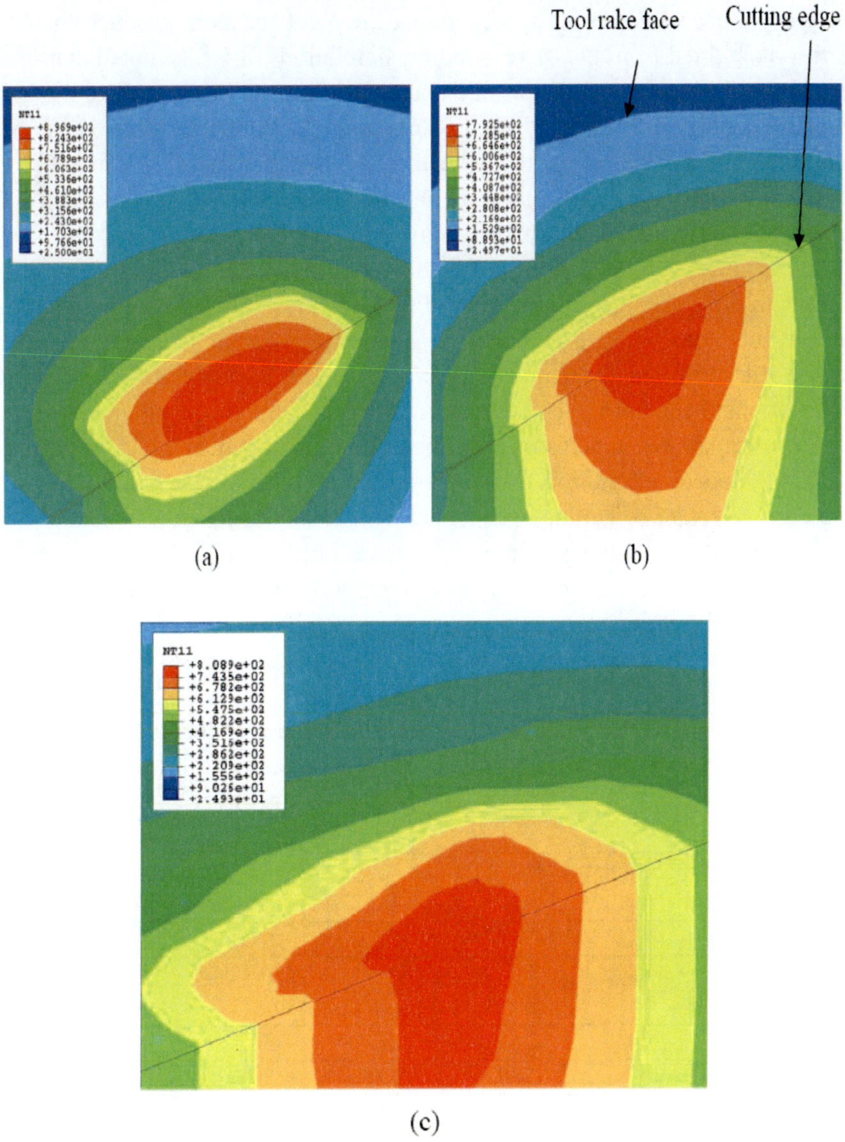

Figure 22. Temperature distribution results at a cutting speed of 565 m/min for (a) uncoated carbide (b) TiN-coated and (c) TiAlN-coated tools.

10.4.1. For Uncoated Carbide Cutting Tool

In the estimation of heat partition into the cutting tool with the hybrid FEM-experimental approach, a slight disagreement was observed between 'single-point' and 'multiple-point' temperature matching for uncoated carbide cutting tools. The tool temperatures obtained form the simulations were compared with the experimentally measured temperatures for uncoated tools. These are presented in the table 8 for three measurements locations (P_1, P_2, and P_3), defined in figure 13. The errors due to location uncertainty in measurements are considered small, as the temperature gradients at these locations are rather insignificant. These results indicate that the tool-chip contact area is more sensitive to heat partition and temperature response than it is to the measurement location. A systematic analysis shows that a difference of 10% in temperature values at the specified measurement locations produces a 16% change in heat partition obtained from the corresponding simulations. It is also noted that temperature reductions obtained from the simulations are between 5 °C to 10 °C, similar to the variation in the experimental measurements. This variation could be due to the fact that when one point is selected for temperature matching, then the thermal plane does not capture the complete variation of the non-uniform heat flux. Furthermore, it is likely that this non-uniform heat flux also varies in the depth of cut direction driven by chip curl.

The finite element thermal analysis of heat partition into the cutting tool is validated by its comparison to the experimentally measured temperatures with thermal imaging camera at three different locations (P_1, P_2, and P_3). A good agreement of temperatures matching is obtained. The difference in the temperature values at P_3 is higher than at the other two measured locations because P_3 is farthest away from the tool-chip contact area zone. Generally, the simulations underestimate temperatures at the beginning of turning at low speeds, but in high-speed machining, the temperatures are overestimated.

Over the entire range of cutting speeds, percentage difference is less than 5%, which is acceptable for three different thermal imaging camera locations (P_1, P_2, and P_3). In summary, a good agreement between experimentally measured and finite element temperatures at three different locations throughout the turning process validates the proposed hybrid FEM-experimental approach to predict heat partition into the cutting tool.

Table 8. Comparison between simulation and experimentally measured tool temperatures for uncoated carbide cutting tool (a_p = 2.5 mm and f = 0.1 mm/rev)

	Simulation temperatures			Experimentally measured temperatures		
	(°C)			(°C)		
Cutting speed (V_c) (m/min)	P_1	P_2	P_3	P_1	P_2	P_3
100	201	266	199	201	259	200
314	207	275	200	207	276	202
565	211	290	199	211	291	200
785	216	305	204	217	305	204
880	222	317	210	222	319	209

Figure 23 compares different analytical heat partition models with the hybrid FEM-experimental approach for uncoated carbide tool. It shows a reduction in the value of heat partition into the cutting tool obtained form the hybrid FEM-experimental approach with an increase in the cutting speed from 100 to 395 m/min, i.e., within the conventional cutting speed region. Heat partition into the cutting tool ($R_{T\,(FEM\text{-}EXP)}$) obtained from the hybrid FEM-experimental approach varies from 41% down to 17% in the conventional (100 to 395 m/min) cutting speed range. This trend within the conventional cutting speed region is influenced by the well-known decrease in the tool-chip contact area (figure 18). The lower value of heat partition ($R_{T\,(FEM\text{-}EXP)}$) at 395 m/min represents a large part of the heat generated at the tool-chip interface to be carried away by the chip. When the cutting speed is increased above 395 m/min, i.e. to the HSM region, the nature of the contact changes and the tool-chip contact length increases, and also the contact area. As a consequence, the fraction of heat flowing into the tool increases gradually with the cutting speed and reaches 23.5% at 880 m/min (figure 23). Thus, the tool-chip contact area is critical in determining heat partition into the cutting tool.

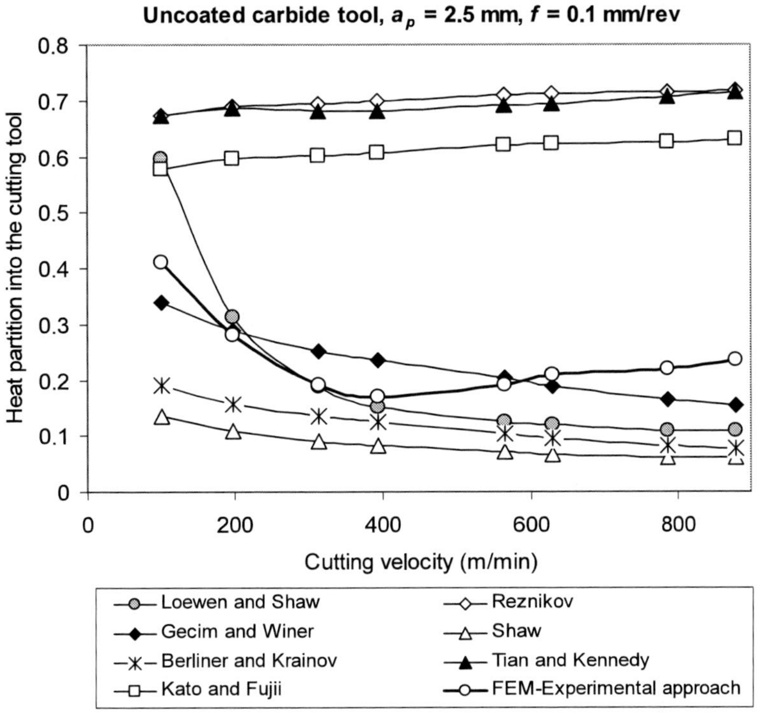

Figure 23. A comparison of heat partition models for conventional to HSM for uncoated carbide tool.

From figure 23, it can be seen that predicted heat partition coefficients by Kato and Fujii (R_{KF}) and Reznikov (R_R) increase with the cutting speed, and the maximum heat going into the tool is 57% and 67%, respectively. These two models depend purely on the thermal properties of the tool and the workpiece materials and do not take into account any contact parameters like the tool-chip contact length, frictional heat intensity, and the chip velocity. Tian and Kennedy model (R_{TK}), which takes into account the influence of Peclet numbers for both the tool and the workpiece, also provides values of heat partition coefficient closer to Kato and Fujii (R_{KF}) and Reznikov (R_R). On the other hand, Loewen and Shaw (R_L) model shows that the heat partition coefficient decreases with an increase in the cutting speed. It is observed that at cutting speed of 100 m/min, the heat partition coefficient is 59.6% dropping to 11% at the maximum cutting speed of 880 m/min. In this case, the heat partition coefficient is modelled as a function of chip velocity, depth of cut, tool-chip contact length, and most importantly, the frictional heat flux. The Berliner and Krainov (R_{BK}) model predicts a heat partition coefficient between 19% down to 7.6% over the range of cutting speeds. The Shaw (R_{SH}) model predicts a heat partition coefficient between 13.6% down to 5.9% which, of all the models, represents the lowest amount of heat going into the tool. The values of heat partition into the tool given by Gecim and Winer (R_{GW}) model, range from 33% down to 15%.

From figure 23, Gecim and Winer (R_{GW}) and Loweon and Shaw (R_L) appears to be the best models for qualitatively estimating the heat partition coefficient in conventional cutting speeds.

In summary, various heat partition models show different trends for the uncoated carbide cutting tool over a wide range of cutting speeds. The hybrid FEM-experimental approach shows that heat partition into the cutting tool strongly depends on the contact phenomenon, thermal properties of the tool and the workpiece materials, and the process variables. In particular, the models proposed by Loweon and Shaw (R_L) and Gecim and Winer (R_{GW}) give good qualitatively estimates of heat partition into the cutting tool for conventional cutting speeds but fail to do so for HSM. Other models either overestimate heat partition (Kato and Fujii (R_{KF}), Reznikov (R_R) and Tian and Kennedy (R_{TK})) or underestimate it (Berliner and Krainov (R_{BK}) and Shaw (R_{SH})).

10.4.2. For TiN-Coated Tool

Figure 24 compares heat partition into the cutting tool for TiN-coated tools following the hybrid FEM-experimental approach and multiple points matching. For TiN-coated tools heat partition varies from 35% down to 9.5%. A continuously decreasing trend for heat partition into the cutting tool is observed over the range of cutting speeds. This implies that the use of single layer TiN-coated tools compared to the uncoated tools can cause a reduction in heat partition into the cutting tool by about 17% at conventional cutting speeds and 60% in the HSM region. The benefits of reduction in heat partition brought by the TiN coating are more pronounced in HSM compared to conventional machining. As presented earlier, table 4 gives temperature-dependent thermal conductivity values for uncoated cemented carbide and TiN cutting tool materials. It can be seen from this table that the thermal conductivity values for TiN coating are lower as compared to those for uncoated tools.

The tool temperatures obtained from the simulations are compared with the experimentally measured temperatures for TiN-coated tools. These are presented in the table 9 for the three measurements locations (P_1, P_2, and P_3). As with uncoated tools, these results also indicate that the tool-chip contact area is more sensitive to heat partition and temperature response than to measurement locations.

Table 9. Comparison between simulation and experimentally measured tool temperatures for TiN-coated cutting tools (a_p = 2.5 mm and f = 0.1 mm/rev)

	Simulation temperatures (°C)			Experimentally measured temperatures (°C)		
Cutting speed (V_c) (m/min)	P_1	P_2	P_3	P_1	P_2	P_3
314	132	248	125	139	256	125
565	115	199	109	122	208	104
785	123	214.5	109	125	224	112
880	131	230	111	130	241	114

From figure 24, it can be seen that the heat partition coefficients predicted by Kato and Fujii (R_{KF}) and Reznikov (R_R) increase with the cutting speed with maximum heat fraction going into the tool to be 45% and 55%, respectively. On the other hand, the Loewen and

Shaw (R_L) model shows that the heat partition coefficient decreases with an increase in the cutting speed. From figure 24, it is also observed that at the cutting speed of 100 m/min, the heat partition coefficient is 18.8%, and at maximum cutting speed of 880 m/min, it is 2%. Berliner and Krainov's (R_{BK}) model predicts the heat partition coefficient between 10.5% down to 3.9% for the range of cutting speeds. Shaw (R_{SH}) model predicts the heat partition coefficient to be between 6.3% and 2%. The values for heat partition into the tool by Gecim and Winer model (R_{GW}) range from 20.59% down to 8.2%, and for the Tian and Kennedy model (R_{TK}), the range is 53.89% down to 48.60%.

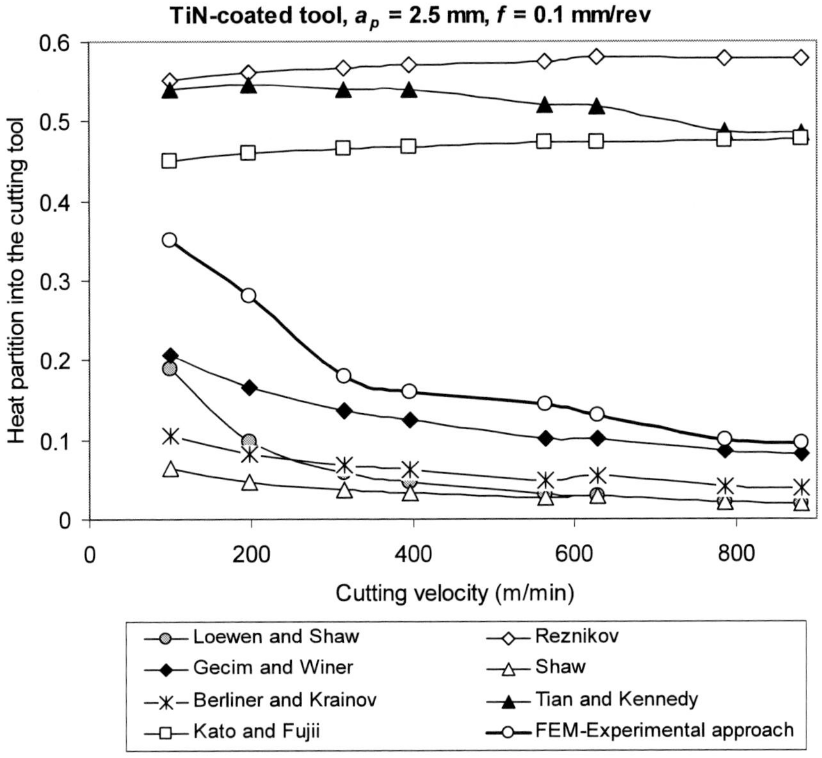

Figure 24. A comparison of heat partition models for conventional to HSM for TiN-coated tool.

In summary, figure 24 shows heat partition data for TiN-coated inserts obtained form the hybrid FEM-experimental approach against various analytical models. The trends from (Kato and Fujii (R_{KF}) and Reznikov (R_R)) models, predicting higher heat partition for increased cutting speeds, are not evident in the FEM-experimental approach. The trend for Loewen and Shaw (R_L), Berliner and Krainov (R_{BK}), and Shaw (R_{SH}) models predicts a lower heat partition with increasing cutting speed. The model used by Gecim and Winer (R_{GW}) is found to be closest to the FEM-experimental approach for TiN-coated tools. These results for TiN-coated tool suggest a reduced dependence of heat partition on the tool-chip contact area (contact phenomena).

10.4.3. For TiAlN-Coated Tool

Figure 25 compares heat partition from different analytical models with the hybrid FEM-experimental approach for TiAlN-coated tools. It shows a reduction from 38% down to 16% in the value of heat partition into the cutting tool obtained from the hybrid FEM-experimental approach, with an increase in cutting speed from 100 to 395 m/min, i.e., within the conventional cutting speed region. This trend within the conventional cutting speed region is influenced by the well-known [30] decrease in the tool-chip contact area (figure 18). When the cutting speed is increased above 395 m/min, i.e., to the HSM region, the nature of the contact changes and the tool-chip contact length, and thus the contact area, increases. As a consequence, the fraction of heat flowing into the tool decreases with the cutting speed and reaches 9.2% at 880 m/min (figure 25).

Tool temperatures for TiAlN-coated tools obtained from the simulations are compared with the experimentally measured temperatures. These are presented in the table 10 for the three measurements locations (P_1, P_2, and P_3). As for TiN-coated tools, these results also indicate that the thermal properties and tool-chip contact area are more sensitive to heat partition and temperature response than to measurement locations.

Heat partition into the cutting tool obtained from the hybrid FEM-experimental approach for uncoated, and TiN- and TiAlN-coated tools show a notable change in trend at 400 m/min. The mechanism for the change in heat partition when the cutting speed approaches 400 m/min needs further investigation. However, the contact area is found to be a dominant influence on heat partition. A cutting speed of 400 m/min represents the HSM transition zone where the trend toward a reduction in the contact length with cutting speed, as experienced in conventional machining [121, 122], changes and starts to increase. This trend is driven by the change in the sliding contact phenomenon and has been reported in an earlier work by Abukhshim et al. [123].

Table 10. Comparison between simulation and experimentally measured tool temperatures for TiAlN-coated cutting tool (a_p = 2.5 mm and f = 0.1 mm/rev)

	Simulation temperatures			Experimentally measured temperatures		
	(°C)			(°C)		
Cutting speed (V_c) (m/min)	P_1	P_2	P_3	P_1	P_2	P_3
100	160	191	136	158	188	130
314	130	240	121	129	240	120
565	120	186	110	125	224	112
785	124	220	107	125	225	111
880	130	240	108	130	241	114

From figure 25, it can be seen that predicted heat partition coefficients by Kato and Fujii (R_{KF}) and Reznikov (R_R) increase with the cutting speed from 100 to 565 m/min and then decrease to 27.9% and 36.8%, respectively. The Tian and Kennedy model (R_{TK}), which takes into account the influence of Peclet numbers for either the tool or the workpiece, also provides values of heat partition coefficient closer to those of Kato and Fujii (R_{KF}) and Reznikov (R_R). The Loewen and Shaw (R_L) model shows that the heat partition coefficient decreases with an increase in the cutting speed. It is observed that at the cutting speed of 100 m/min, the heat partition coefficient is 17.4% dropping to 1.4% at the maximum cutting speed of 880 m/min. In this case, the heat partition coefficient is modelled as a function of chip velocity, depth of cut, tool-chip contact length, and most importantly, the frictional heat flux. The Berliner and Krainov (R_{BK}) model predicts a heat partition coefficient between 6.1% down to 2.5% over the range of cutting speeds. Shaw's (R_{SH}) model predicts a heat partition coefficient between 3.8% down to 1.3%, which, of all the models, represents the lowest amount of heat going into the tool. The values of heat partition into the tool in the Gecim and Winer (R_{GW}) model range from 12.5% down to 5.4%. From figure 25, the Loweon and Shaw (R_L) model appears to be the best for qualitatively estimating the heat partition coefficient for a wide range of cutting speeds.

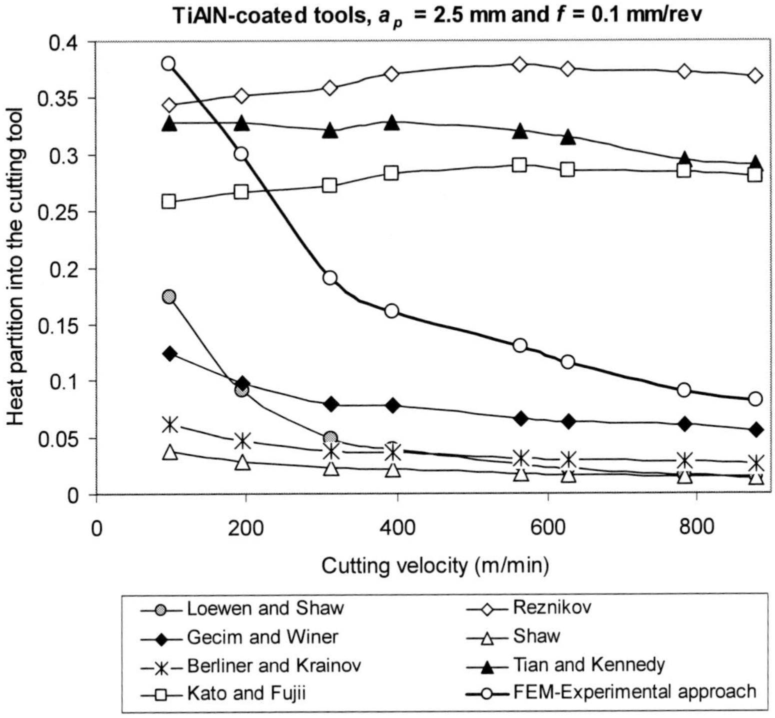

Figure 25. A comparison of heat partition models for conventional to HSM for TiAlN-coated tool.

11. Conclusion

Accurate prediction of heat partition into the cutting tool is a major challenge in metal cutting. This is due to numerous practical difficulties involved in the process. Review of the works on analytical modelling of heat partition shows that these models have limitations and suffer mainly due to their simplified assumptions. Moreover, due to the complexity and the severity of deformation in the HSM process, it is clear that purely analytical approaches are severely restrictive in predicting heat partition.

This chapter presents results of heat partition for uncoated, and TiN- and TiAlN-coated tools over a wide range of cutting speeds. These results clearly show that coatings influence the contact area and thus also influence heat flux in metal cutting. In general, existing analytical models of heat partition into the cutting tool do not model the effect of the contact area. These models do not yield agreeable results or provide an accurate quantitative estimate of the fraction of heat flowing into the tool. The best quantitative estimates of heat partition were obtained using the Loewen and Shaw (R_L) and Gecim and Winer (R_{GW}) models for uncoated carbide tools, derived from the Gecim and Winer (R_{GW}) model for TiN-coated tool and from the Loewen and Shaw (R_L) model for TiAlN-coated tool. These models suggest that heat partition is influenced by the frictional heat source, tool-chip contact length, chip velocity, area shape factor, and the thermal properties of the tool and the workpiece materials. Whilst most numerical modelling work assumes a uniform heat flux, the evidence from this investigation suggests that the heat flux may be uniform or non-uniform depending on the sticking and sliding lengths. This factor should be taken into account in the FE model. Results show that the TiN and TiAlN coatings on the cutting tool reduce the percentage of heat entering into the cutting tool compared to the uncoated tool. If the tool coatings are compared, it is evident that the use of TiAlN reduces heat partition into the cutting tool in the high-speed cutting region. Additionally, this chapter provides heat partition values as a function of cutting speeds, which are useful in the selection of optimal machining parameters.

The analytical models by Kato and Fujii (R_{KF}) and Reznikov (R_R), which are based solely on the thermal properties of the tool and workpiece, appear to overestimate the percentage of heat that goes into the tool for most cutting speeds except the lowest speed, for all the tools. The Kato and Fujii model (R_{KF}) is similar to Reznikov model (R_R) but incorporates a scaling factor. The Tian and Kennedy model (R_{TK}), despite additionally considering the Peclet numbers, follows the same trend as Kato and Fujii (R_{KF}) and Reznikov (R_R) for uncoated tools. For low cutting speeds, the models developed by Loewen and Shaw (R_L) and Gecim and Winer (R_{GW}), which consider contact phenomena and chip velocity in their formulations, appear to be in closer agreement for estimates of uncoated tools on heat partition with the hybrid experimental-FEM approach. The Loewen and Shaw model (R_L) neglects the increase in the amount of heat going into the tool in the HSM region. Berliner and Krainov (R_{BK}) and Shaw (R_{SH}) follow a similar trend, but with lower accuracy in quantitative prediction.

It can be argued that existing analytical models for the quantitative prediction of heat partition are inadequate. New models are required, especially for coated tools used at high-speed cutting. From the analysis presented in this chapter, it can be concluded that the area shape factor, mean temperature rise due to plastic deformation in the primary deformation zone (shear plane temperature), frictional heat flux, and contact area, in addition to the temperature-dependent thermal properties of tool and workpiece materials, should all be used in developing an accurate heat partition model.

REFERENCES

[1] Trent, E. M. (1984). *Metal Cutting.* Second ed., Butterworth & Co, London.
[2] Tlusty, J. (1993). High-speed machining. *Annals of the CIRP*, 42(2), 733-738.

[3] Dawes, R. C. and Aspinwall, D. K. (1997). A review of ultra high speed milling of hardened steels. *Journal of Materials Processing Technology*, 69, 1-17.

[4] Akbar, F., Mativenga, P. T. and Sheikh, M. A. (2008). An evaluation of heat partition in high-speed turning of AISI/SAE 4140 steel with uncoated and TiN-coated tools. *Proceedings of the Institution of Mechanical Engineers, Part B: Journal of Engineering Manufacture*, 222, 759-771.

[5] Fallbohmer, P., Rodrguez, C. A., Ozel, T. and Altan, T. (2000). High-speed machining of cast iron and alloy steels for die and mold manufacturing. *Journal of Materials Processing Technology*, 98, 104-115.

[6] Schulz, H. and Moriwaki, T. (1992). High-speed machining. *Annals of the CIRP*, 41(2), 637-643.

[7] Nakayama, K. and Ogawa, M. (1987). Burr formation in metal cutting. *Ann. CIRP* 36(1), 33-36.

[8] Flom, D. G., Komanduri, R. and Lee, M. (1984). HIGH-SPEED MACHINING OF METALS. *Ann. Rev. Mater. Sci.*, 14, 231-278.

[9] Schulz, H. (1996). *High-Speed Machining.* (Carl Hanser, Munich, Germany).

[10] Denkena, B., Amor, R. B., Leon-Garcia, L. and Dege, J. (2007). Material specific definition of the high speed cutting range. *International journal of machining and machinability of materials*, 2(2), 176-185.

[11] Kramer, B. M. (1987). On tool materials for high speed machining. *J. Eng. Ind. Trans. ASME*, 109(2), 87-91.

[12] Prengel, H. G., Pfouts, W. R. and Santhanam, A. T. (1998). State of the art in hard coatings for cutting tools. *Surface and Coatings Technology*, 102, 183-190.

[13] Klocke, F. and Krieg, T. (1999). Coated tools for metal cutting-features and applications. *Annals of the ClRP*, 48(2), 515-525.

[14] Wallen, P. and Hogmark, S. (1989). Influence of TiN coating on wear of high speed steel at elevated temperature. *Wear*, 130, 123-135.

[15] Akbar, F., Mativenga, P. T. and Sheikh, M. A. (2007). An investigation of the tool-chip interface temperature and heat partition in high-speed machining of AISI/SAE 4140 steel with TiN-coated tool. *Proceedings of the 35th International MATADOR Conference*, Springer London, 215-218.

[16] Zieberta, C. and Ulrich, S. (2006). Hard multilayer coatings containing TiN and/or ZrN: A review and recent progress in their nanoscale characterization. *J. Vac. Sci. Technol*, 24(3), 554-583.

[17] Grzesik, W. (2003). Friction behaviour of heat isolating coatings in machining: mechanical, thermal and energy-based considerations. *International Journal of Machine Tools & Manufacture*, 43, 145-150.

[18] Rech, J. and Djouadi, M. A. (2001). High speed gear hobbing. *Wear*, 250, 45-53.

[19] Klocke, F., Krieg, T., Gerschwiler, K., Fritsch, R., Zinkann, V., Pohls, M. and Eisenblatter, G. (1998). Improved cutting processes with adapted coating systems. *Annals of the ClRP*, 47(1), 65-68.

[20] Klocke, F. and Eisenblatter, G. (1997). Dry Cutting. *Annals of the ClRP*, 46(2), 519-526.

[21] Hogmark, H., Hedenqvist, P. and Jacobson, S. (1997). Tribological properties of thin hard coatings: demands and evaluation. *Surface and Coatings Technology*, 60, 247-257.

[22] Grzesik, W. and Nieslony, P. (2003). Thermophysical-Property-Based Selection of Tool Protective Coatings for Dry Machining of Steels. *ASME, Journal of Manufacturing Science and Engineering*, 125, 689-695.

[23] Pfouts, W. R. (2000). Cutting Edge Coatings. *Manuf. Eng.*, 125(1), 98-107.

[24] Yen, Y.-C., Jain, A., Chigurupati, P., Wu, W.-T. and Altan, T. (2004). Computer simulation of orthogonal cutting using a tool with multiple coatings. *Machining Science and Technology*, 8(2), 305-326.

[25] Balaji, A. K. and Mohan, V. S. (2002). An 'effective cutting tool thermal conductivity' based model for tool-chip contact in machining with multi-layer coated cutting tools. *J. Mach. Sci. Technol.*, 6(3), 415-436.

[26] Grzesik, W. (2000). The influence of thin hard coatings on frictional behaviour in the orthogonal cutting process. *Tribology International*, 33, 131-140.

[27] Grzesik, W. (1998). The role of coatings in controlling the cutting process when turning with coated indexable inserts. *Journal of Materials Processing Technology*, 79, 133-143.

[28] Grzesik, W. (2000). An integrated approach to evaluating the tribo-contact for coated cutting inserts. *Wear*, 240, 9-18.

[29] Childs, T., Maekawa, K., Obikawa, T. and Yamane, Y. (2000). *Metal machining theory and applications.* Arnold, a member of Hodder and Headline group, London.

[30] Akbar, F., Mativenga, P. T. and Sheikh, M. A. (2009). On the Heat Partition Properties of TiAlN Compared to TiN Coating in High-Speed Machining. *Proceedings of the Institution of Mechanical Engineers, Part B: Journal of Engineering Manufacture*, 223, 363-375.

[31] Jawahir, I. S. and Van Luttervelt, C. A. (1993). Recent developments in chip control research and applications. *Annals of the CIRP*, 42(2), 659-693.

[32] Da Silva, M. B. and Wallbank, J. (1999). Cutting temperature: prediction and measurement methods-a review. *Journal of Materials Processing Technology* 88, 195-202.

[33] Trent, E. M. (1988). Metal cutting and the tribology of seizure I: Seizure in metal cutting. *Wear*, 128, 29-45.

[34] Trent, E. M. (1988). Metal cutting and the tribology of seizure, II: Movement of work material over the tool in metal cutting. *Wear*, 128, 46-64.

[35] Akbar, F., Mativenga, P. T. and Sheikh, M. A. (2008). An evaluation of heat partition models in high-speed machining of AISI/SAE 4140 steel. *The 6th International Conference on Manufacturing Research (ICMR08),* 1, 13-21.

[36] Kalpakjian, S. (1997). Manufacturing Processes for Engineering Materials. Addison-Wesley.

[37] Childs, T. H. C., Maekawa, K. and Maulik, P. (1988). Effects of coolant on temperature distribution in metal machining. *Mater. Sci. Technol.*, 4, 1006-1019.

[38] Wright, P. K., McCormick, S. P. and Miller, T. R. (1980). Effect of rake face design on cutting tool temperature distributions. *ASME Journal of Engineering for Industry*, 102(2), 123-128.

[39] Tay, A. E., Stevenson, M. G. and DeVahl Davis, G. (1974). Using the finite element method to determine temperature distribution in orthogonal machining. *Proceedings of the Institution of Mechanical Engineers*, 188, 627-638.

[40] Huang, Y. and Liang, S. Y. (2005). Cutting temperature modeling based on non-uniform heat intensity and partition ratio. *Machining Science and Technology*, 9, 301-323.

[41] Blok, H. (1938). Theoretical study of temperature rise at surfaces of actual contact under oiliness lubricating conditions. *Proceedings of General Discussion on Lubrication and Lubricants, Institute of Mechanical Engineers London*, 222-235.

[42] Chao, B. T. and Trigger, K. J. (1955). Temperature distribution at the tool-chip interface in metal cutting. *Transactions of the ASME*, 72, 1107-1121.

[43] Komanduri, R. and Hou, Z. B. (2001). Thermal modeling of the metal cutting process Part II: temperature rise distribution due to frictional heat source at the tool-chip interface. *International Journal of Mechanical Sciences*, 43, 57-88.

[44] Liu, J. and Chou, Y. K. (2007). On temperatures and tool wear in machining hypereutectic Al–Si alloys with vortex-tube cooling. *International Journal of Machine Tools & Manufacture*, 47, 635-645.

[45] Liu, J. and Chou, Y. K. (2007). Cutting Tool Temperature Analysis in Heat-Pipe Assisted Composite Machining. *Transactions of the ASME*, 129, 902-910.

[46] Yvonnet, J., Umbrello, D., Chinesta, F. and Micari, F. (2006). A simple inverse procedure to determine heat flux on the tool in orthogonal cut. *International Journal of Machine Tools & Manufacture*, 46, 820-827.

[47] Karpat, Y. and Özel, T. (2006). Predictive analytical and thermal modeling of orthogonal cutting process - Part I: Predictions of tool forces, stresses, and temperature distributions. *Transactions of the ASME, Journal of Manufacturing Science and Engineering*, 128, 435-444.

[48] Karpat, Y. and Özel, T. (2006). Predictive analytical and thermal modeling of orthogonal cutting process - Part II: Effect of tool flank wear on tool forces, stresses, and temperature distributions. *Transactions of the ASME, Journal of Manufacturing Science and Engineering*, 128, 445-453.

[49] Francis, H. A. (1970). Interfacial temperature distribution within a sliding Hertzian contact. *ASLE Trans.*, 14, 41-54.

[50] Bos, J. and Moes, H. (1995). Frictional heating of tribological contacts. *Trans.ASME J. Tribol.*, 117, 171-177.

[51] Komanduri, R. and Hou, Z. B. (2001). Analysis of heat partition and temperature distribution in sliding systems. *Wear*, 251, 925-938.

[52] Gecim, B. and Winer, W. O. (1985). Transient temperatures in the vicinity of an asperity contact. *Transactions of the ASME Journal of Tribology*, 107, 333-342.

[53] Loewen, E. G. and Shaw, M. C. (1954). On the analysis of cutting tool temperatures. *Transactions of ASME*, 71, 217-231.

[54] Takeuchi, Y., Sakamoto, M. and Sata, T. (1982). Improvement in the working accuracy of an NC lathe by compensating for thermal expansion. *Precision Engineering*, 4(1), 19-24.

[55] Ghani, M. U., Abukhshim, N. A. and Sheikh, M. A. (2008). An investigation of heat partition and tool wear in hard turning of H13 tool steel with CBN cutting tools. *Int J Adv Manuf Technol*, 39, 874-888.

[56] Trigger, K. J. and Chao, B. T. (1951). An analytical evaluation of metal cutting temperature. *Transactions of the ASME*, 73, 57-68.

[57] Stephenson, D. A. and Ali, A. (1992). Tool temperatures in interrupted metal cutting. *ASME J Eng Ind*, 14(2), 127-136.

[58] Trent, E. M. (1991). *Metal Cutting.* Third edition, Butterworth & Co, London.

[59] Wang, J. Y. and Liu, C. R. (1999). The effect of tool flank wear on the heat transfer, thermal damage and cutting mechanics in finish hard tuning. *Annals of the ClRP*, 48(1), 53-58.

[60] Casto, L. S., Valvo, L. E. and Micari, F. (1989). Measurement of temperature distribution within tool in metal cutting. Experimental tests and numerical analysis. *Journal of Mechanical Working Technology*, 20, 35-46.

[61] Abukhshim, N. A., Mativenga, P. T. and Sheikh, M. A. (2005). Investigation of heat partition in high speed turning of high strength alloy steel. *International Journal of Machine Tools and Manufacture*, 45, 1687-1695.

[62] Grzesik, W. and Nieslony, P. (2003). A computational approach to evaluate temperature and heat partition in machining with multilayer coated tools. *International Journal of Machine Tools and Manufacture*, 43, 1311-1317.

[63] Cotterell, M. and Byrne, G. (2008). Characterisation of chip formation during orthogonal cutting of titanium alloy Ti–6Al–4V. *CIRP Journal of Manufacturing Science and Technology*, 1, 81-85.

[64] Shih, A. J. (1996). Finite element analysis of orthogonal metal cutting mechanics. *Int. J. Mach. Tools Manufact.*, 36, 255-273.

[65] Lei, S., Shin, Y. C. and Incropera, F. P. (1999). Thermo-mechanical modeling of orthogonal machining process by finite element analysis. *International Journal of Machine Tools & Manufacture*, 39, 731-750.

[66] Xie, L.-J., Schmidt, J., Schmidt, C. and Biesinger, F. (2005). 2D FEM estimate of tool wear in turning operation. *Wear 258*, 1479-1490.

[67] Arrazola, P. J., Ugarte, D. and Dominguez, X. (2008). A new approach for the friction identification during machining through the use of finite element modeling. *International Journal of Machine Tools and Manufacture*, 48, 173-183.

[68] Arrazola, P. J., Villar, A., Ugarte, D. and Marya, S. (2007). Serrated chip prediction in finite element modeling of the chip formation process. *Machining Science and Technology*, 11, 367-390.

[69] Reginaldo, T. C., Ng, E.-G. and Elbestawi, M. A. (2007). Tool wear when turning hardened AISI 4340 with coated PCBN tools using finishing cutting conditions. *International Journal of Machine Tools and Manufacture*, 47, 263-272.

[70] Shet, C. and Deng, X. (2000). Finite element analysis of the orthogonal metal cutting process. *Journal of Materials Processing Technology*, 105, 95-109.

[71] Shet, C., Deng, X. and Bayoumi, A. E. (2003). Finite element simulation of high-pressure water-jet assisted metal cutting. *International Journal of Mechanical Sciences* 45, 1201-1228.

[72] Shi, G., Deng, X. and Shet, C. (2002). A fnite element study of the efect of friction in orthogonal metal cutting. *Finite Elements in Analysis and Design*, 38, 863-883.

[73] Shet, C. and Deng, X. (2003). Residual stresses and strains in orthogonal metal cutting. *International Journal of Machine Tools and Manufacture*, 43, 573-587.

[74] Shi, J. and Liu, C. R. (2006). On predicting chip morphology and phase transformation in hard machining. *International Journal of Advanced Manufacturing Technology*, 27, 645-654.

[75] Shi, J. and Liu, C. R. (2005). On predicting softening effects in hard turned surfaces - Part II: Finite Element Modeling and verification. *Transactions of the ASME*, 127, 484-491.
[76] Kai, L. and Melkote, S. N. (2006). Material strengthening mechanisms and their contribution to size effect in micro-cutting. *Transactions of the ASME*, 128, 730-738.
[77] Shi, J. and Liu, C. R. (2004). The influence of material models on finite element simulation of machining. *Journal of Manufacturing Science and Engineering*, 126, 849-857.
[78] Li, K., Gao, X.-L. and Sutherland, J. W. (2002). Finite element simulation of the orthogonal metal cutting process for qualitative understanding of the effects of crater wear on the chip formation process. *Journal of Materials Processing Technology* 127, 309-324.
[79] Mabrouki, T. and Rigal, J.-F. (2006). A contribution to a qualitative understanding of thermo-mechanical effects during chip formation in hard turning. *Journal of Materials Processing Technology*, 176, 214-221.
[80] Ng, E. G., Aspinwall, D. K., Brazil, D. and Monaghan, J. (1999). Modelling of temperature and forces when orthogonally machining hardened steel. *International Journal of Machine Tools & Manufacture*, 39, 885–903.
[81] Zong, W. J., Li, D., Cheng, K., Sun, T. and Liang, Y. C. (2007). Finite element optimization of diamond tool geometry and cutting-process parameters based on surface residual stresses. *International Journal of Advanced Manufacturing Technology*, 32, 666-674.
[82] Mohamed, A. N., Ng, E. G. and Elbestawi, M. A. (2007). Modelling the effects of tool-edge radius on residual stresses when orthogonal cutting AISI 316L. *International Journal of Machine Tools and Manufacture*, 47, 401-411.
[83] Ghani, M. U., Abukhshim, N. A. and Sheikh, M. A. (2007). An investigation of heat partition and tool wear in hard turning of H13 tool steel with CBN cutting tools. *Int J Adv Manuf Technol*, In press.
[84] Jaeger, J. C. (1942). Moving sources of heat and the temperature at sliding contacts. *Proceedings Royal Society of NSW*, 76, 203-224.
[85] Silin, S. S. (1981). *Similarity Methods in Metal Cutting.* Mashinostroenie, Moscow.
[86] Grzesik, W. and Nieslony, P. (2004). Physics based modelling of interface temperatures in machining with multilayer coated tools at moderate cutting speeds. *International Journal of Machine Tools & Manufacture*, 44, 889–901.
[87] Oxley, P. L. B. (1989). Mechanics of Machining. An Analytical Approach to Assessing Machinability. Ellis Horwood, New York
[88] Boothroyd, G. and Knight, W. A. (1989). *Fundametals of Machining and Machine Tools.* Second edition, Marcel Dekker, New York and Basel.
[89] Akbar, F., Mativenga, P. T. and Sheikh, M. A. (2009). Predictive modelling of average heat partition in high-speed machining of AISI/SAE 4140 steel. *International journal of machining and machinability of materials*, 5(2/3), 135-154.
[90] Reznikov, A. N. (1981). *Thermophysical Aspects of Metal Cutting Processes.* Mashinostroenie, Moscow.
[91] Gecim B. and Winer, W. O. (1985). Transient temperatures in the vicinity of an asperity contact. *Transactions of the ASME Journal of Tribology*, 107, 333-342.
[92] Shaw, M. C. (1989). *Metal Cutting Principles.* Clarendon Press, Oxford.

[93] Berliner, E. M. and Krainov, V. P. (1991). Analytic calculations of the temperature field and heat flows on the tool surface in metal cutting due to sliding friction. *Wear* 143, 379-395.

[94] Li, R. and Shih, A. J. (2007). Tool Temperature in Titanium Drilling. *Transactions of the ASME*, 129, 740-749.

[95] Tian, X. and Kennedy, F. E. (1994). Maximum and average flash temperatures in sliding contact. *Transactions of the ASME Journal of Tribology*, 116.

[96] Kato, T. and Fujii, H. (1999). Energy partition in conventional surface grinding. *Transactions of the ASME Journal of Manufacturing Science and Engineering*, 121, 393-398.

[97] (1997). Hibbitt, Karlsson, Sorensen. ABAQUS/Explicit User's Manual.

[98] Yen, Y.-C., Jain, A., Chigurupati, P., Wu, W.-T. and Altan, T. (2003). Computer simulation of orthogonal cutting using a tool with multiple coatings. *Proceedings of the 6th CIRP International Workshop on Modeling of Machining Operation*, McMaster University, Hamilton, Ontario, Canada, 119-130.

[99] Gale, W. F. and Totemeier, T. C. (2004). *Smithells metals Reference book.* Elsevier, London.

[100] Lengauer, W., Binder, S., Aigner, K., Ettmayer, P., Guillou, A., Debuigne, J. and Groboth, G. (1995). Solid state properties of group IVb carbonitrides. *Journal of Alloys and Compounds*, 217, 137-147.

[101] Barsoum, M. W., Rawn, C. J., El-Raghy, T., Procopio, A. T., Porter, W. D., Wang, H. and Hubbard, C. R. (2000). Thermal properties of Ti_4AlN_3. *Journal of Applied Physics*, 87(12), 8407-8414.

[102] Finkel, P., Barsoum, M. W. and El-Raghy, T. (2000). Low temperature dependencies of the elastic properties of Ti_4AlN_3, $Ti_3Al_{1.1}C_{1.8}$, and Ti_3SiC_2. *Journal of Applied Physics*, 87(4), 1701-1703.

[103] Woolman, J. and Mottram, R. A. (1964). *The mechanical and physical properties of the British standard En steels.* 1, Pergamon press, London.

[104] AY, H. and Yang, W. (1998). Heat transfer and life of metal cutting tools in turning. *Int. J. Heat Mass Transfer.* , 41(3), 613-623.

[105] Komanduri, R. and Hou, Z. B. (2001). A review of the experimental techniques for the measurement of heat and temperature generated in some manufacturing processes and tribology. *Tribol. Int.*, 34, 653-682.

[106] O'Sullivan, D. and Cotterell, M. (2001). Temperature measurement in single point turning. *J. Mater. Process. Technol.*, 118, 301-308.

[107] Sutter, G., Faure, L., Molinari, A., Ranc, N. and Pina, V. (2003). An experimental technique for the measurement of temperature fields for the orthogonal cutting in high speed machining. *Int. J. Mach. Tools Manufac*, 43, 671-678.

[108] Boothroyd, G. (1963). Temperatures in orthogonal metal cutting. *Proceedings of the Institution of Mechanical Engineers*, 177(29), 789-810.

[109] Wanigarathne, P. C., Kardekar, A.S., Dillon, O.W., Poulachon, G. and Jawahir, I.S. (2005). Progressive tool-wear in machining with coated grooved tools and its correlation with cutting temperature. *Wear*, 259, 1215-1224.

[110] Kwon, P., Schiemann, T. and Kountanya, R. (2001). An inverse estimation scheme to measure steady-state tool-chip interface temperatures using an infrared camera. *International Journal of Machine Tools & Manufacture*, 41, 1015-1030.

[111] Davies, M. A., Cao, Q, Cooke, A. L. and Ivester, R. (2003). On the measurement and prediction of the temperature fields in machining. *Annals of the CIRP*, 52, 77-80.

[112] M'Saoubi, R., Le Calvez, C., Changeux, B. and Lebrun, J. L. (2002). Thermal and microstructural analysis of orthogonal cutting of a low alloyed carbon steel using an infrared-charge-coupled device camera technique. *Proceedings of the Institution of Mechanical Engineers*, 216, 153-165.

[113] M'Saoubi, R. and Chandrasekaran, H. (2004). Investigation of the effects of tool micro-geometry and coating on tool temperature during orthogonal turning of quenched and tempered steel. *International Journal of Machine Tools and Manufacture*, 44, 213-224.

[114] Davies, M. A., Cooke, A. L. and Larsen, E. R. (2005). High Bandwidth Thermal Microscopy of Machining AISI 1045 Steel. *Annals of the CIRP*, 54(1), 63-66.

[115] Davies, M. A., Ueda, T., M'Saoubi, R., Mullany, B. and Cooke, A. L. (2007). On The Measurement of Temperature in Material Removal Processes. *Annals of the CIRP*, 56(2), 581-604.

[116] Filice, L., Umbrello, D., Beccari, S. and Micari, F. (2006). On the FE codes capability for tool temperature calculation in machining processes. *Journal of Materials Processing Technology*, 174, 286-292.

[117] M'Saoubi, R., Le Calvez, C., Changeux, B. and Lebrun, J. L. (2002). Thermal and microstructural analysis of orthogonal cutting of a low alloyed carbon steel using an infrared-charge-coupled device camera technique. *Proceedings of the Institution of Mechanical Engineers Part B*, 216, 153-165.

[118] Shaw, M. C. (2005). *Metal Cutting Principles.* Oxford University Press.

[119] Bailry, J. A. (1975). Friction in metal machining-mechanical aspects. 31, 243-275.

[120] Gekonde, H. O. and Subramanian, S. V. (2002). Tribology of tool-chip interface and tool wear mechanisms. *Surface and Coatings Technology*, 149, 151-160.

[121] Gad, G. S., Armarego, J. A. and Smith, A. J. R. (1992). Tool-chip contact length in orthogonal machining and its importance in tool temperature predictions. *International Journal of Production Research*, 30(3), 485-501.

[122] Sadik, M. I. and Lindstrom, B. (1993). The role of tool-chip contact length in metal cutting. *Journal of Materials Processing Technology*, 37, 613-627.

[123] Abukhshim, N. A., Mativenga, P. T. and Sheikh, M. A. (2004). An investigation of the tool-chip contact length and wear in high-speed turning of EN19 steel. *Proceedings of the Institution of Mechanical Engineers, Part B: Journal of Engineering Manufacture*, 218, 889-903.

In: Metal Cutting: Research Advances
Editor: J. Paulo Davim pp. 77-110

ISBN: 978-1-60876-207-1

Chapter 3

MECHANICAL AND THERMAL EXPERIMENTS IN CUTTING PROCESS FOR NEW BEHAVIOUR LAW

Olivier Cahuc*[*], *Philippe Darnis, and Raynald Laheurte
Université de Bordeaux, 351 cours de la Libération,
33405 Talence-cedex, France.

ABSTRACT

High speed machining has been improved thanks to considerable advancement in tools, in machined materials, and finally in the machines. However at high-speed loading velocities, the cutting process generates high strain and high-speed strain, which cause various complex and irreversible phenomena in plasticity. These phenomena are comprehended through the complete measurement of mechanical actions using a six-component dynamometer and flux and temperature measurements at the tip of the cutting tool. Balanced energy assessments are the starting point of our reflection on machining modelling. The modelling of these phenomena and the material behaviour under this type of loading requires a suitable theoretical approach. The main points of the strain gradient theory are developed. An experimental approach is proposed to reproduce shearing strains generated during machining. A friction load similar to the secondary shear zone loading is generated and analysed in detail. Then, the first points of a phenomenon modelling are developed using the strain gradient theory. A theoretical behaviour law adapted to the cutting phenomena is then expressed.

1. INTRODUCTION

Evolution and Revolution in the Scientific Approach to the Cutting Process

In the highly competitive industrial sector of machining, it has now become necessary to control the production costs. Costs are primarily due to the duration of machining, the type of machine used, and the consumption of tools and lubricants. For a number of years,

[*] Corresponding author, E-mail: olivier.cahuc@u-bordeaux1.fr

technological advances such as tool materials and numerically controlled machines have enabled industrialists to maintain a linear reduction in production costs without having to resort to scientific methods.

Thus, a series of a few tests could easily solve a particular problem and give satisfaction to the customer via a material or a specific tool geometry.

Currently, this approach is undergoing changes; and industrialists have few alternative solutions other than High Speed Machining, which many see as a radical technological step.. As a matter of fact, improvements in machining performance in the last ten years relate more to increased performance in data processing than to an actual optimisation of tool-matter interaction.

Nevertheless, the computing capacities of C.A.M. software or of the numerical controls are restricted to the physical limits imposed by a non-optimised chip formation.

These sorts of problems arise in many advanced machining industries, which have already reached these limits, and tool and machine manufacturers are looking for clear solutions to meet the strong demand for increased productivity.

Thus, to satisfy the industry's expectations, it seems essential to deal with the problem as a whole. The modelling of the cutting process is a response to the problems raised by this technology. However, this type of modeling requires a comprehensive understanding of the phenomena.

In this respect, the difficulty of apprehending the global geometry of the process can be better highlighted by passing from a 3D to a plane representation, illustrated in Figure 1 (according to A view) to identify the characteristic zones of machining. In fact, most of the scientific community dealing with this subject is aware of the importance of three-dimensional analysis. Indeed, simple observation of chip formation in machining tests is convincing evidence of the non-relevance of two-dimensional, or assimilated, aspect of the starting assumptions (figure 2).

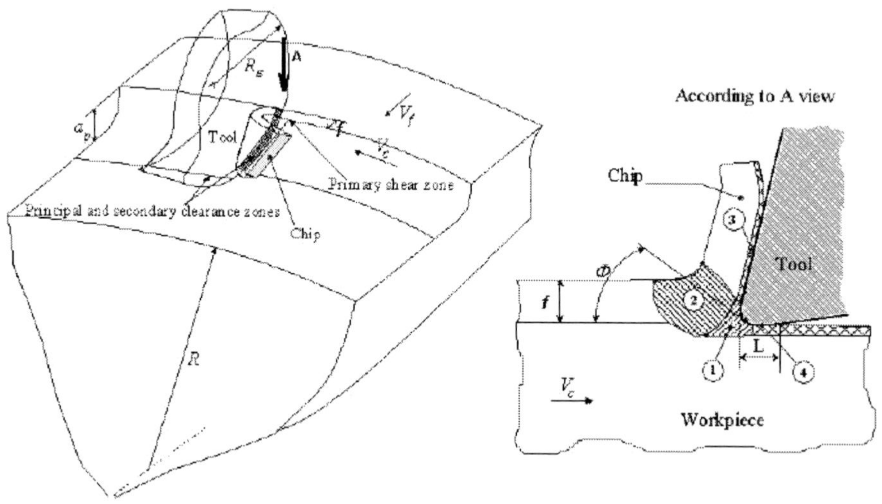

Figure 1. Main machining zones.

Experimental results of turning (Toulouse, 1998) were obtained at the Laboratory of Physical Mechanics (LMP) (University Bordeaux using a six-component dynamometer

(Couétard, 1993), which allowed COUÉTARD (Couétard, 2000) to inspect the cutting process as a truly three-dimensional phenomenon. For this purpose, he used the whole range of the mechanic's tools and focused more particularly on the action of a body on another body through a force vector and a moment vector.

We strongly advise users of the cutting process to treat it in a three-dimensional space in order to avoid making assumptions that might reduce the study to an orthogonal (or oblique) case, which would fail to reflect the real machining process in turning, drilling, or milling. Such assumptions are too restrictive in high speed or very high speed machining.

Cutting speeds are always increasing due to the development of new tools and materials. Thus, the very high deformation velocities in the cutting process involve high strain velocities and strain (Moufski et al. 2000).

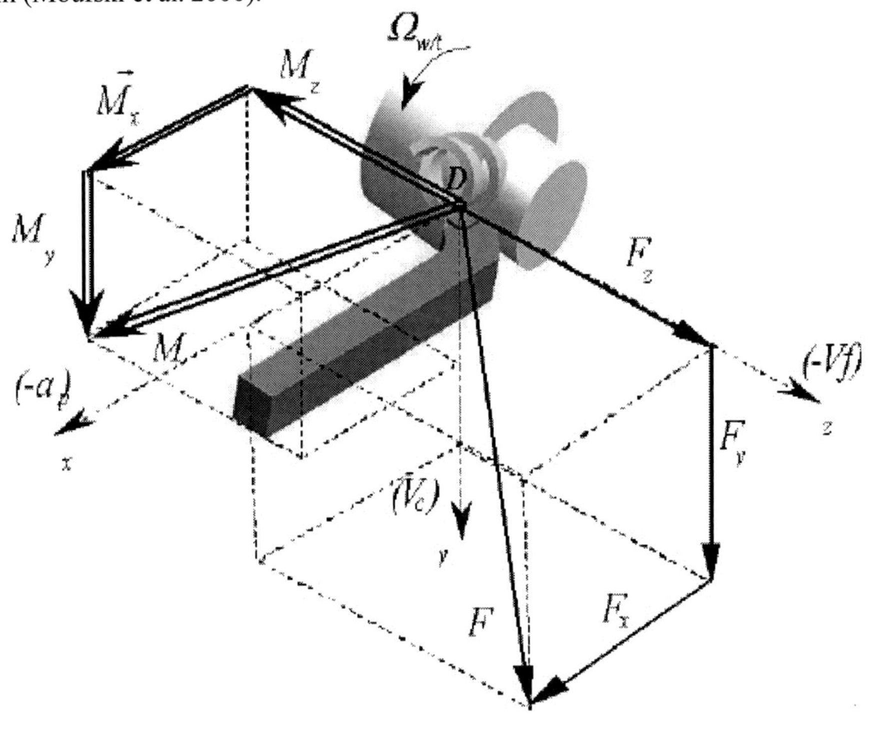

Figure 2. Three-dimensional cut in turning (kinematics and cutting actions).

The behaviour of the materials under such conditions is still unknown, and it seems possible that a better expression of the material load could be obtained by improving the behavioural laws of the material.

Friction and temperature (or flux) conditions at the interfaces between the tool and the chip are not easy to characterize or model.

Older mechanical models are too simplistic to explain and predict the physical phenomena occurring in the cutting process. Therefore, more complete analytical or numerical models must be developed for the cutting process to be representative of fit today's industrial reality.

As many companies are adopting HSM, it is also necessary for them to change from two-dimensional models to a three-dimensional models, in which the tool-chip contact is not be regarded as a punctual contact.

Some of the things that various existing models lack include the following:

- The requirement to take into account the 3D aspect of the mechanical actions exerted on the material,
- The requirement to produce better analysis of thermal aspects, their measurement, and their modelling,
- The requirement to develop experimental devices recreating the phenomena occurring at the tool-chip contact and to analyse those phenomena,
- The requirement to develop new, more complete behavioural laws adapted to the complex transformation of material.

This paper addresses these considerations.. The results presented here are the synthesis of our work toward developing a new approach to the cutting process. The test and model results are presented for various cutting conditions and machined material (steel, hard steel, and aluminium alloy) and for turning, hard turning, and milling configurations.

2. Moments at the Tip of the Tool

The synthesis of experimental statements in turning indicates the presence of six mechanical actions between the workpiece and the tool (Cahuc et al. 2001; Toulouse et al. 1997).

The existence of two (orthogonal cut) or even three (oblique cut) force components is beyond question. Conversely, the presence of a vector moment directly applied to the tip of the tool and not directly resulting from the preceding force vector is debatable as this concept goes against all theories based only on force vectors with mechanical effects on stress and strain states in the various zones identified.

Nevertheless, this approach makes it possible for the first time to obtain a balanced energy assessment, which means that an initial power input into a machine tool is thoroughly transformed into mechanical spindle power, while at the same time being degraded into thermal energy through the cutting process (Cahuc et al. 2001).

2.1. Experimental Devices

The experimental devices used in this test allow measurement of the 6 components of the mechanical actions and also of the thermal flux at the tip of the tool during machining.

The 6 components of the mechanical action are measured with the 6 component dynamometer (Couétard 1993) (figure 3).

Figure 3. The 6-component turning dynamometer, left-to right: on conventional lathe, on CNC lathe, and on milling.

The principles and the procedures of calibration that were consulted can be found in COUETARD's Thesis (Couétard 2000).

The dynamometer accuracy is about $\pm 8N$ for forces and $\pm 0.5Nm$ for moments in this loading case. Thus, six contact actions may be obtained in real time during the cutting process.

2.2. Examples of Experimental Mechanical Results

2.2.1. Turning Mechanical Results

Sample result obtained during an operation of hard turning on a cemented steel is given now. The tool insert used is a CBN TNMA 16 04 08 from SANDVIK COROMANT.

The vector moment is expressed at the tip of the tool (figure 4).

2.2.2. Milling Mechanical Results

High speed milling tests (N=17000 rpm $^{-1}$ - Vf=4080 mm.min $^{-1}$ - ap=6 mm - ar=1 mm) were carried out on an aluminum alloy piece with a double-edged milling tool under the configuration shown in figure 5. Figure 6 shows the mechanical actions measured during this test.

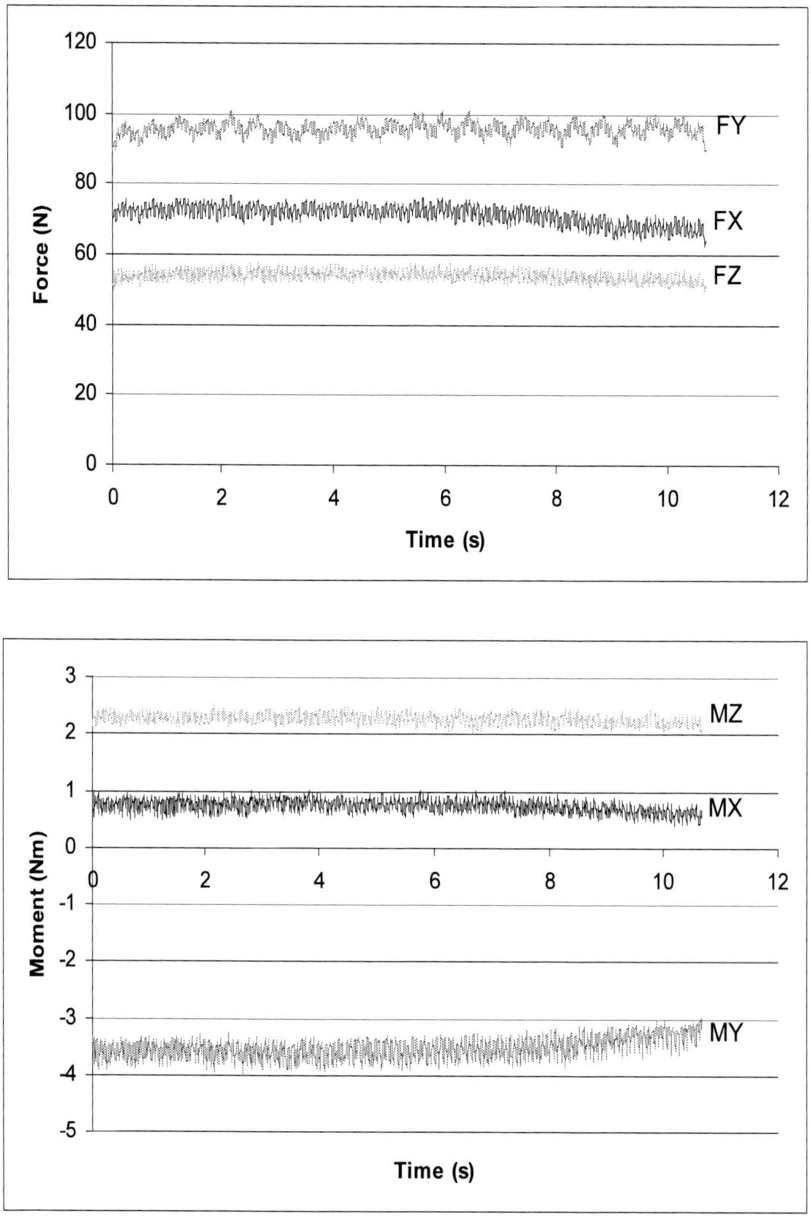

Figure 4. Measurement of the forces Fx, Fy, Fz; and of the moments Mx, My, Mz during an operation of turning (65 HRC steel); cutting conditions: Vc=173 $m.mn^{-1}$ - f=0.1 mm/rev. - ap=0.185 mm - machined diameter 69.45mm, N=793 tr/min).

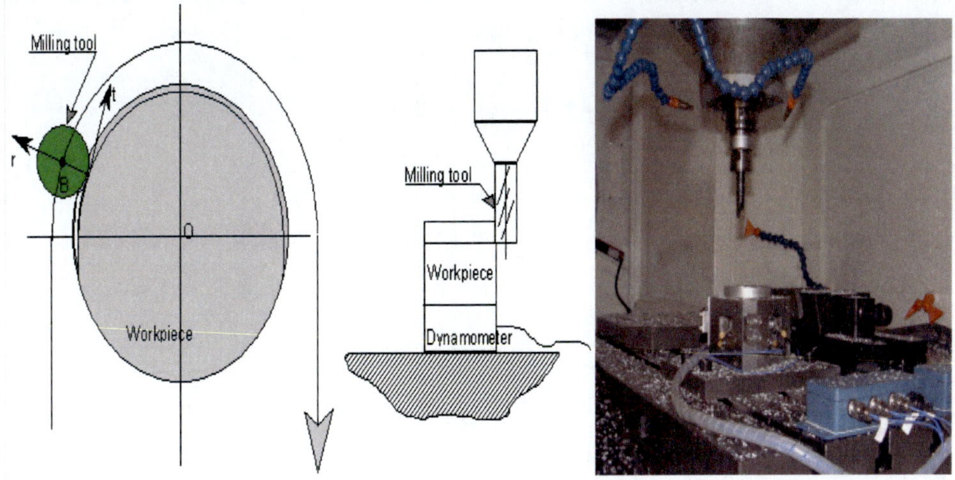

Figure 5. Left to right: High Speed Milling; Experimental device (illustration); experimental device (photograph).

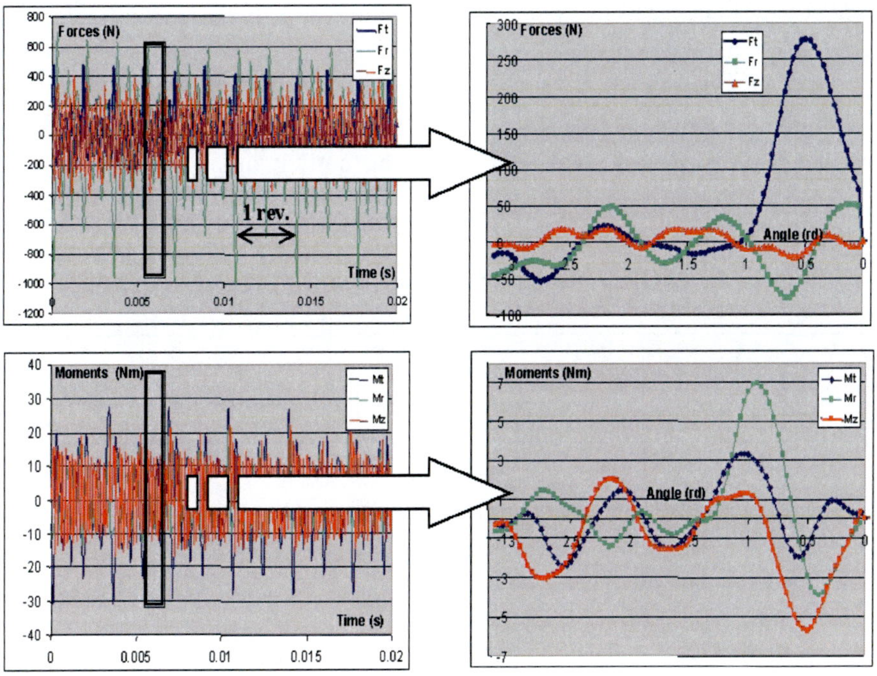

Figure 6. Extracted acquisition file; Results for a half revolution.

2.3. Analysis of the Mechanical Experimental Results

2.3.1. First Part - Energy Considerations

The presence of all the components has a direct effect on the total power consumed by the process. Indeed, mechanical power can be expressed with the static torsor T and the kinematic torsor C in a reference axis system related to the tool. Its origin D is a point located, by assumption, in the middle of the cutting edge.

$$P_m = \{T_{\text{Tool}\to\text{Workpièce}}\}_{D,x,y,z} \cdot \{C_{\text{Workpièce}\to\text{Tool}}\}_{D,x,y,z} = F_X V_X + F_Y V_Y + F_Z V_Z + M_X \Omega_X + M_Y \Omega_Y + M_Z \Omega_Z \quad (1)$$

During a longitudinal turning test in the direction $-\vec{z}$ (figure 1), the kinematic torsor has 3 non-null components V_Y, V_Z et Ω_Z. The total mechanical power can be expressed as

$$P_m = -F_Y V_Y - F_Z V_Z - M_Z \Omega_Z \quad (2)$$

where V_Y and V_Z are, respectively, the cutting speed and feed velocity expressed in m.s^{-1} and Ω_Z is the spindle speed expressed in rad.s^{-1}.

The feed velocity used for the tests was very weak, and the term $F_Z V_Z$ is only about 2% of the total consumed power. Thus, it may be ignored.

In practice, many users take into account only the first term ($P_m = F_Y . V_Y$), as they assume that the tool-chip contact is almost punctual . The torsor between the two bodies is then summarized with a two-component force vector. Note that this assumption was acceptable in 1945, and it served as a reference for reference models (Merchant 1945a; b). However, it is much too simplistic today. For the hard turning operations described above, we have shown (Couétard et al. 2001; Darnis et al. 2000) that the portion of power consumed ($F_y V_c$) accounts for only 60% of the total power (table 1).

Table 1. Mechanical power in hard turning

$F_y V_c$	$F_z V_f$	$M_z \omega$	P_m
277.5 ± 23.5 W	-0.07 ± 0.01 W	190 ± 42 W	467 ± 65 W
60%	□ 0%	40%	100%

Similarly, the results obtained during the high-speed milling tests show that it is important to integrate all the components of the cutting action torsor to obtain a better analysis of the phenomena involved (table 2).

Table 2. Estimate of mechanical power in high-speed milling

$F_t V_c$	$M_z \omega$	P_m
3966 W	10254 W	14220 W
27.9%	72.1%	100 %

2.3.2. Second Part - Mechanical Considerations

One of the mechanical ways to determine the existence of this moment vector consists of searching for the position of the central axis (i.e., the place of minimum moment value) in experimental results of turning (figure 7).

If this central axis passes by the tip of the tool, then the torsor is just a force vector. If the position of the central axis is several millimetres (even centimetres) off the tip of the tool point, then the presence of a complete cutting action torsor, and therefore of a vector moment, is undoubtedly validated.

Figure 8 shows the position of the central axis in the (O, X, Z) for various experimental cases, with error bars representing the measurement dispersion of the dynamometer (for these tests ± 8N on the resultant and ±0.5Nm on the moments). Figure 9 shows that the position of the central axis does not coincide with the tool's tip D, but instead, shows the presence of pure torque at the tool tip.

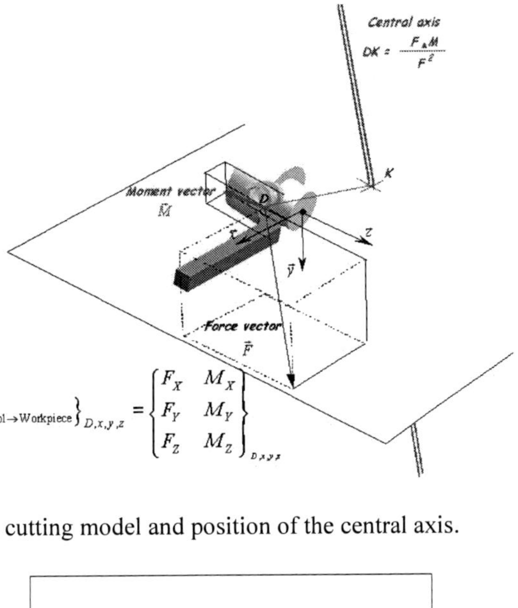

$$\{T_{Tool \rightarrow Workpiece}\}_{D,x,y,z} = \begin{Bmatrix} F_X & M_X \\ F_Y & M_Y \\ F_Z & M_Z \end{Bmatrix}_{D,x,y,z}$$

Figure 7. Three-dimensional cutting model and position of the central axis.

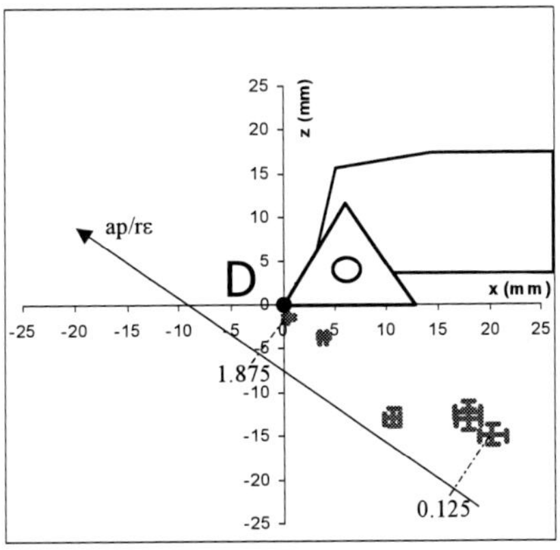

Figure 8. Example of the central axis positions (with accuracy bars) compared to the tip of the tool D in the oxz plan for ap/rε evolution tests.

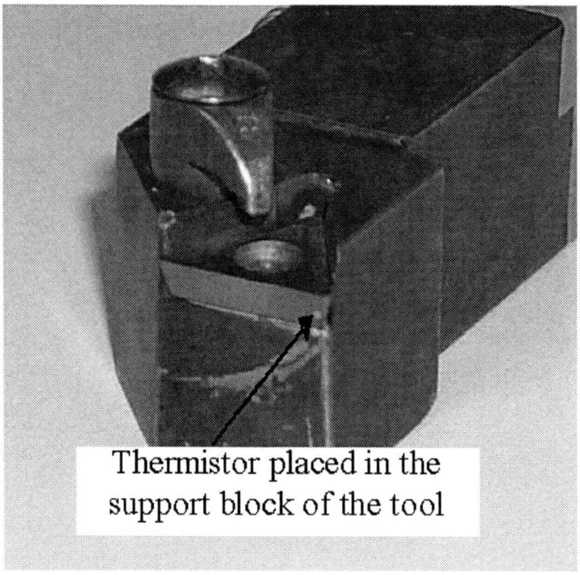

Figure 9. Installation of the temperature sensor (thermistor chip) on the tool.

The results presented in figure 8 show the variation in position of the central axis when the ratio ap/rε (cutting depth/nose radius of the tool) varies and Vc and f are constant.

Moreover, it appears that the position of the central axis is not influenced by the precision of the dynamometer, which validates the concept of a complete torsor.

2.4. Conclusion

This experimental work shows that the moments at the tip of the tool are real and that their presence has an influence on: the total cutting power consumed, and the geometry of the chip, its evacuation, the machined surface quality, the tool, and its wear.

The reliable information we obtained on the mechanical actions led us subsequently to search for experimental fluxes and temperatures generated by the process in order to chart its energy consumption, described below.

3. Temperature Measurement Device at the Tip of the Tool

To measure the temperature at the tip of the cutting tool, a thermistor chip on is affixed to the cutting tool as illustrated in figure 9. The presence of this chip in front of the cutting device limits the use of any other existing commercial solution (pyrometer, infra-red camera, and the like) to determine temperature at the tip of the tool.

In order to correlate the heat flux on the cutting edge with the temperature, a model of the heat transfer occurring in the tool is realized using an inverse method, by means of an instrument to measure the temperature (shown in figure 9) at one or more points far away from the cutting edge (figure 10).

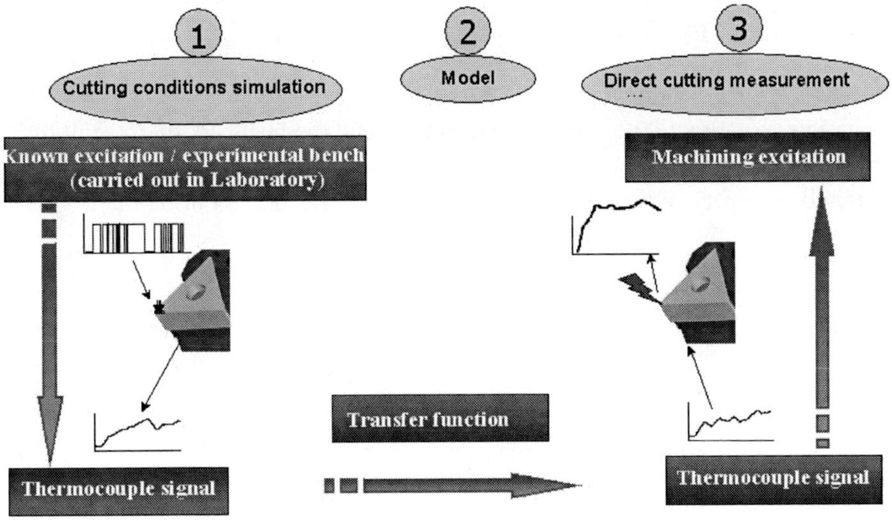

Figure 10. Procedure for measuring temperature at the tip of the cutting tool.

This model includes the following three phases:

Phase ①: Tool characterization. The parameters of a non-integer model of the cutting tool behaviour are estimated from measurements obtained using an experimental device, which reproduces the thermal conditions of machining on the cutting edge of the tool.

Phase ②: Inversion procedure. The procedure of inversion is carried out based on temperature characterization and tool sensor measurements indicated in the model.

Phase ③: Evaluation of fluxes and tool temperature. Fluxes and tool temperature are evaluated during machining with the transfer function using an exogenous autoregressive model (ARX) (Ljung 1987).

In this case, which uses a mono-variable system, one supposes the existence of a link between the variable of entry $q(t)$ (here the thermal solicitation imposed on the cutting edge: flow or average temperature) and the output variable (in our case the temperature at the tip of the tool $T(t)$). This relationship can be expressed as follow :

$$\sum_{i=0}^{M} a_i\, D^i T(x,t) = \sum_{i=0}^{L} b_i\, D^i q(t), \quad D^i = \frac{d^i}{dt^i}, \quad a_0 = 1 \tag{3}$$

where a_i and b_i are the identification parameters of the thermal behaviour model of the tool.

The number of parameters occurring in the non-integer identified model is relatively small, which makes enables a very fast simulation of the behaviour at the measurement point and a timely estimate of the imposed solicitation, in real time, on the cutting edge of the tool.

The complete work (Battaglia 2002; Battaglia et al. 1999; Battaglia et al. 2000a; Battaglia et al. 2000b; Puigsegur 2002; Puigsegur et al. 2001) allows further examination of the methods used.

3.1. Examples of Experimental Results

3.1.1. Thermal Measurement Results in Turning

It is possible to estimate the average temperature and the heat flux on the cutting edge of the tool by using the inversion procedure and the temperature measurement.

The signals measured at the sensor and calculated by the model during the machining test are presented in figure 11.

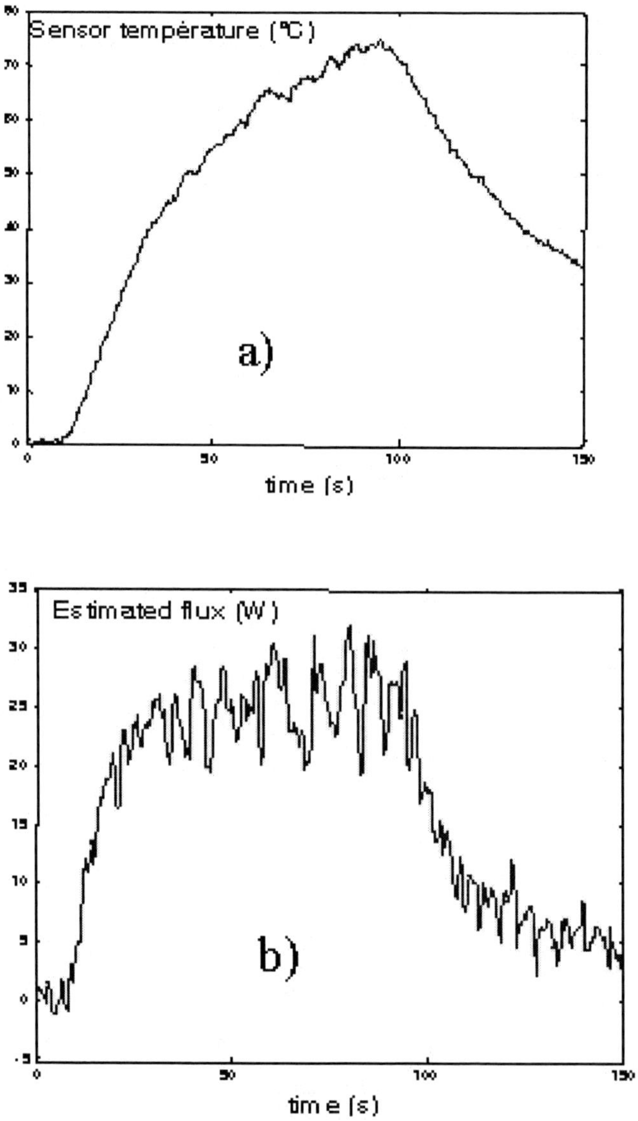

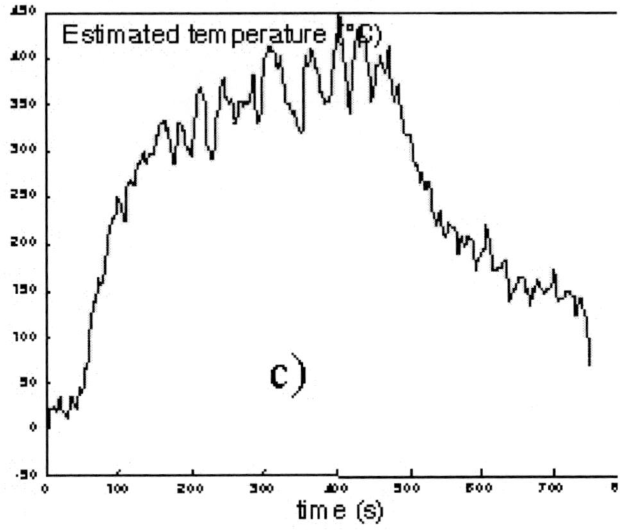

Figure 11. a) Temperature measured as at the sensor; b) estimated heat flux and c) estimated average temperature at the tip of the tool.

Finally, this experimental work on the thermal aspect of machining has facilitated a new experimental methodology for measuring the temperature and the flux at the tip of the tool. Other works have allowed us to obtain the temperature on the workpiece surface during machining (Battaglia et al. 2005). This methodology allows to improve existing models (Komanduri and Hou 2000; 2001a; b) of heat transfers in the workpiece, the chip and the tool without having to consider the contact resistance value between the tool and the chip.

4. Mechanical / Thermal Energy Balance

Concluding the experimental testing, some turning tests allow analysis of the three types of power used: P_E , power produced by the spin of the turning, mechanical power, P_m, established using the six-component measurement of the cutting actions, and finally thermal power, P_{th}, calculated using the temperature measurement at the tip of the tool. For the first time a combined mechanical/thermal energy assessment (Cahuc et al. 2001) is obtained using these three powers.

Two series of tests are carried out under traditional turning conditions in two configurations: (1) on a CC35 steel with a TNMM 16 04 08 insert and (2) under hard turning conditions on a cemented steel (65 HRC hardness) with a TNMM 16 04 08 CBN insert. The cutting parameters and necessary data for these tests are shown in table 3.

Table 4 shows the distribution of the mechanical power and dispersions related to the measuring devices, which indicates that the moments could not be neglected during these tests.

Table 5 shows the results obtained for the three types of power by various experimental methods described.

Table 3. Cutting parameters and data

First configuration						
D_i (mm)	N (rev/mn)	f (mm/rev)	a_p (mm)	γ (°)	λ (°)	α_t (°)
69.86	697	0.1	1	-6	24	6
V_c $(m s$	V_{chip} $(m s^{-1})$	V_f $(m s^{-1})$	ρC_p $(J m^3 K)$	Φ (°)	S (mm^2)	L (mm)
2.47	1.53	0.0011	$3.58 10^6$	30	0.1	0.173
Second configuration						
D_i (mm)	N (rev/mn)	f (mm/rev)	a_p (mm)	□□	□□	α_t □
69.45	840	0.1	0.185	-6	24	6
V_c $(m s$	V_{chip} $(m s^{-1})$	V_f $(m s^{-1})$	ρC_p $(J m^3 K)$	Φ (°)	S (mm^2)	L (mm)
2.89	1.79	0.0013	$3.75 10^6$	30	0.0185	0.173

Table 4. Distribution of the mechanical power

	$F_y V_c$	$F_z V_f$	$M_z \omega$	P_m
1st configuration	735 ± 20 W	0.21 ± 0.01 W	459 ± 36 W	1194 ± 56 W
2d configuration	277.5 ± 23.5 W	-0.07 ± 0.01 W	190 ± 42 W	467 ± 65 W

Table 5. Comparison of the three types of power: mechanical (Pm), thermal (P th), and spin power *(P_E)*

First configuration			Second configuration		
P_m	P_{th}	P_E	P_m	P_{th}	P_E
1194 W	1283 W	1139 W	467 W	477 W	439 W

Small variations (*5%* to *6.5%* between the power provided to the spin power P_E and the mechanical power P_m, and *7.5%* to *8.6%* between the spin power P_E and the thermal power P_{th}) point to a complete energy assessment for turning (Cahuc et al. 2001).

These results point to the necessity to consider the combined mechanical tool and material aspects for studying and modeling cutting phenomena.

5. Experimental Conclusions

The development of new mechanical and thermal experimental devices make it possible to obtain reliable results, which ascertain the presence of a complete action torsor at the tip of the cutting tool during machining.

This torsor alone does not bring a direct solution to the problems of process optimisation posed to manufacturers. But by analysing the torsor evolution during machining or comparing it with two other machining processes, together with other parameters such as tool wear or roughness, allows the improved comprehension of these phenomena, as well as substantial productivity gains .

These various developments have opened new fields of investigation for correctly analysing and modeling these phenomena. And further, this work allows us to understand that the phenomena identified cannot be modelled according to existing behavioural approaches, but require the development of adapted behavioural laws.

6. A Three-Dimensional Semi-Analytical Thermo Mechanical Cutting Model

The modelling of the complex phenomena during a machining operation must be performed with consideration for previous results. A model developed by TOULOUSE (Toulouse, 1998) and other following models that evolved from it evolved in the direction of energy aspects of the phenomena realize.

The main characteristic of this type of model is the predictive use of the presence of forces and moments at the tip of the tools during chip formation, in order to minimize the power consumed by the cutting process.

6.1. Model Description

A three-dimensional view of the initial model deviates from previous planar models and to deal with sets of the following requisite zones: two shear zones (primary ② and secondary ③) and two rake zones (primary ④ and secondary ⑤) Zone ① is the common zone shared by the other four zones and allowing their connection.

The principal assumptions relative to this model are:

- Permanent rate of flow
- Homogeneous, isotropic medium with Von Mises plastic threshold
- Stationary and incompressible plastic flow
- Constant shear stress along the average primary shearing line

The principal developments and results of this initial model are presented in Cahuc et al. 2001; and Toulouse 1998.

The significant modification of the previous modelling was applied to the geometrical description of the tool and thus also to the tool/chip/workpiece contact.

Figure 12 presents the pressure distributions currently used on the contact zones between the tool and the workpiece. The choice of the new pressure distribution on the cutting face is based on MOLINARI and DUDZINSKI's work (Moufski et al., 1998). A sensitivity study shows that the distribution used by TOULOUSE (Toulouse, 1998) over-estimates the pressure level. In addition, it is assumed that shear stresses in the secondary shear zone follow the same law. On the two rake faces, the shear pressure and stress distributions are assumed to decrease linearly (Toulouse, 1998).

The equilibrium of both the tool and the chip is expressed to determine the cutting actions applied to them (figure 12).

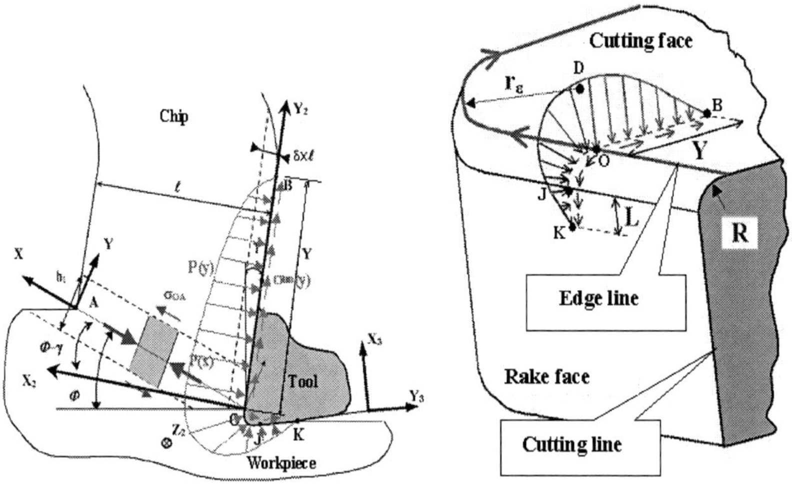

Figure 12. Loading representation and geometrical description.

The tool/workpiece contact zone is considered a continuous surface (figure 12). Both the cutting face and the rake face are included in this zone. The connection of these surfaces is made by the tool radius r_ε and the tool edge acuity R. This approach is in accordance with modern cutting tools, such as insert tools, which generally have a large radius. This approach also allows a more complete definition of the tool/chip/workpiece contact interface.

This three-dimensional contact surface is defined by decoupling the surface, starting from an edge line and a draft line (or cutting line).

The edge line defines the contact surface between the tool and the workpiece and is made up of a rectilinear zone and a curvilinear zone. The dimensions of these two zones vary according to (1) the geometrical parameters of the tool, such as the edge angle κ_r and the tool radius r_ε, and (2) kinematic parameters, such as the depth of cut a_p and the feed rate f.

The definition of the tool/workpiece contact surface is completed with the cutting line. The cutting line is defined by three zones (figure 12) which delimit it: the zone BO (or cutting face), the edge acuity zone OJ, and the zone JK (rake face). Geometrically, the cutting line is defined by (1) Y, the length between B and O points; and (2) secondly by L, the length between J and K points. The acuity radius R determines the connection between these two lines.

The length Y is unknown in the problem and will be evaluated using balance considerations. The tool/workpiece contact length L is estimated with the hydrostatic pressure P_0 in O, the Young modulus E of the workpiece, the cutting speed V_c, the speed velocity V_f, and the edge acuity R (Toulouse 1998).

In this new model, the assumption and the difficulty arise in taking into account the edge acuity radius R and the tool radius r_ε, which creates locally a torus surface, where a pressure and shear stress are applied. It results in a curvilinear integral equation:

$$\vec{R} = \int_S p.ds.\vec{n} + \int_S \tau.ds.\vec{t} \qquad (4)$$

All the components of the mechanical actions of the chip/tool system can now be expressed.

The resolution of the total energy problem consists in evaluating the unknown factors in two stages. In stage 1, the unknown kinematics factors (h_1 et Y) and the thermal factors are determined from the balance of the chip. In the second stage, the cutting actions are obtained by establishing the tool balance.

This work allows an improvement in tool geometrical modelling and chip kinematics. The whole of this work provides better knowledge of the influence of the parameters (such as the rate of engagement ap/r_ε) on machining and thus on the experimental models. The model was confirmed by video acquisition (Laheurte 2004; Laheurte et al. 2003) in order to consider the chip orientation during ejection.

6.2. Thermomechanical Model of the Cutting Process

This section describes an analytical thermal model of the phenomena occurring during a turning operation in order to integrate it into the preceding model. Existing models use empirical results (Boothroyd, 1963) for the primary shear zone, and these do not always allow determination of divisions of flux in the various zones (Tay et al., 1976).

It is assumed that the mechanical energy coming from plasticity, viscous dissipation of machined material, and frictions of the tool on the workpiece and on the chip is degraded into thermal energy (Cahuc et al., 2001). The physical phenomena in the cutting zone contribute to generating a heat flux in the tool, the chip, and the workpiece. The connection between the variations of these thermo-mechanical phenomena and the cutting parameters is fundamental to a total grasp of the process.

6.2.1. Thermal Transfers in Cutting Zones

To characterize thermal transfers in cutting zones, we undertook a study with analytical models and identification of non-integer systems of opposite problem resolution.

Modelling of thermal transfers in the workpiece and chip was carried out analytically, making it possible to partition the heat flux and to establish a thermal mapping of the workpiece and chip.

The modelling of thermal transfers in the cutting tool was performed differently. Tool geometry is complex and variable; and heat transfer is three-dimensional. The tool is composed of several elements whose thermo-physical properties are unknown due to their interrelationships.

Analysis of the thermal transfer between the chip and the tool is complex. The tribology phenomenon between the chip and the tool has been modelled by both a Coulomb-type and a "stick-slip" approach. Our approach, however, introduces the concept of contact thermal resistance and friction from the thermal point of view. But the analysis of contact thermal resistance is difficult because of the complexity of interaction of its components, that is, the mechanical actions (e.g., stresses, strains, and high-strain speeds), the intrinsic nature and state of the materials in the secondary shear zone, the nature of chip surfaces, and the surface quality of the tool. Furthermore, as these factors interact during machining, the actual thermal contact between chip and tool is imperfect, as shown in figure 13.

In the secondary shear zone, some analytical thermo-mechanical models (Lowen and Shaw 1954; Trigger and Chao 1951) of chip and tool, using the theory of mobile sources (Jaeger 1942), determine the distribution of the heat flux between the two solids and the surface temperatures of those solids. For these models, the tool-chip contact is assumed to be perfect and the coefficient of the flux division between the two elements is given as a function of the thermo-physical parameters of materials.

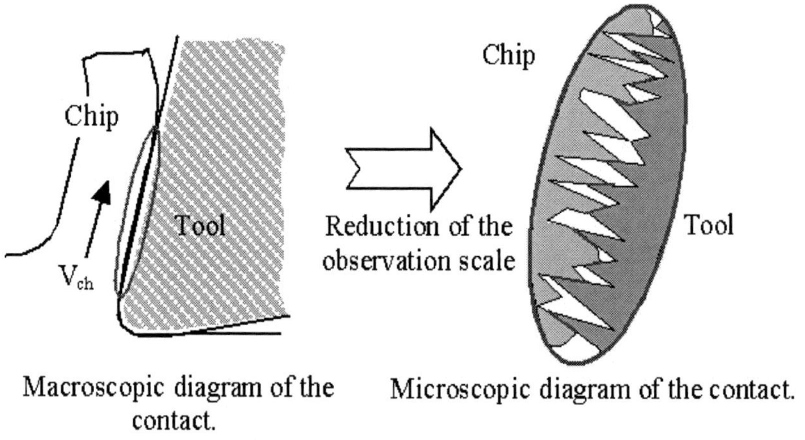

Figure 13. Tool-chip contact.

However, experimental tribology measurements (Laraqi, 1992) highlight the existence of considerable temperature variation between surfaces of the two solids. During friction, the surface asperities form a thermal barrier which disturbs exchange mechanisms to the interface. This in turn causes constriction of the lines of flux and a temperature discontinuity between surfaces of the two solids. This phenomenon is modelled by a thermal contact resistance or, if one of the solids is mobile, by slipping constriction resistance .

The non-stationary thermal behaviour of the tool is obtained using a non-integer identification system. The model of the tool we identify connects the heat flux or the average temperature on the cutting edge with the temperature at a distant point from the cutting edge. The average temperature is measured by a sensor placed on the tool (figure 9). The estimate of the thermal requests imposed on the tool are then carried out in real time, by resolution of the thermal inverse problem (Battaglia 2002; Battaglia et al. 2000a; Battaglia et al. 2000b). This method integrates the tool complexities (its several parts and its unknown thermal contact resistance factor) through a specific experimental device and a specific calibration.

6.2.2. Behaviour Laws in the Shear Zones

Changeux (Changeux 2001) identifies the parameters that influence the material behaviour during cutting tests on austenitic steels. In the primary shear zone, the author highlights the sensitivity of the mechanical behaviour of the machined material to deformation velocity and temperature. A Johnson-Cook law used to translate the influence of all these phenomena on the material behaviour correctly is expressed as

$$\sigma_{eq} = \left(A + B\varepsilon_{eq}{}^{n}\right)\left[1 + C\ln\left(\frac{\dot{\varepsilon}_{eq}}{\dot{\varepsilon}_{eq}^{0}}\right)\right]\left[1 - \left(\frac{T - T_{amb}}{T_{fus} - T_{amb}}\right)^{m}\right] \quad (5)$$

In our model, however, for precision reasons, the initial NORTON HOFF law is replaced by two different JOHNSON COOK laws (Joyot, 1994) one for primary shear zone and the other, for a secondary shear zone.

The plastic transformation that takes place around the primary shear zone involves a rise in temperature of about 100°C to 200°C in the configurations studied (i.e., for the tool-material couple and cutting conditions observed). In the depth of the primary shear zone, the temperature is then high but it does not vary during the abrupt direction change of the material. One generally speaks about adiabatic shearing. Thus, in the primary shear zone, the influence of the strain and strain velocity is predominant over that of temperature.

We chose to use the modified JOHNSON-COOK law (Joyot 1994), which does not take into account the temperature dependence of the mechanical behaviour of material on the material flow characterization in this zone

$$\sigma_{eq} = \left(A + B\varepsilon_{eq}{}^{n}\right)\left[1 + C\ln\left(\frac{\dot{\varepsilon}_{eq}}{\dot{\varepsilon}_{eq}^{0}}\right)\right] \quad (6)$$

The volume source term is then equal to the mechanical power consumed

$$Q_1 = \sigma_{OA}\dot{\varepsilon}_{OA} \quad (7)$$

In the secondary shear zone, the shear stress along (OB) is calculated with a complete Johnson Cook behaviour law:

$$\sigma_{OB} = \left(A + B\varepsilon_{OB}^{n}\right)\left[1 + C\ln\left(\frac{\dot{\varepsilon}_{OB}}{\dot{\varepsilon}_{0}}\right)\right]\left[1 - \left(\frac{T_{OB} - T_{amb}}{T_{fus} - T_{amb}}\right)^{m}\right] \quad (8)$$

This expression of σ_{OB} is first calculated with the temperature at point O. T_{OB} is calculated starting from the model of the thermal transfers in the chip due to the thermal influence of the primary shear zone, for the first iteration. Then, the σ_{OB} value is obtained using the average temperature on the segment $[OB]$ (tool-chip length contact calculated starting from the model of the heat transfers in the secondary shear zone).

The volume source term generated in the secondary shear zone is the sum of the shearing and friction terms

$$Q_2 = \sigma_{OB}\dot{\varepsilon}_{OB} + Q_{f2} \quad (9)$$

The source term induced by friction is written in the following way

$$Q_{f2} = \mu_2 N_{OB} V_{g2} / s_2 \tag{10}$$

with μ_2 the tool-chip friction coefficient, N_{OB} the normal force applied to (OB), V_{g2} the slip speed, and the tool-chip contact surface.

Thermo-mechanical coupling also participates in the expression (9). The temperature takes its place in the secondary shear zone stress calculation, along the tool-chip contact. As explained above, σ_{OB} is found after several iterations between the mechanical part (stress calculation) and the thermal part (calculation of the temperature along (OB)).

In the clearance zone, one supposes that the material has an elastic behaviour and that the only power consumed is that of the tool/workpiece friction.

The source term induced by the friction in the clearance zone is written in the following way

$$Q_3 = \mu_3 N_{JK} V_{g3} / s_3 \tag{11}$$

where μ_3 is the tool/workpiece friction coefficient in the clearance zone, N_{JK} is the normal force applied to (JK), $V_{g3} = V_c / \cos\alpha$ (α the clearance angle) and s_3 the tool-workpiece contact surface.

Results of an additional study (Puigsegur, 2002) then enabled us to show the sensitivity of the previous various source terms to the JOHNSON COOK law parameters. It is important to note that the combination of approximations of all the various parameters of the behaviour law and the friction law can lead to erroneous values for the sources terms. For this reason, we chose to combine the most influential parameters: h_1 for the source term in the primary shear zone, m and δ for the source term in the secondary shear zone, to obtain temperature values in the cutting zone, in coherence with the average temperature estimated at the cutting edge of the tool.

As we explained above, a global modelling must take into account the existence of the slipping resistance thermal contacts in both the secondary shear zone and the clearance zone. Their characterization is extremely delicate, and we show elsewhere (Puigsegur, 2002) that the thermal phenomena in the cutting zone can be uncoupled and treated separately. It is therefore possible to model the workpiece, the chip, and the tool independently, without the value of the slipping resistance thermal contact.

6.2.3. Thermal Balance Sheet

The thermal balance sheet in the cutting zone is thus carried out zone by zone:

- we use the results of KOMANDURI and HOU (Komanduri and Hou 2000; 2001a; b) in the primary shear zone,
- in the secondary shear zone, we use the model of the chip developed by KOMANDURI and HOU
- we extend the precedent model to the clearance zone,

Together, these models enable us to carry out the modelling of the thermal transfers in the cutting zone through a partition of the complete system, as illustrated in figure 14.

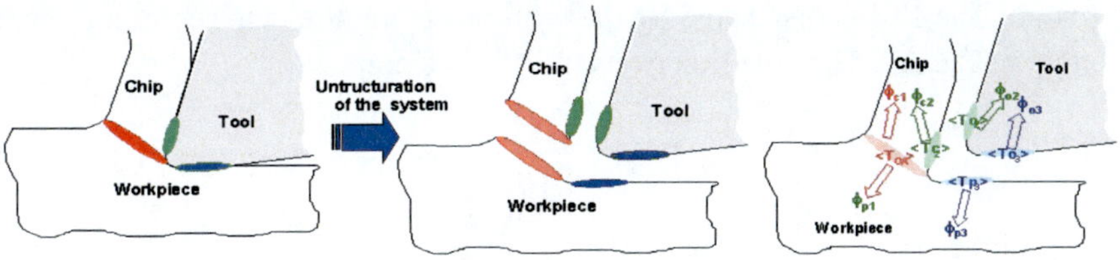

Figure 14. Partition of the cutting zone and modelling of thermal transfers in the three zones.

This last point constrains our modelling to the use of experimental results from the flux and temperature measurement at the tip of the tool during machining. Thus, to obtain the thermo-mechanical assessment of the cutting zone, we carry out a machining test, using a slide lathe, with a TNMA 16 04 08 tool insert with steel alloy of AISI4142.

Table 6 shows the parameters used for machining and calculations.

Table 6. Parameters used for calculations

V_c (m.min-1)	100	m	0,0003
f (mm.rev-1)	0,1	h_1 (mm)	0,5
a_p (mm)	1	δ	0,025
γ (°)	-6	ℓ (mm)	0,17
L (mm)	0,2	T_{amb} (K)	293
Φ (°)	29,15	T_{fus} (K)	1810
V_{cop} (m.min-1)	59,6	λ (W.m-1.K-1)	33
Y (mm)	1	ρC_p (J.m-3.K-1)	5313000
A (MPa)	598	α (m2.s-1)	$6{,}21.10^{-6}$
B (MPa)	768	N_{OB} (Pa)	$1{,}4530.10^{3}$
C	0,0137	μ_2	0,4
n	0,2092	φ_2 (W.m-2)	$2{,}62.10^{8}$
$\dot{\varepsilon}_0$ (s-1)	0,001	φ_o (W.m-2)	2.107
φ_1 (W.m-2)	7,59.10 8		

The results provide a complete heat assessment of the cutting zone, presented in figure 15, that is, the distribution of the flux in the tool, the workpiece and the chip, the thermal mapping of the workpiece and the chip, as well as the estimated average temperature on the cutting edge of the tool.

The results show that the chip evacuates the greatest part of the thermal energy created in the cutting zone. The fluxes dissipated by the workpiece and the chip represent, respectively, approximately 8.5% and 7% of the flux evacuated by the chip.

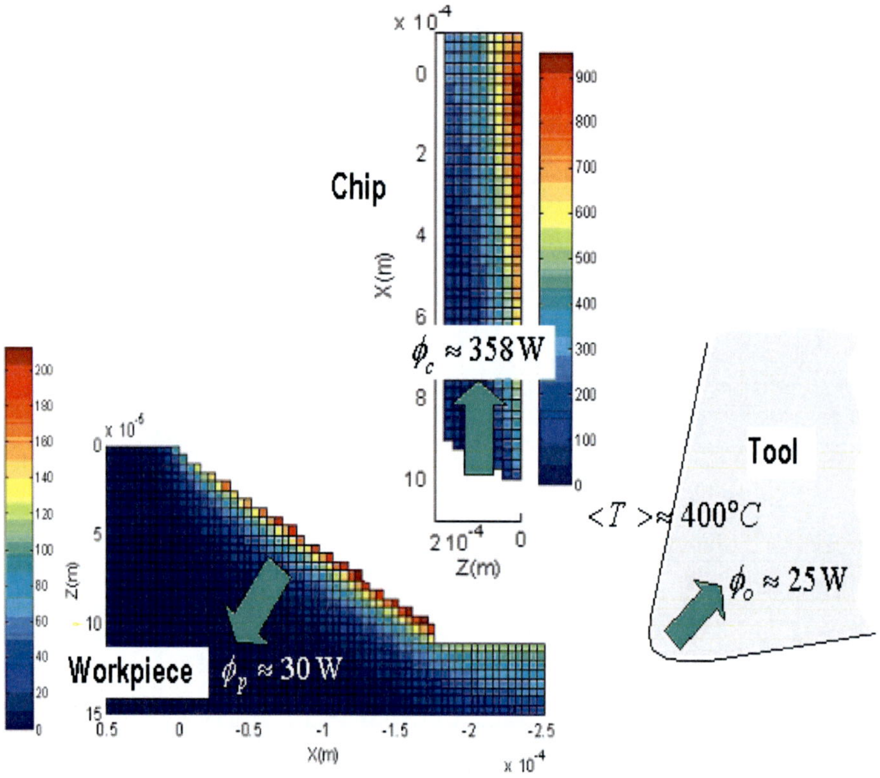

Figure 15. Distribution of flux and thermal cartography of the cutting.

6.2.4. Analysis

It should be noted that the results obtained from the thermal part of this model depend strongly on the coefficients used in the behaviour laws and on the friction and geometrical parameters (h_1, Φ, Y, δ). We used the parameter values of the behaviour laws given in the literature (Joyot 1994). In the same way, the value of the friction coefficient results from (Joyot 1994; Toulouse 1998) and the geometrical parameters were fixed on the basis of physical considerations derived from observations of the tool inserts.

It is obvious that the thermal mappings obtained for the chip and the workpiece using the values of these various parameters have an unknown amplitude error. The thermal and mechanical models are coupled through the behaviour law used for the secondary shear zone, and the error is redundant for the chip thermal mappings, since it is calculated from the primary shear zone results.

In the case of the primary shear zone, the most influential factor is its thickness h_1. A 5% variation of h_1 involves a 50% variation of the Q_1 source term in the primary shear zone. The

h_1 values included (for our sensitivity study), between 1.10^{-5} m and 1.10^{-3} m, indicate that an error can easily be made on this value parameter and thus on the source term value.

It is equally true for the geometrical parameters Y and δ in the secondary shear zone. However, the parameter that most influences the source term due to shearing is the thermal softening coefficient m. It translates the temperature influence on the structural properties of the material and thus on its flow. This parameter has an important influence also for lower-than-1 values. Beyond these values, the term raised with the power m is so small that it becomes negligible against 1 and thermal softening has no more influence. When m is lower than 1, the dependence of the source term due to shearing is very strong. A 5% variation of m involves a 67% variation of Q_2.

The coefficient of the friction law μ_2 and the slip velocity V_{g2} are also significant since the source term due to friction Q_{f2} depends on it linearly.

It is thus significant to note the importance of the estimate of the behaviour laws and friction parameters. We are currently focusing all our attention on this point. The installation of experiments that highlight the relevant parameters related to the rewriting of new behaviour laws is no easy matter.

As these new lawsare developed, the thermal work described here is been integrated into the mechanical model and makes it possible to improve modelling of the thermo mechanical phenomena occurring in a turning operation.

An example of comparison between experimental results and simulation in 2D and 3D cases is shown in table 7.

Table 7. Compared results between simulation and experimental results in 2D and 3D cases

	Fx (N)	Fy (N)	Fz (N)	Mx (Nm)	My (Nm)	Mz (Nm)	Computed power (W)
3D Model	369	992	436	- 4.85	-11.6	9,47	2614
3D experimental	391	1006	485	-10,84	-11,96	6,72	2458650
2D Merchant	***	985	438	***	***	***	1969
Oxley model	44	933	480	***	***	***	1866
2D experimental	***	1006	485	***	***	***	2010650

The cutting conditions are as follow :

Nuance of machined steel AISI4142, Vc = 120 m/min, f= 0.2 mm/tr, ap = 2 mm, cutting angle $\gamma = 7°$, clearance angle $\gamma = 7°$, workpiece diameter D= 59 mm.

Table 7 shows that the realized model must take into account all the mechanical components of the cutting actions to be representative of the real process.

7. Analysis and Discussion on the Validity of the Behaviour Laws Used Classically in Cutting Models

Our modelling is currently built around a modified law (Joyot, 1994) characterizing material flow in the primary shear zone, and the complete JOHNSON-COOK law in the secondary shear zone. In this zone, the contact between the tool and the chip must be modelized.

Our most recent work concerns the validity of the behaviour laws in these two zones, particularly in the more complex secondary shear zone. It was selected because the present phenomena are more easily reproduced on an experimental device independent of the machining operation. Thus, the material behaviour could be identified through new forms of behaviour laws.

7.1. Reproduction of Complex

Contact Tool-Chip Phenomena

7.1.1. Experimental Device

The experimental system used proposes to reproduce the friction strain observed during machining in the secondary shear zone.

The experimental system is based on the slug-disk system (figure 16) widely used in tribology. It functions like this: A normal load presses the two elements, the AISI4042 disc being in rotation compared to the tool insert (TNMA 16 04 R_{ε}).

The use of a six-component dynamometer on a tribometer assembled on a CNC turning machine (figure 16) makes it possible to take measurements during the loading phase.

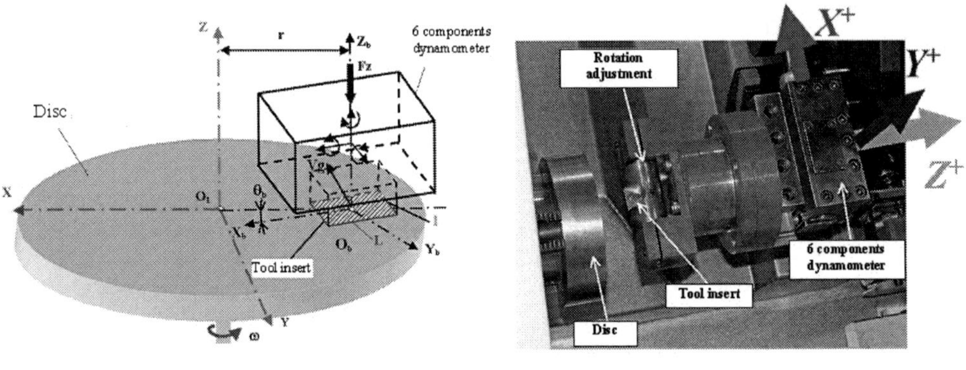

Figure 16. Tribometer (slug-disk system) and assembly on the CNC turning machine.

The moment's measurement on the friction surface highlights complex phenomena. For this experimental study, the complementary analyses (such as micrographic observations, rubbing surface measurements) show the non-linear character of the sub-surface strain evolution due to hard friction between the two bodies.

These observations reveal that the material subjected to intense shearing is subject to the rotation strain which can be modelled by a strain gradient. Therefore, modelling of this type of phenomena requires a fine mechanical analysis. The classical theory of the continuous media is no longer sufficiently representative. The second gradient theory, which integrates the material strain, is representative of the observed phenomena.

7.1.2. Results of the Friction Forces

Figure 17 represents the measurements results of the forces applied by the tool insert on the disc in the reference measurement coordinate system (O, X, Y, Z) (figure 16).

These measurements of forces make it possible to quantify the Coulomb friction coefficient (μ_2 = Fy/Fz).

The friction surface evaluation *S* (figure 18) allows definition of the pressure applied on the friction surface. The evolution of this friction coefficient compared to pressure is also analysed (figure 17).

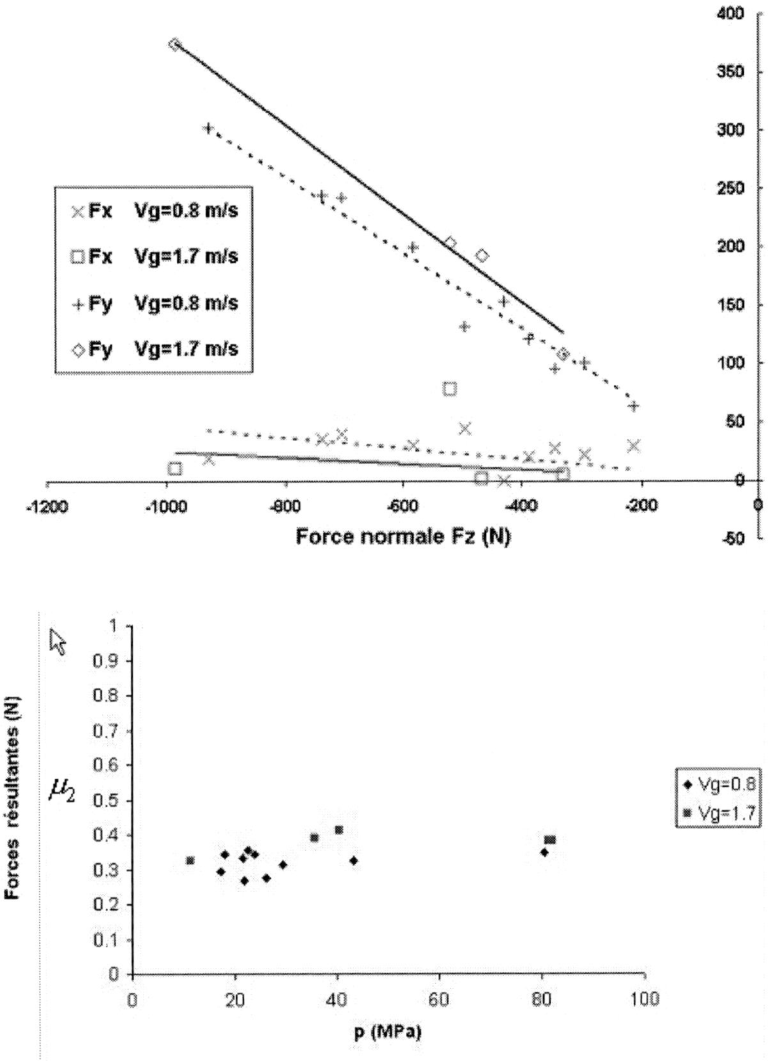

Figure 17. Friction forces and evolution of the friction coefficient.

Figure 18. Real friction surface.

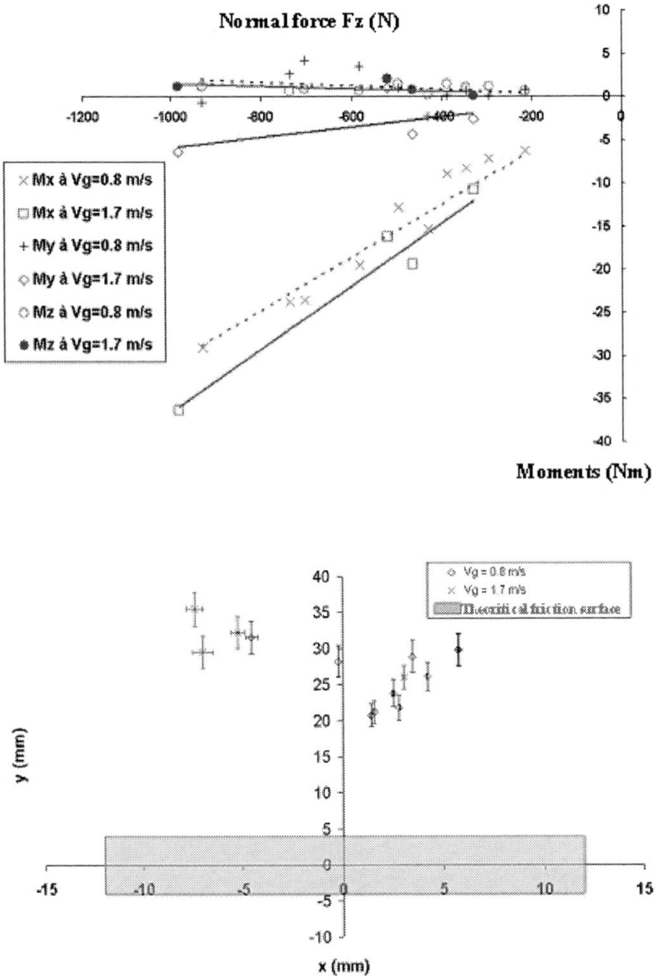

Figure 19. Moments in the centre of the friction surface and positions of the central axis compared to the friction surface.

Other tests carried out under similar conditions (Meiller 2002; Ryckelynck and Meiller 2002) corroborate these results and thus validate the experimental device.

7.1.3. Results of the Friction Moments

The moments are measured in the dynamometer centre point and are then expressed in the centre O_b of the theoretical friction surface. These moments represent the tool-insert actions on the disc in the reference coordinate system (O, X, Y, Z) and are expressed in figure 19.

If the Coulomb assumption is verified, it is possible to consider that the moments in the centre of the friction surface are equal to zero. The moments around the axes Y and Z cancel each other out, and the non-null values can be assimilated to an orientation default of the disc introduced during the assembly into the rotation adjusting device. The moment around the X-axis is not null and takes important values, which contradicts the Coulomb assumption. The moment's measurement highlights a phenomenon not taken into account in the Coulomb models, and which cannot be assimilated to a bearing, or revolving, moment such as the ones defined in the literature (Zhuravlev, 1998).

This moment may be due to the effect of a sub-surface strain (figure 20) with strong gradient.

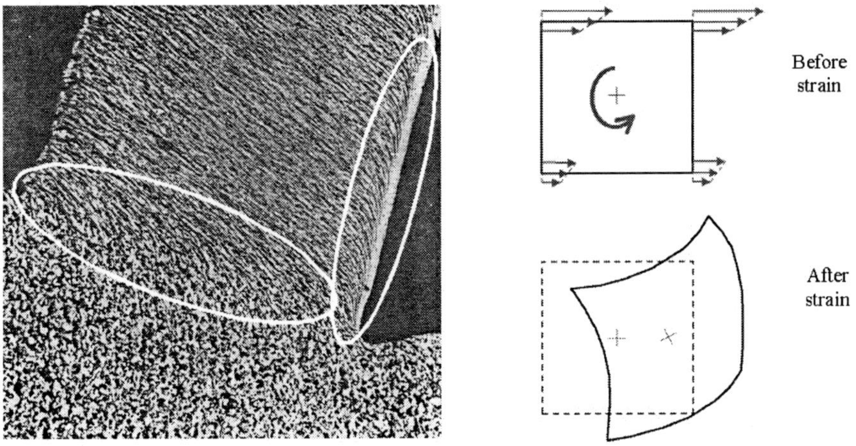

Figure 20. Micrograph of the strain gradient zones and 2d gradient modeling.

7.2. Second Gradient Theory – New Behaviour Law Forms for the Secondary Shear Zone

An intense shearing load generates the strain observed under the friction surface, and the non-linear evolution of the strain imposes a modelling that represents the strain gradient and the stress gradient, which generate the moments measured (figure 21).

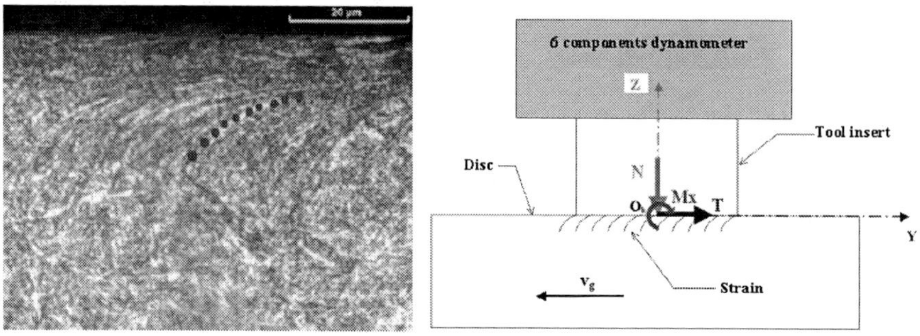

Figure 21. Micrograph of the friction-deformed material - Diagram of the experimental system.

The most adequate generalized theory is the second gradient theory. It is a generalization of the classical theory of the continuous media (Germain, 1973a), and the stress couple theory can be developed from this theory (Germain, 1973b) with some specific considerations. The description of the movement allows introduction of the material strain, as in the classical theory of the continuous media, as well as the strain gradient (table 8).

Table 8. Equations for the first and the second gradient theories

First gradient theory
$\mathcal{P}_i + \mathcal{P}_c + \mathcal{P}_d = \mathcal{P}_a$
$\mathcal{P}_i = -\int_{\mathcal{D}} \sigma_{ij}\dot{\varepsilon}_{ij}\,d\upsilon$
$\mathcal{P}_i = \int_{\mathcal{D}} \sigma_{ij,j}U_i\,d\upsilon + \int_{\partial\mathcal{D}} \sigma_{ij}n_jU_i\,dS$
$\mathcal{P}_c = \int_{\partial\mathcal{D}} T_iU_i\,dS$
Second gradient theory
$\mathcal{P}_i + \mathcal{P}_c + \mathcal{P}_d = \mathcal{P}_a$
$\mathcal{P}_i = -\int_{\mathcal{D}} \left(\sigma_{ij}\dot{\varepsilon}_{ij} + \mu_{ijk}\dot{\eta}_{ijk}\right)d\upsilon$
$\mathcal{P}_i = \int_{\mathcal{D}} \mathcal{F}_iU_i\,d\upsilon + \int_{\partial\mathcal{D}} \left(\mathcal{T}_iU_i + \mathcal{M}_i\omega_i + \mathcal{N}_iD_{\underline{nn}}\right)dS + \int_{\Gamma} \mathcal{R}_iU_i\,ds$
$\mathcal{P}_c = \int_{\partial\mathcal{D}} \left(T_iU_i + M_i\omega_i\right)dS + \int_{\Gamma} R_i\left(DU_i\right)ds$

The movement description then becomes more precise and the strain gradients are in agreement with the rotation strain phenomena modelling. This method was developed in the 1960s by MINDLIN (Mindlin and Eshel, 1968), TOUPIN (Toupin, 1962), and GERMAIN (Germain, 1973b). These works define the bases of the second gradient theory, but only in the case of linear elasticity. The case of plasticity was treated in the later work of FLECK and

HUTCHINSON (Fleck and Hutchinson, 1997); finally, the case of visco-plasticity was treated later by GURTIN (Gurtin, 2003).

The loading is imposed in the (y,z) plan; therefore, the assumption of plane stresses is used. The assumption of the plan problem and the plane stresses imply (Tomasi, 1994):

- the non evolution according to x ; the derived terms compared to x are null,
- the first order planes stresses σ_{ix} are null $\forall i$,
- the second order stresses μ_{ij} are null $\forall i,j$ except μ_{yx} and μ_{zx},
- the second order stresses μ_{ijk} are null $\forall i,j,k=x$.

Application of the preceding assumptions and use of the developments resulting from LAHEURTE (Laheurte 2004) generates the following equilibrium equations:

$$\sigma_{yy,y}+\sigma_{yz,z}-\mu_{yyy,yy}-\mu_{yzy,yz}-\mu_{yyz,zy}-\mu_{yzz,zz}=0 \tag{12}$$

$$\sigma_{zy,y}+\sigma_{zz,z}-\mu_{zyy,yy}-\mu_{zzy,yz}-\mu_{zyz,zy}-\mu_{zzz,zz}=0 \tag{13}$$

Application of the same assumptions to the boundary conditions gives us the following equations:

for the boundary conditions of couple surface density

$$M_{z1}=\mu_{zx}(y,0),\quad \forall y, \tag{14}$$

for the boundary conditions of forces surface density

$$\begin{aligned} T_{z2}&=\sigma_{zy}(y,0)-\mu_{zyy,y}(y,0)-\mu_{zyz,z}(y,0),\quad \forall y,\\ T_{z3}&=\sigma_{zz}(y,0)-\mu_{zzy,y}(y,0)-\mu_{zzz,z}(y,0),\quad \forall y. \end{aligned} \tag{15}$$

These new formulations obtained for the boundary condition and equilibrium equations indicate the new elements which must appear in the behaviour laws. These laws are defined by the following equations:

$$\dot{\varepsilon}^p_{ij}=\frac{3}{2}H(f)g'(\Sigma_{eq})\frac{\langle\dot{\Sigma}_{eq}\rangle}{\Sigma_{eq}}\sigma'_{ij} \tag{16}$$

$$\dot{\eta}^p_{ijk}=H(f)g'(\Sigma_{eq})\frac{\langle\dot{\Sigma}_{eq}\rangle}{\Sigma_{eq}}\sum_{I=1}^{3}\left[\ell_I^{-2}\mu'^{(I)}_{ijk}\right] \tag{17}$$

The overall effective stress tensor Σ_{eq} is then defined by:

$$\Sigma_{eq}{}^{2} = \sigma_{eq}^{2} + \overset{=2}{\mu}_{eq} + \sum_{I=1}^{3}\left[\ell_{I}^{-2}\overset{\equiv 2}{\mu}{}^{I}_{eq}\right] \tag{18}$$

After application of the orthogonal decomposition (Laheurte, 2004) , it becomes:

$$\sigma_{eq}^{2} = \frac{3}{2}\sigma'_{ij}\sigma'_{ij} = 3.\left(\sigma_{zy}\right)^{2} \tag{19}$$

$$\overset{=2}{\mu}_{eq} = \frac{3}{2}\mu_{ij}\mu_{ij} = \frac{3}{2}.\left(\mu_{zx}\right)^{2} \tag{20}$$

$$\left(\overset{\equiv 1}{\mu}_{eq}\right)^{2} = \mu^{1}_{ijk}\mu^{1}_{ijk} = \frac{12}{5}\left(\mu_{yzz}\right)^{2} + \frac{12}{5}\left(\mu_{yyz}\right)^{2} \tag{21}$$

$$\left(\overset{\equiv 2}{\mu}_{eq}\right)^{2} = \mu^{2}_{ijk}\mu^{2}_{ijk} = \frac{1}{576}\left(\mu_{yzz}\right)^{2} \tag{22}$$

$$\left(\overset{\equiv 3}{\mu}_{eq}\right)^{2} = \mu^{3}_{ijk}\mu^{3}_{ijk} = \frac{53}{576}\left(\mu_{yzz}\right)^{2} + \frac{13}{90}\left(\mu_{yyz}\right)^{2} \tag{23}$$

These developments are deferred in the relation (16) by taking a cubic definition of the characteristic volume of material as an assumption. The form of the overall effective stress tensor is written as:

$$\begin{aligned} \Sigma_{eq} &= \sqrt{\sigma_{eq}^{2} + \overset{=2}{\mu}_{eq} + \sum_{I=1}^{3}\left[\ell_{I}^{-2}\overset{\equiv 2}{\mu}{}^{I}_{eq}\right]}, \\ &= \sqrt{3.\left(\sigma_{zy}\right)^{2} + \frac{3}{2}.\left(\mu_{zx}\right)^{2} + \frac{1}{\ell_{c}}\left(\frac{399}{160}\left(\mu_{yzz}\right)^{2} + \frac{229}{90}\left(\mu_{yyz}\right)^{2}\right)}. \end{aligned} \tag{24}$$

The definition of the form of the work hardening function $g'(\Sigma_{eq})$ characterizing the plasticity state in relations (16) and (17), requires the evaluation of the field of strain velocity. A first approach to measurement of this field within the material was elaborate. The micrograph (figure 20) reveals the trajectory of a material particle. From this trajectory, it is possible to determine the evolution of the displacement field rates and, indirectly, also the field of strain velocity. The observation is carried out after loading, which does not allow estimation of the duration of displacements during the test: the current test duration being 10 seconds, it is impossible to determine if the strain state was obtained at the end of 2 seconds, 5 seconds or

10 seconds. A complementary study is needed to remove this ambiguity and to obtain more information on the evolution of the strain velocity field.

8. Conclusion

Machining processes introduce complex phenomena during chip formation, relating to various fields such as deformable solid mechanics or thermal phenomena, which make it difficult to control energy parameters. Measurements carried out using a six-component dynamometer allow a better understanding of these phenomena. They reveal the presence of moments at the tip of the tool, which the traditional cutting models fail to evaluate.

A first three-dimensional cutting modelling integrating this concept of moment and based on the theory of the stresses couples was adopted in 1995, but it did not seem totally adapted to real cutting phenomena. As a matter of fact, observations of the chip strain show that the deformations have a non-linear evolution, which leads to use of the stress gradient concept. The second gradient theory is the best suited to describe the cutting phenomena occurring in the secondary shear zone.

Thus, the principal developments of the second gradient theory were developed. Works resulting from the literature (Germain, 1973a; b) present a method theory wherein all the terms that might be encountered are defined. This theory can thus be adapted to a great number of cases. In our case, the developments were adapted in order to be directly applied to machining.

Recent works of FLECK and HUTCHINSON (Fleck and Hutchinson, 1997) have made it possible to establish the general form of the behaviour laws using the second gradient theory for cases of plasticity. Actually, works dealing with the cases of high strain and high strain velocity do not exist, as yet.

Theoretical elements were presented, but the goal of this work is to provide behaviour laws adapted to the cutting phenomena in the two shear zones. The analytical and numerical resolution of the presented equations has not yet been achieved. This resolution will determine the evolution of the stresses and stress gradient fields applied to the material. The microscopic observations carried out after loading, as well as an analysis of the strain state, will provide the strain and the strain gradient fields. Then, the behaviour law will define the connection between stresses and strain as well as their gradients. Complementary experiments must be carried out to completely determine the evolution of the strain and strain gradient fields.

To validate these works, experiments are being developed to reproduce the shearing strains generated by a "stick and slip" loading similar to the secondary shearing ones. Our results show the relevance of this modelling theory. Significant experimental and theoretical works remain to be carried out, first to determine the values of the coefficients of the behaviour law in the secondary shear zone and also to show that this theory can also find its application in the primary shear zone.

ACKNOWLEDGMENTS

This article has been previously published in International Journal of Forming Processes, LAVOISIER, Volume 10, N°2-2007, pp 235-269. With the agreement of Jean-Marc Bocabeille Editorial manager of Lavoisier SAS reviews - 14 rue de Provigny 94236 CACHAN CEDEX

The authors would like to thank the Centre National de la Recherche Scientifique (CNRS) UMR 5469 for the financial support, the publisher LAVOISIER for its permission to reuse papers cited above.

REFERENCES

Battaglia, J. L. (2002). "Méthodes d'identification de modèles à dérivées d'ordre non entier et de réduction modale – Application à la résolution de problèmes thermiques inverses dans des systèmes industriels," Bordeaux 1, France.

Battaglia, J. L., Cahuc, O., Batsale, J. C., and Toulouse, D. (1999). "Métrologie des transferts thermiques dans un bec d'outil en condition d'usinage." *Mécanique Industrielle et Matériaux*, 52(1), 155-161.

Battaglia, J. L., Cois, O., Puigsegur, L., and Oustaloup, A. (2000a). "Solving an inverse heat conduction problem using a non-integer identified model." *International Journal of Heat and Mass Transfer*, 14(44), 2671-2680.

Battaglia, J. L., Le Lay, L., Batsale, J. C., Oustaloup, A., and Cois, O. (2000b). "Heat flow estimation through inverted non integer identification models." *International Journal of Thermal Science*, 39(3), 374-389.

Battaglia, J. L., Puigsegur, L., and Cahuc, O. (2005). "Estimated temperature on a machined surface using an inverse approach." *Int J. Experimental Heat Transfer* 18, 13-32.

Boothroyd, G. (1963). "Temperatures in orthogonal metal cutting. GB IME LONDON " *Proceedings of the Institution of the Mechanical Engineers*, 177(29), 789-802.

Cahuc, O., Darnis, P., Gérard, A., and Battaglia, J. L. (2001). "Experimental and analytical balance sheet in turning applications." *The International Journal of Advanced Manufacturing Technology*, 18(9), 648-656.

Changeux, B. (2001). "Loi de comportement pour l'usinage. Localisation de la déformation et aspects microstructuraux," Ecole Nationale Supérieure des Arts et Métiers centre de Paris, France.

Couétard, Y. (1993). "Capteur de forces à deux voies et application à la mesure d'un torseur de forces."

Couétard, Y. (2000). "Caractérisation et étalonnage des dynamomètres à six composantes pour torseur associé à un système de forces," Université Bordeaux 1, France.

Couétard, Y., Cahuc, O., and Darnis, P. (2001). "Mesure des 6 actions de coupe en fraisage grande vitesse." *Third International Conference on Metal Cutting and high speed machining – CIRP – AFM, Metz - France*.

Darnis, P., Cahuc, O., and Couétard, Y. (2000). "Mesure des actions mécaniques nécessaires à l'établissement du bilan énergétique complet lors d'une opération de tournage." *International Seminar on Improving Machine Tool Performance, La baule*, La baule.

Fleck, N. A., and Hutchinson, J. W. (1997). "Strain gradient plasticity." *Advances in applied mechenics*, 33, 295-361.

Germain, P. (1973a). "La méthode des puissances virtuelles en mécanique des milieux continus. Première partie: Théorie du second gradient." *Journal de la Mécanique*, 12(2), 235-274.

Germain, P. (1973b). "The method of virtual power in continuum mechanics. Part 2: Microstructure." *SIAM J. Appl. Math*, 25(3), 556-575.

Gurtin, M. E. (2003). "On a framework for small-deformation viscoplasticity: free energy, microforces, strain gradients." *International Journal of Plasticity*, 19(1), 47-90.

Jaeger, J.-C. (1942). "Moving sources of heat and the temperature at sliding contacts." *Proceedings of the Royal Society of New South Wales*, 203-224.

Joyot, P. (1994). "Modélisation numérique et expérimentale de l'enlèvement de matière, Application à la coupe orthogonale," Université de Bordeaux 1, France.

Komanduri, R., and Hou, Z. B. (2000). "Thermal modelling of the metal cutting process - Part I : Temperature rise distribution due to shear plane source." *International Journal of Mechanical Sciences*, 42, 1715-1752.

Komanduri, R., and Hou, Z. B. (2001a). "Thermal modelling of the metal cutting process - Part II : Temperature rise distribution due to frictional heat source at the tool-chip interface." *International Journal of Mechanical Sciences*, 43, 57-88.

Komanduri, R., and Hou, Z. B. (2001b). "Thermal modelling of the metal cutting process - Part III : Temperature rise distribution due to the combined effects of shear plane and at tool-chip interface frictional heat source." *International Journal of Mechanical Sciences*, 43, 89-107.

Laheurte, R. (2004). "Application de la théorie du second gradient à la coupe des matériaux ", Université de Bordeaux 1, France.

Laheurte, R., Darnis, P., and Cahuc, O. (2003). "Energy assessment and cutting tool behavior in machining." *Recent Advances in Integrated Design and Manufacturing in Mechanical Engineering*, ISBN 1-4020-1163-6, 281-290.

Laraqi, N. (1992). "Température de contact et coefficient de partage de flux généré par frottement sec entre deux solides. Approche nouvelle de la génération de flux." *International Journal of Heat and Mass Transfer*, 35(11), 3131-3139.

Ljung, L. (1987). *System identification: theory for the user.*

Lowen, E. G., and Shaw, M. C. (1954). "On the analysis of cutting-tool temperatures." *Transaction of ASME*, 71, 217-231.

Meiller, M. (2002). "Etude expérimentale du frottement outil/pièce en coupe orthogonale," Ecole Nationnale Supérieure d'Arts et Métiers Centre de Cluny, France.

Merchant, E. (1945a). "Mechanics of the metal cutting process. I. Orthogonal cutting and a type 2 chip." *Journal of Applied Physics*, 16, 267-275.

Merchant, E. (1945b). "Mechanics of the metal cutting process. II. Plasticity conditions in orthogonal cutting." *Journal of Applied Physics*, 16, 318-324.

Mindlin, R. D., and Eshel, N. N. (1968). "On first strain-gradient theories in linear elasticity." *Int. J. Solids Structures*, 4, 109-124.

Moufski, A., Dudzinski, D., and Molinari, A. (1998). "Modelling of oblique cutting. Application to bar turning and face milling." *International Colloqium on the Machinability and Damage Mechanisms of Cutting Tools. CEM, Saint Etienne, 10.1-10.8*, Saint Etienne.

Moufski, A., Molinari, A., Dudzinski, D., and Rausch, R. (2000). "Thermoviscoplastic modelling of oblique cutting : forces and chip flow predictions." *International Journal of Mechanical Sciences*, 42, 1205-1232.

Puigsegur, L. (2002). "Caractérisation thermique d'un procédé d'usinage par tournage. Approche analytique et par identification de systèmes non entiers," Université de Bordeaux 1, France.

Puigsegur, L., Cois, O., and Battaglia, J.-L. (2001). "Utilisation de modèles d'identification non entiers pour la résolution de problèmes inverses en thermique " *Congrès Français de Thermique, Editions Elsevier*, 745-750.

Ryckelynck, D., and Meiller, M. (2002). "Modélisation du frottement outil-pièce pour la simulation de la coupe par la méthode des éléments finis: Friction modelisation of tool workpiece contact for the finite element simulation of cutting process." *Mécanique & Industries*, 3, 323-332.

Tay, A. O., Stevenson, M. G., De Vahl Davis, G., and Oxley, P. L. B. (1976). "A numerical method for calculating temperature distributions in machining, from force and shear angle measurements." *International Journal of Machining Tool Design Research*, 16, 335-349.

Tomasi, G. (1994). "Comportement d'une poutre stratifiee avec glissement interlaminaire." *Mechanics Research Communications*, 21(4), 317-327.

Toulouse, D. (1998). "Contribution à la modélisation et à la métrologie de la coupe dans le cas d'un usinage tridimensionnel," Université Bordeaux 1, France.

Toulouse, D., Couétard, Y., Cahuc, O., and Gérard, A. (1997). "An experimental method for the cutting process in three dimensions." *Actes des conférences DYMAT 97, TOLEDO, SPAIN*, C3-21-26.

Toupin, R. A. (1962). "Elastic material with couple-stresses." *Archs ration.Mech. Analysis*, 11, 385-414.

Trigger, K. J., and Chao, B. T. (1951). "An analytical evaluation of metal cutting temperatures." *Transaction of ASME*, 73, 57-68.

Zhuravlev, V. G. (1998). "The model of dry friction in the problem of the rolling of rigid bodies." *Journal of Applied Mathematics and Mechanics*, 62, 705-710.

In: Metal Cutting: Research Advances
Editor: J. Paulo Davim pp. 111-133

ISBN: 978-1-60876-207-1

Chapter 4

FINITE ELEMENT MODELLING OF MACHINING ALUMINIUM ALLOY (AL 7075) AND EXPERIMENTAL VALIDATION

C. Maranhão and J. Paulo Davim*
Department of Mechanical Engineering, University of Aveiro,
Campus Santiago, 3810-193, Aveiro, Portugal.

ABSTRACT

The objective of this chapter is to study the thermo-mechanical behavior of an aluminium alloy when machined with polycrystalline diamond (PCD) cutting tools; and to determine the influence of the friction coefficient in the tool chip interface on cutting and feed forces, cutting temperature, plastic strain, plastic strain rate, and maximum shear stress. An experimental validation of the cutting process was conducted in order to verify the simulated results, and comparison shows that friction modeling at the tool-chip interface significantly influences the final results.

Therefore, it can be concluded that the friction coefficient has a significant effect on the cutting process and is crucial to obtaining valuable predictions when machining with the finite element modeling (FEM) method.

1. INTRODUCTION

Machining aluminium with traditional cutting tools shortens the tool life and has an adverse effect on surface and edge quality, because of the formation of built-up edges and burrs. This is significant because aluminium parts require deburring and finishing processes after cutting, raising the cost of the workpiece. Burrs are damaging even during machining because they hit the cutting tool edge and cause groove wear in the tool; groove wear, in turn, accelerates burr formation. Therefore, machining technology that does not form built-up edges and produces a good surface is much in demand, and use of polycrystalline diamond

* Corresponding author: pdavim@ua.pt

(PCD) machining appears to be a likely technology. A degraded tool that is not identified early, may lead to a significant drop in the quality of the workpiece [1-4]. The overall surface quality is influenced by tool geometry, chip flow, temperature generation, heat flow, and tool wear [5].

PCD cutting tools are considered noble material (i.e., highly resistant to oxidation and corrosion) and show substantially higher performance in machining aluminium alloys when compared to traditional cutting tool materials [6]. With diamond tools, the turning process is a technique of precision machining for automobile, aerospace, computer, optic components, and for products of the moulding and die industry. Along with other advancements science and technology and requirements for increased machining accuracy, optimization of machining techniques for accurate cutting parameters and predictable machined surface quality is essential [7]. Optimized processing factors and tool geometry, predictable final surface residual stress, cutting temperature, and thermal deformation together can improve the accuracy and integrity of machined surfaces [8].

Aluminium alloys have good machinability among metals due to outputting lower cutting forces, lower cutting temperatures, and lower tool wear rates. Machinability advantages concerned with aluminium alloys include tool life, chip characteristics, chip disposal, recycling potential, and surface finish [9-10]. Dry machining of aluminium alloys is especially difficult because aluminium exhibits high friction and therefore has a strong tendency to yield a built-up edge (BUE) formation. Cutting tools used for machining aluminium should possess low affinity to aluminium, low friction coefficient, and high hardness, which makes PCD tools the suitable choice due to their ability to significantly reduce the built-up edge formation and hence improve cutting performance [11-12].

For some alloys, the high tendency of aluminium to adhere during cutting presents a significant risk that can lead to tool breakage. The reason for this is that long, ductile chips that complicate the machining process form during cutting regardless of the cutting tool geometry chosen. From thermal and deformation points of view, aluminium alloys have many advantages, such as allowing complex shapes with power requirements (due to lower flow strengths), thus permitting the use of smaller equipment [13].

Therefore, in dry machining aluminium alloy (Al 7075), a PCD cutting tool was used.

From previous simulations and experimental work, it can be concluded that finite element modelling is a valid approach to predicting the thermo-mechanical behaviour of the stainless steel. Parameters like cutting and feed forces, cutting power, maximum cutting temperature, plastic strain, plastic strain rate, and maximum shear stress were studied.

2. FEM Analysis

The choice of finite element modelling software is an important factor in determining the quality and scope of analysis that can be performed. When dealing with metal cutting, input parameters like friction coefficient and flow stress curves greatly influence the reliability and precision of the results. Finite element software (specific for machining operations) was chosen to simulate the metal cutting process, therefore, Advantedge^{TM} (provided by Third Wave Systems) was used to aid this study. In figure 1, an example of a mesh in 2D in the tool, chip and workpiece can be seen.

This commercial software was built with machining operations in mind and allows simulation of turning, drilling, milling, micro-machining, and related operations in two or three dimensions. It also uses adaptive meshing to help to improve quality and precision of the results and supports a wide variety of workpiece materials. The solver is actually controlled by the software, and its clean software interface allows fast setups for multiple simulations . AdvantEdge™, on the other hand, does not give the user much flexibility in configuring the controls of the solver. While this limitation may be preferable in some cases, it means that the user is restricted to the preset controls of the software.

Third Wave Advantedge™ allows improvement and optimization of machining processes and makes it possible to determine optimal machining parameters and tool configurations that allow lower cutting and feed forces and temperature, without spending time and money on experimental processes.

In the end, Advantedge™ allows for very fast setups with its accompanying tool geometry libraries provided. It also includes an extensive material model library, with support for a wide range of materials, aerospace alloys included. It uses adaptive meshing, but the controls cannot be modified. In the overall control of the simulations, very minimal changes are allowed. Figure 2 illustrates some of the input parameters necessary for a two-dimensional turning operation.

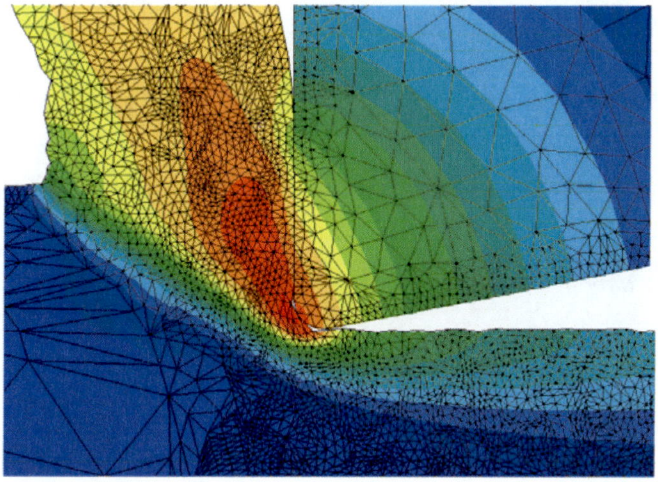

Figure 1. Example of a mesh using ThirdWave Advantedge™ commercial package.

To model the thermal-visco plastic behavior of the workpiece material, ThirdWave Advantedge™ software uses a constitutive equation, the Johnson-Cook law, which can be represented by the following formula:

$$\sigma_{eq} = (A + B\varepsilon^n)(1 + C\ln(\frac{\dot{\varepsilon}}{\dot{\varepsilon}_0}))(1 - (\frac{T - T_{room}}{T_m - T_{room}})^m) \tag{1}$$

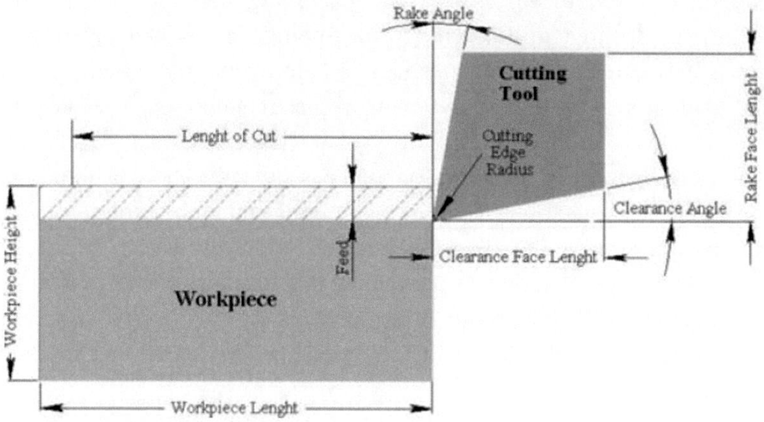

Figure 2. Software input parameters (Advantedge™).

where ε is the plastic strain, $\dot{\varepsilon}$ is the plastic strain rate (s^{-1}), ε_0 is the reference plastic strain rate (s^{-1}), T is the temperature of the workpiece material (°C), Tm is the melting temperature of the workpiece material (in °C), and T_{room} is the room temperature (°C). Coefficient A is the yield strength (MPa), B is the hardening modulus (MPa), and C is the strain rate sensitivity coefficient; n is the hardening coefficient and m is the thermal softening coefficient.

Experimental work was carried out in order to obtain the friction coefficient to use as an input in the FEM simulation. The friction coefficient was calculated using the Coulomb model through the following equation:

$$u = \frac{Ff + Fc \times tan\gamma}{Fc - Ff \times tan\gamma} \qquad (2)$$

where Ff represents the feed force, Fc the cutting force, and γ is the tool rake angle. Both cutting forces were obtained experimentally.

3. Methodology

Experimental Procedure

Table 1 presents the properties of the aluminium alloy used in this study.

The turning tests were performed with a high rigidity CNC lathe Kingsbury 50 with 18kW spindle power and a maximum spindle speed of 4500 rpm. In the machining of the aluminium alloy, a PCD cutting tool (TPUN 110304FP) was tested. It has 0° of rake angle and 7° of relief angle with 0.4 mm of nose radius. A 91°-edge major tool cutting angle and 0°-cutting-edge inclination angle were given by CTGPL 2020K11 (ISO) tool holder.

Table 1. Chemical composition and mechanical properties of the aluminium alloy Al 7075

Chemical composition (%)	
Zn	5.6
Mg	2.5
Cu	1.6
Fe	0.5
Si	0.4
Mn	0.3
Cr	0.23
Ti	0.2
Mechanical properties	
Tensile Strength, Yield	260 MPa
Tensile Strength, Ultimate	360 MPa
Elongation at Break	Max 10 %
Modulus of Elasticity	70 GPa
Density	2.77 g/cm^3

The acquisition of the cutting, feed, and depth forces (the latter being negligible due to the cut being orthogonal) was made with a Kistler® piezoelectric dynamometer. The cutting and feed forces were continuously monitored and acquired throughout the test by using a charge amplifier with three independent channels.

The cutting parameters in the machining of the aluminium alloy were as follows: cutting speed (V_c) of 500 m/min, feed rate (f) of 0.05 - 0.10 and 0.2 mm/rev, and constant depth of cut (a_p) of 2 mm. Work bars of Al 7075 with 100 mm of diameter and 200 mm of length were tested.

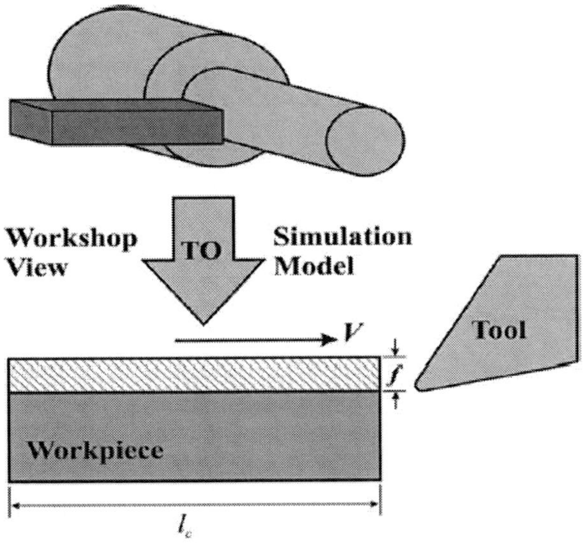

Figure 3. Schematic on how the experimental work was conducted and comparison with the finite element approach *(Bil et al, 2004)*.

The orthogonal machining was conducted under Merchant theory [14] as shown in figure 3.

FEM Machining Input Parameters

The simulations were carried out in order to meet what was made experimentally, allowing for an experimental validation. The input parameters of the machining operation for the aluminium alloy are shown in table 2.

Table 2. Aluminium alloy input parameters

Workpiece	
Workpiece Length	6 mm
Workpiece Height	3 mm
Workpiece Material	Al 7075
Tool	
Rake Angle	0°
Rake Face Length	3 mm
Relief Angle	7 °
Relief Face Length	3 mm
Cutting Edge Radius	0.02 mm
Material	PCD
Process	
Depth of Cut	2 mm
Length of Cut	2 mm
Feed	0.05, 0.1, 0.2 mm/rev
Cutting Speed	500 m/min
Initial Temperature	20 °C
Friction Coefficient	specific for each case
Coolant	not used
Simulation	
Maximum Number of Nodes	12000
Maximum Element Size	0.1 mm
Minimum Element Size	0.02 mm

Experimental Validation

The forces obtained in the experimental validation using the dynamometer are directly compared to the mean forces obtained in the simulation; and the same can be said for the cutting power. A more elaborate method is necessary to find the maximum cutting temperature. Metal cutting generates high temperatures; thus, the cutting temperature is carefully calculated using Boothroyd analytic model [15].

4. Results and Discussion

Table 3 presents experimental values measured in the machining of the aluminium alloy with the PCD cutting tool as a function of feed rate. Cutting and feed forces represent an average of two different experimental tests.

Table 3. Experimental cutting and feed forces for the machining of the aluminium alloy with the PCD cutting tool with Vc= 500 m/min; f= 0.05 – 0.1 – 0.2 mm/rev; a_p= 2 mm

Feed rate [mm/rev]	Cutting force [N]	Feed force [N]
0.05	94.7	46.4
0.1	171.3	53.3
0.2	317.3	52.8

FEM Analysis Validation with Coulomb Friction Coefficient

Table 4 presents the Coulomb friction coefficients obtained using equation (2) as a function of the feed rate for the dry machining of the aluminum alloy.

Table 4. Coulomb friction coefficients for aluminium alloy as a function of the feed rate

Feed rate [mm/rev]	Al 7075
0.05	0.31
0.1	0.25
0.2	0.17

By comparing the obtained Coulomb friction coefficients with the friction coefficients found in the literature, it is possible to conclude that an agreement is established. Typically, aluminium alloys have friction coefficients lower than 0.5.

To provide an overview of the differences obtained between experimental work and the simulation using the Coulomb friction coefficient, a single case study is presented. An experimental validation on the cutting and feed forces, cutting power, maximum cutting temperature, and plastic strain is addressed in order to verify if the simulation values met experimental values. The differences in these values were calculated using the following formula:

$$difference = \frac{|\,Experimental\ \ value - Simulated\ \ value\,|}{Experimental\ \ value} \times 100\% \qquad (3)$$

Figure 4 presents the results obtained (for cutting and feed forces, cutting power, maximum cutting temperature, and plastic strain) for a feed rate of 0.1 mm/rev and the respective Coulomb friction coefficient (0.25),. It is important to note that all values are an average of two tests in the experimental work and an average of one simulation in the numerical simulation.

In figure 4, the experimental cutting force value is 171.3 N, while the feed force value is 53.3 N. The numerical results of the experiment are superior to the simulated results. In the latter case, the cutting force value is 197.9 N and the feed force is 81.2 N.

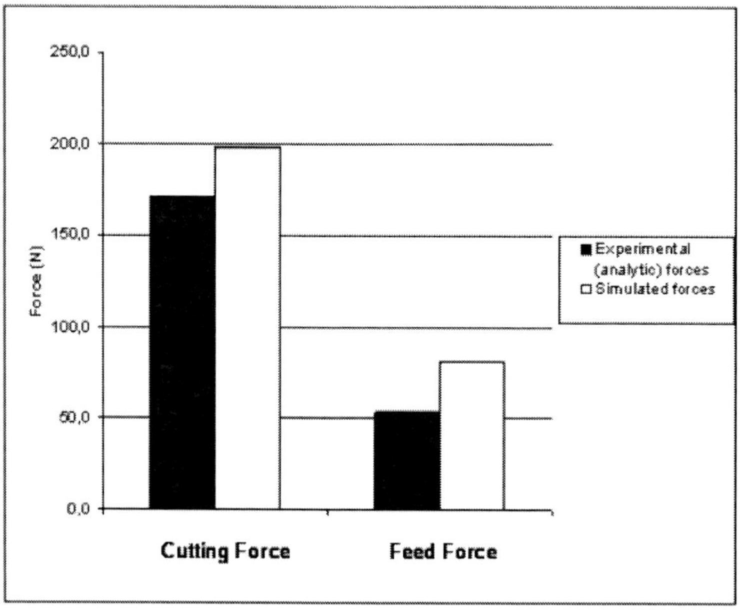

Figure 4. Comparison between experimental and simulated cutting and feed forces for machining aluminium alloy with a feed rate of 0.1 mm/rev and the Coulomb friction coefficient of (0.25).

Figure 5 a) presents comparative values for cutting power in experimental (1427.5 W) and numerical (1641) approaches.—

The maximum cutting temperature has a minimum difference between experimentation and simulation when using the Coulomb friction coefficient as can be seen in figure 5 b): 268.2 ºC in the analytical models and 271.4 ºC in the numerical simulation.

Figure 5 c) shows plastic strain values of 2.65 deformation in the experimental work and about 2.85 in the numerical simulation.

Table 5 presents values showing the differences in percentage between the experimental and simulated results.

The differences shown are reasonable for every case, except the feed force. This values is due to the use of the Coulomb friction coefficient in the numerical simulation.

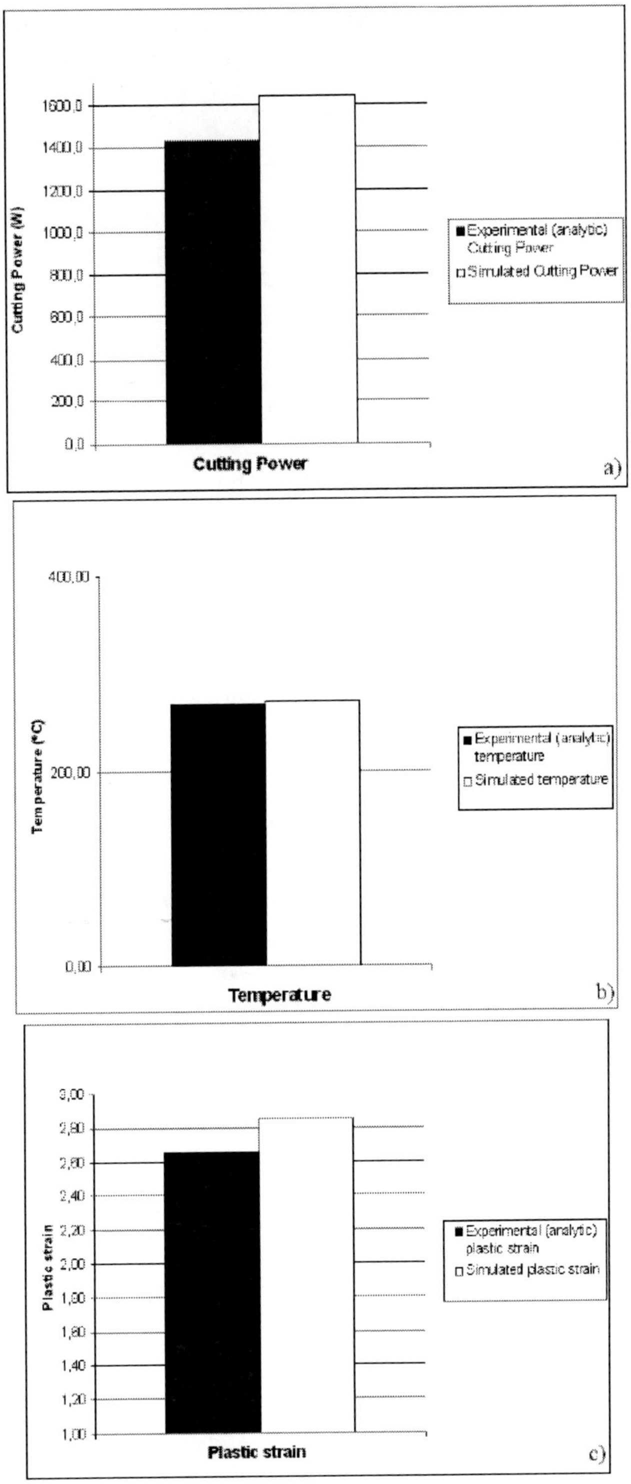

Figure 5. Comparative values for experimental and simulated data for machining aluminium alloy with a feed rate of 0.1 mm/rev and a Coulomb friction coefficient (0.25) showing: a) cutting power, b) maximum cutting temperature, and c) plastic strain.

Table 5. Differences between experimental and simulated data for the machining of Al 7075 with the PCD cutting tool with a feed rate of 0.1 mm/rev

Differences [%]	
Cutting force (Fc)	15.5
Feed force (Fa)	52.3
Cutting power (Pc)	15
Maximum cutting temperature (T)	1.2
Plastic strain (ε)	7.4

FEM Analysis Validation with Coefficient Friction Adjustment

As noted, the friction coefficient obtained through the Coulomb model fails to provide a solution that would output results close to the those obtained using analytical models for any of the materials. While the Coulomb model is a useful starting point for establishing the friction value, it is clear that several iterations must be carried to produce satisfactory results. The Coulomb model is able to determine whether the friction coefficient is bigger or smaller than the standard friction coefficient (0.5) in the software. The stopping criterion is found when the cutting and feed forces obtained through numerical simulation are comparable with the results found in the experimental work.

As stated, the Advantedge^{TM} software has a limited capability for changing Johnson-Cook coefficients, so all simulations were conducted using the software database flow stress. Even if the Johnson-Cook coefficient values were improved to fit the experimental material used in the experiments, the numerical and analytical results would likely differ due to the critical friction coefficient. Besides, changing the friction coefficient is a fast and reliable way to minimize differences between experimentation and simulation., Industry is always pushing towards faster results, making the cost of testing each material prohibitive. Moreover, for each lot of materials, new parameters must be found and several tests done on the material in order to find the new constitutive law constants. Thus, the software database flow stress value was used and the friction coefficient was iterated instead.

Table 6 shows the optimal friction coefficient in function of the feed rate for the several simulations made in order to adjust the friction coefficient for each case.

Table 6. Iterated friction coefficients as a function of the feed rate

Feed rate [mm/rev]	Al 7075
0.05	0.1
0.1	0.1
0.2	0.06

Experimental Validation

Experimental validations were conducted using formula 3 on cutting and feed forces, cutting power, maximum cutting temperature, and plastic strain. Figure 6 shows a comparison between the experimental and simulated cutting forces as a function of the feed rate.

For a feed rate of 0.05 mm/rev, the iterated friction coefficient is 0.1 while the Coulomb friction coefficient is 0.31. Figure 6, also shows the experimental cutting force value of 94.7 N and a simulated cutting force value of 97.8 N.

For a feed rate of 0.1 mm/rev, the iterated friction coefficient is 0.1 and Coulomb friction coefficient is 0.25. Once again, the experimental results are comparable to the numerical results, as presented in figure 6. The experimental cutting force value is 171.3 N while the simulated cutting force value is 171.8 N.

For a feed rate of 0.2 mm/rev, the iterated friction coefficient is 0.06 and the Coulomb friction coefficient is 0.17. Once again, the difference between numerical and experimental values are minimal as shown in figure 6. The experimental cutting force value is 317.3 N and the simulated cutting force value is 315.2 N.

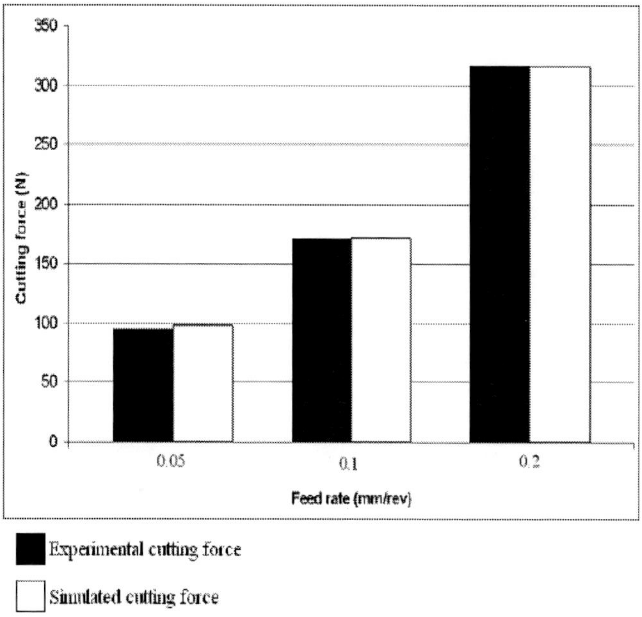

Figure 6. Comparison of experimental and simulated cutting forces when machining aluminium alloy with an iterated friction coefficient as a function of the feed rate, a constant cutting speed of 500 m/min, and a fixed depth of cut of 2 mm.

Figure 7 shows a comparison of feed force for both methods as a function of the feed rate. For a feed of 0.05 mm/rev, the experimental feed force value is 46.4 N and the simulated feed force value is 47.9 N. For a feed rate of 0.1 mm/rev, while the experimental feed force value reached 53.3 N, the simulated feed force value was 57.8 N. Last but not least, for a feed rate of 0.2 mm/rev, experimental feed force reached 52.8 N and the simulation value, 63.1 N.

Cutting power values are also very similar for experimental and numerical results. Figure 8 shows the cutting power as a function of the feed rate. For a feed of 0.05 mm/rev, the analytical model value was 789.2 and the value in the simulation, about 825.4 W. For a 0.01 mm/rev feed rate, the experimental cutting power value reached 1427.5 W, while the simulation value was 1433.9 W. For 0.2 mm/rev, the analytical model cutting power value reached 2644.2 W, while the simulation value was 2550.8 W.

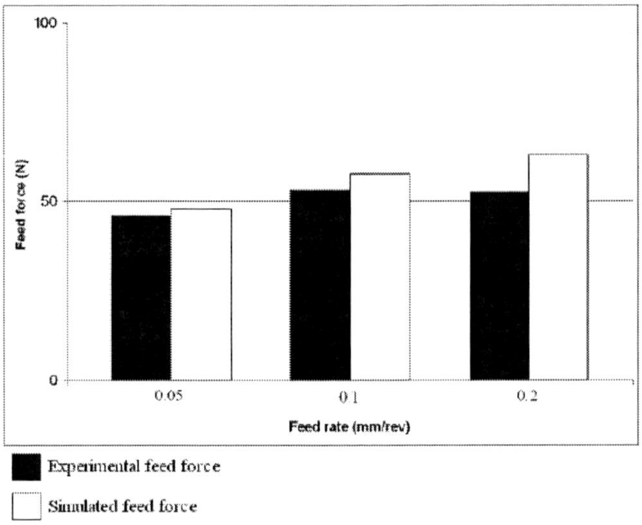

Figure 7. Comparison of experimental and simulated feed forces when machining aluminium alloy with an iterated friction coefficient as a function of the feed rate, a constant cutting speed of 500 m/min, and a fixed depth of cut of 2 mm.

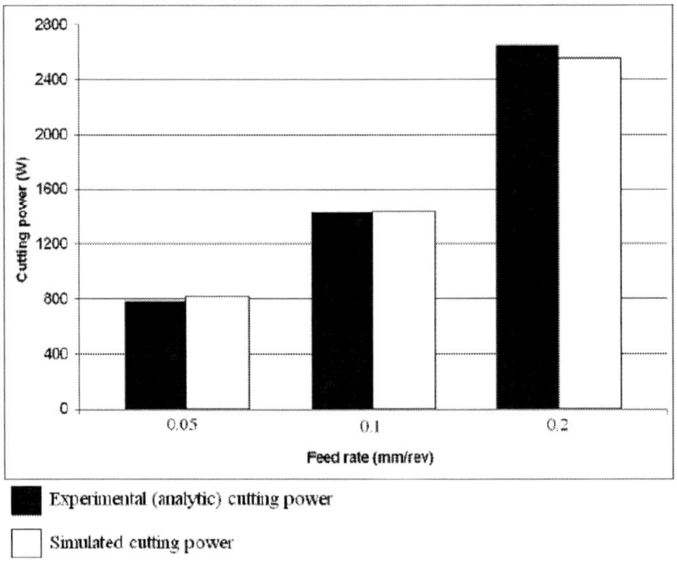

Figure 8. Comparison of experimental and simulated cutting power when machining aluminium alloy with the iterated friction coefficient as a function of the feed rate. a constant cutting speed of 500 m/min, and a fixed depth of cut of 2 mm.

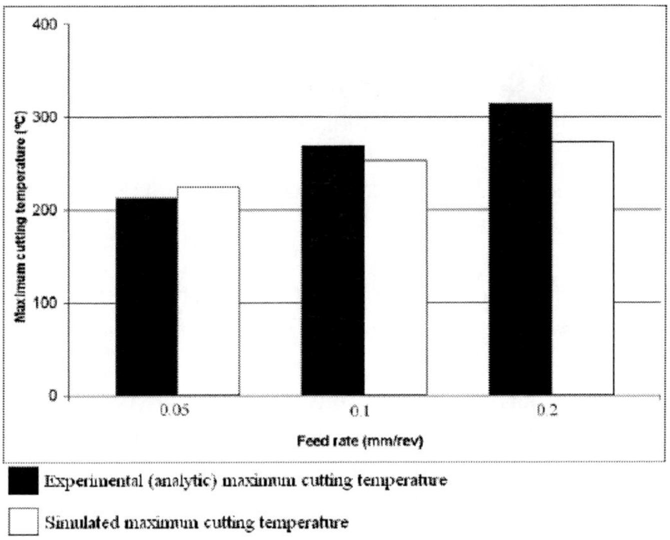

Figure 9. Comparison of experimental and simulated maximum cutting temperature when machining aluminium alloy with the iterated friction coefficient as a function of the feed rate, a constant cutting speed of 500 m/min, and a fixed depth of cut of 2 mm.

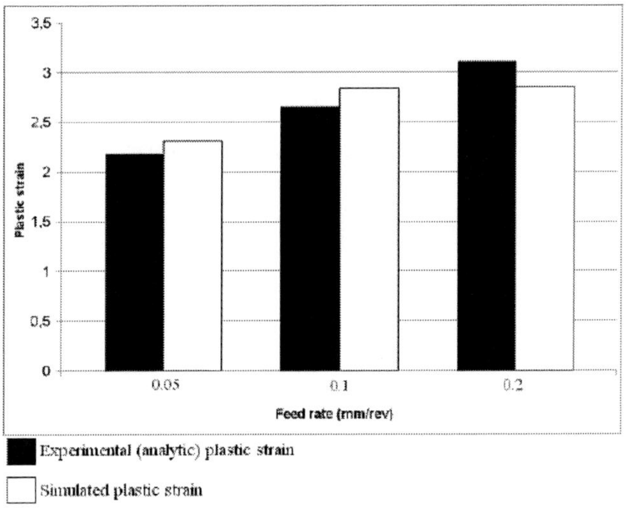

Figure 10. Comparison of experimental and simulated plastic strain when machining aluminium alloy with the iterated friction coefficient as a function of the feed rate, a constant cutting speed of 500 m/min, and a fixed depth of cut of 2 mm.

Regarding the maximum cutting temperature, once again the results of the methods are similar, as figure 9 suggests. Figure 9 presents a comparison of the experimental (analytical) maximum cutting temperature values with the numerical simulation maximum cutting temperature values as a function of the feed rate. Values of about 213.8 ºC were obtained using analytical models and 224.2 ºC using numerical simulation, both using a feed rate of 0.05 mm/rev. A feed rate of 0.1 mm/rev yielded a cutting temperature value of about 268.2 ºC in the analytical models and 252.5 ºC in the numerical simulation. A feed rate of 0.2 mm/rev yielded a cutting temperature of about 314.8 °C in the analytical models and 272.1 °C in the numerical simulation.

Figure 10 presents plastic strain values, as a function of the feed rate, of 2.18 deformation in the experimental (analytical) work and about 2.31 in the numerical simulation, for a feed rate of 0.05 mm/rev in both cases. For 0.1 mm/rev feed rate, the plastic strain reached 2.65 in the experimental work and about 2.84 through numerical simulation. Finally, for a feed rate of 0.2 mm/rev, the plastic strain reached 3.11 in the experimental work and about 2.85 through numerical simulation.

Table 7 shows the differences in percentage between the experimental results and simulated values as a function of the feed rate in the machining of the aluminium alloy.

Table 7. Differences in experimental and simulated data for the machining of Al 7075 with the PCD cutting tool with a variable feed rate

	Differences (%)				
Feed [mm/rev]	Cutting force	Feed force	Cutting Power	Maximum Cutting Temperature	Plastic strain
0.05	0.29	8.4	0.45	5.9	6.9
0.1	0.3	8.4	0.4	5.9	7.0
0.2	0.7	19.5	3.5	13.6	8.5

As shown, when the friction coefficient is iterated, very precise results are obtained for almost all cases. The most critical results were obtained in the simulation with a feed rate of 0.2 mm/rev. In this particular case, the feed force is 19.5%, but in general these results can be considered very satisfactory and much more precise in comparison to the Coulomb friction results for all cases of study.

Modelling and Prediction

The following sections present a more detailed analysis of cutting and feed forces, cutting power, maximum cutting temperature, plastic strain, plastic strain rate, and maximum shear stress applied to the machining of aluminium alloys.

Cutting and Feed Forces, Cutting Power, and Maximum Cutting Temperature

Figure 11 illustrates patterns for both experimental results and expected (theoretical) results representing the evolution of cutting and feed forces with the variation of feed rate.. Notice that as the feed increases, the feed force also increases, although the increase in the cutting force is much more obvious. This difference because as the feed increases, the amount of material removed also increases.

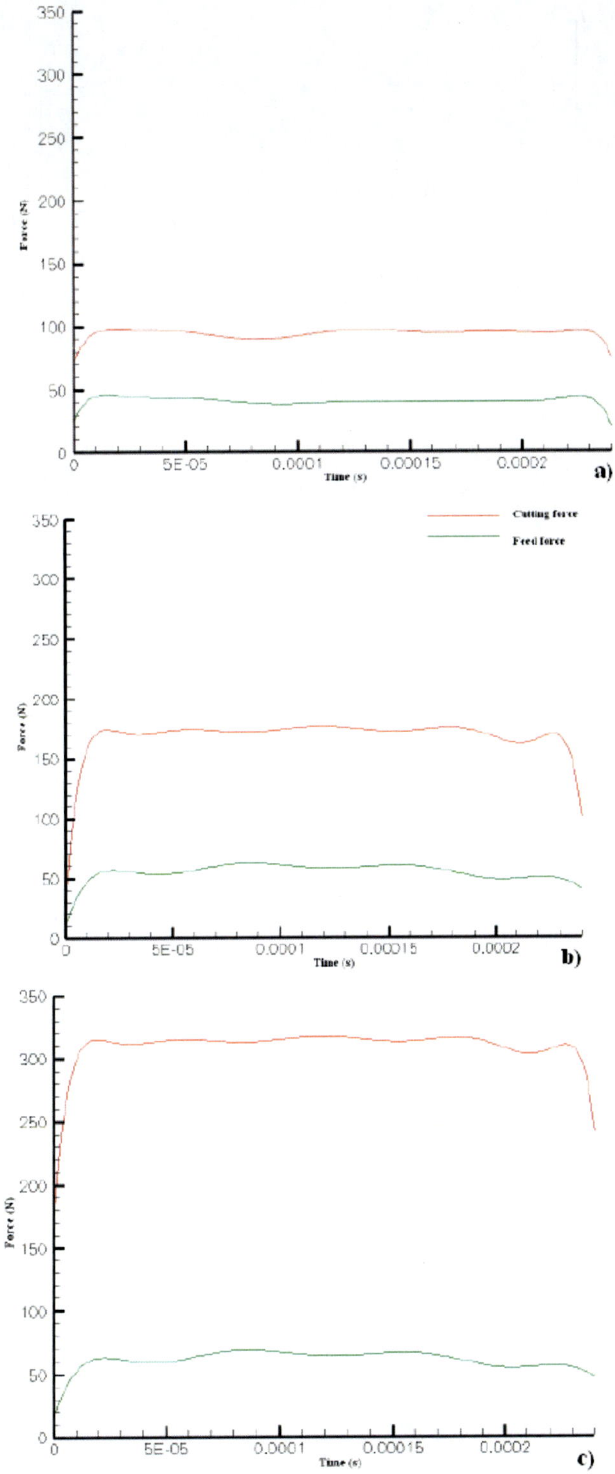

Figure 11. Comparison of cutting force and feed force, through time in machining aluminium alloy with a cutting speed of 500 m/min, a depth of cut of 2 mm, and the following variable feed rates: a) 0.05 mm/rev, b) 0.1 mm/rev, and c) 0.2 mm/rev.

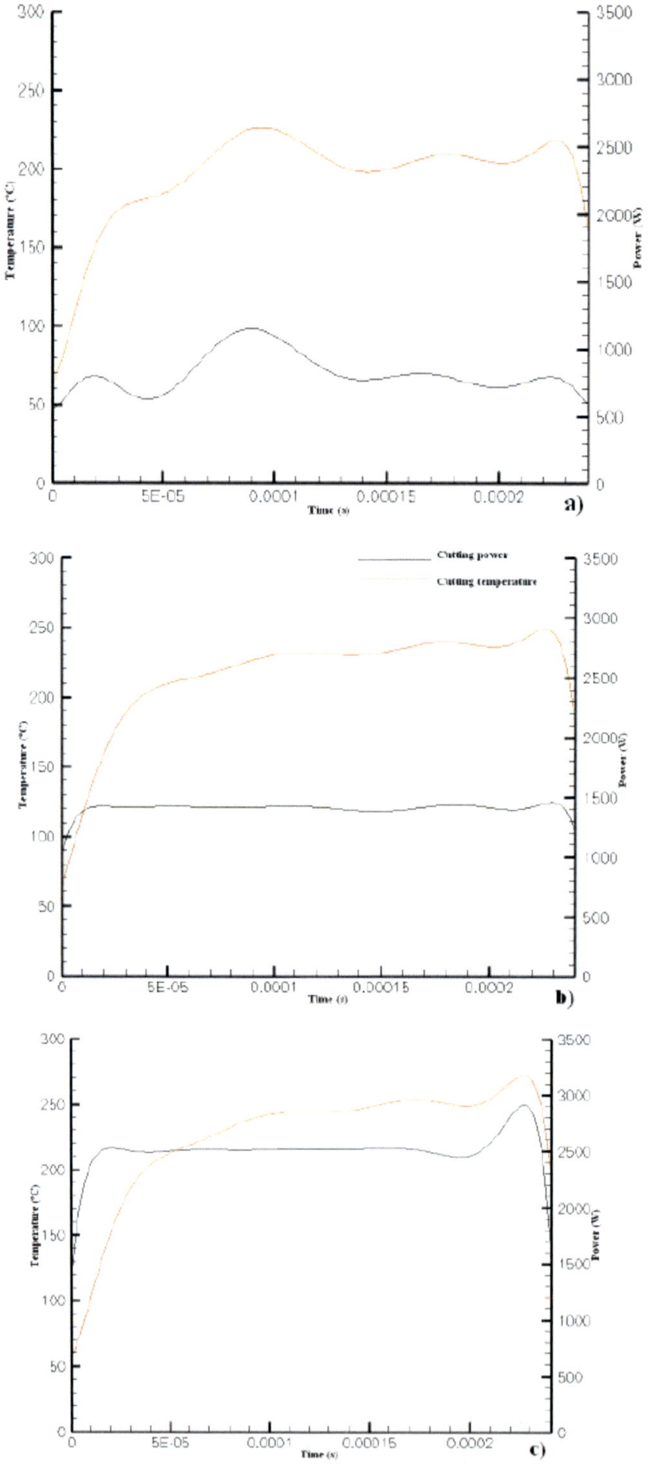

Figure 12. Maximum cutting temperature and cutting power over time in machining aluminium alloy with cutting speed of 500 m/min, depth of cut of 2 mm, and the following variable feed rates: a) 0.05 mm/rev, b) 0.1 mm/rev, and c) 0.2 mm/rev.

Figure 12 illustrates cutting power and maximum cutting temperature (which is felt in the tool), as a function of cutting time. Once again, the experimental results agree with initially expected results. Cutting power, which is function of cutting force, grows as the feed rate increases.

The maximum cutting temperature increases also with variation in feed rate, but not by a considerable margin. Yet even this minimal increase in temperature can cause a decrease in tool life. The increase in temperature shown might be due increases in the cutting and feed forces with the increase in feed rate. Because PCD also has good thermal conductivity, part of the generated heat is conducted through the cutting tool. Significantly, PCD transforms into graphite at about 700°C. Figure 12 shows that none of the feed rates achieves this critical temperature.

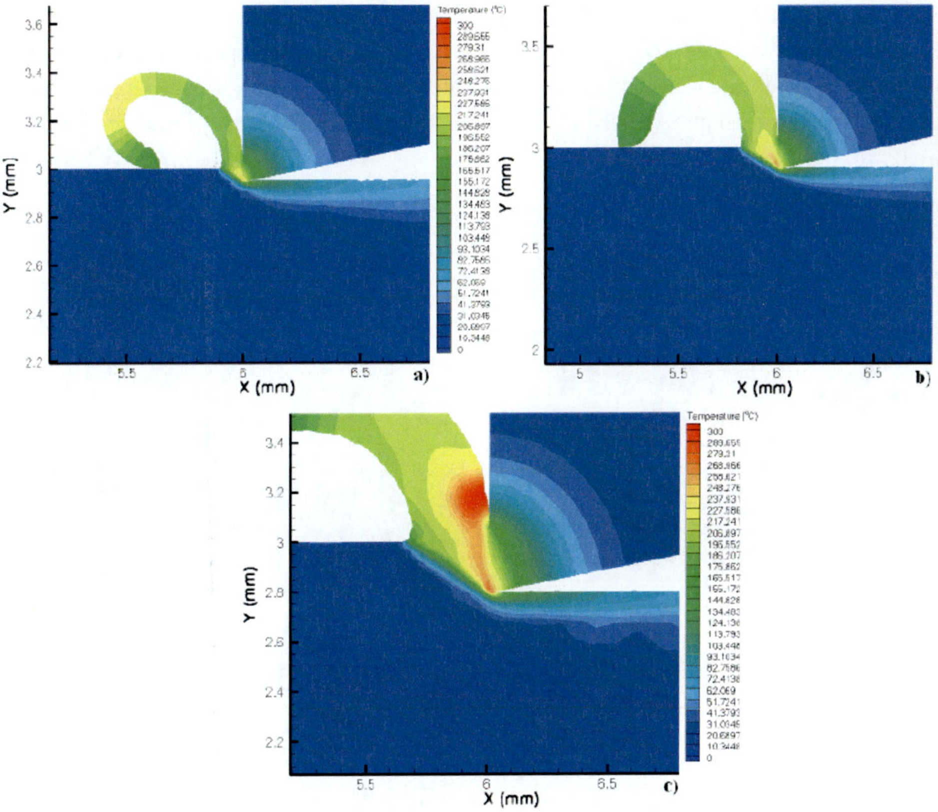

Figure 13. Detailed comparison of maximum cutting temperature in the tool, workpiece, and chip, at the end of the length of cut, in the machining of aluminium alloy with a cutting speed of 500 m/min, a depth of cut of 2 mm, and the following variable feed rates: a) 0.05 mm/rev, b) 0.1 mm/rev, c) 0.2 mm/rev.

Figure 13 presents a detailed view of the distribution of cutting temperature in tool, workpiece, and chip. The figure shows temperature at the end of the length of cut where higher temperatures and steady state conditions occur. The maximum temperature occurs in the primary shear zone for all cases (although for c], a feed rate of 0.2 mm/rev, the chip has a noticeable hotter zone).As illustrated, cutting temperature increases with feed rate, and heat disperses from the cutting zone into the chip, where the outside of the chip curling is hotter than the inside). Also note that both the aluminium and PCD cutting tool have good heat conductivity release the heat principally through the chip. This distribution of heat is beneficial in achieving a longer tool life. However, additional heat is generated in a thermal affected layer in the workpiece and increases with feed rate.

Plastic Strain, Plastic Strain Rate, and Maximum Shear Stress

Figure 14 illustrates in detail the plastic strain distribution in the workpiece and chip at the end of length of cut, where steady-state conditions exist. At first glance, differences in the plastic strain may seem not to correlate with variations in feed rate, as a bigger deformed zone is visible in the chip with a feed rate of 0.05 mm/rev. However, on closer examination, if the primary shear zone is considered, the plastic strain rate exhibits a trend to slightly increase with increases in feed rate. This phenomenon can be explained by a similar increase in maximum shear stress with an increase in feed rate. A larger area (layer) of plastic strain also appears with higher feed rates.

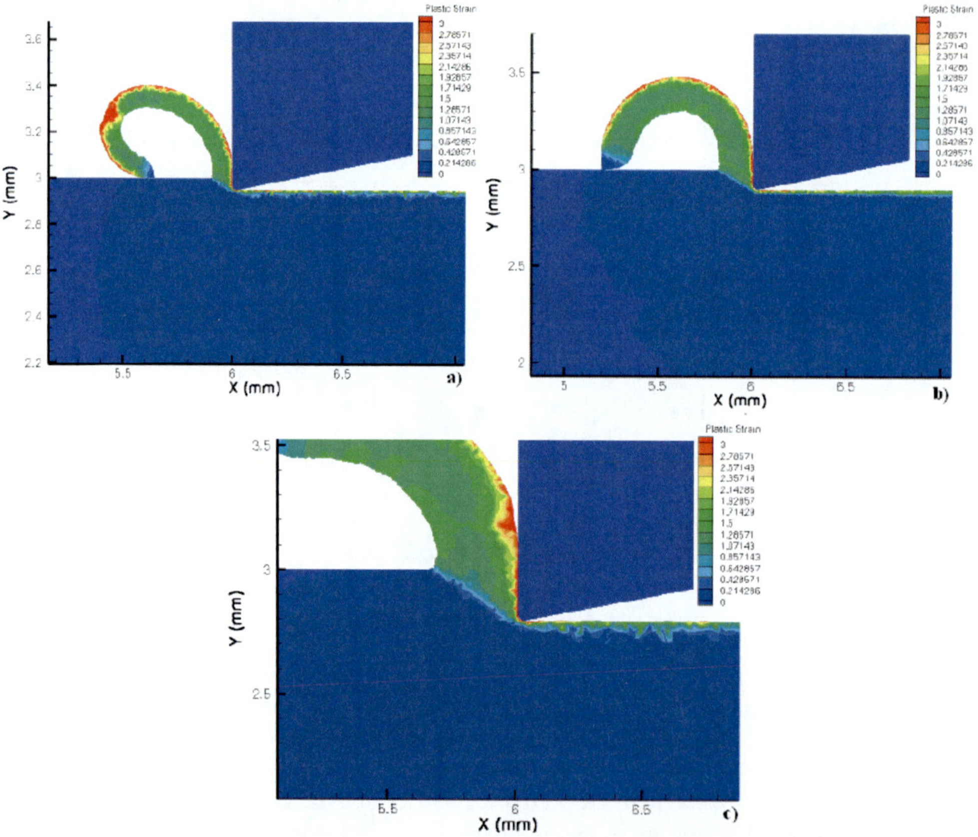

Figure 14. Detailed comparison of plastic strain in tool, workpiece, and chip at the end of cut length in machining of aluminium alloy with cutting speed of 500 m/min, depth of cut of 2 mm, and the following variable feed rates: a) 0.05 mm/rev, b) 0.1 mm/rev, and c) 0.2 mm/rev.

Figure 15 shows a detailed distribution of plastic strain rate in the workpiece and chip shown at the end of the cut length, where steady state conditions exist. The cutting parameter that most affects the plastic strain rate is the cutting speed. However, here, the cutting speed is kept constant in all cases, with only the feed rate as a variable. It is evident that the area affected by plastic strain is larger at higher feed rates. However, peaks in strain rate occur during machining at lower feed rates.

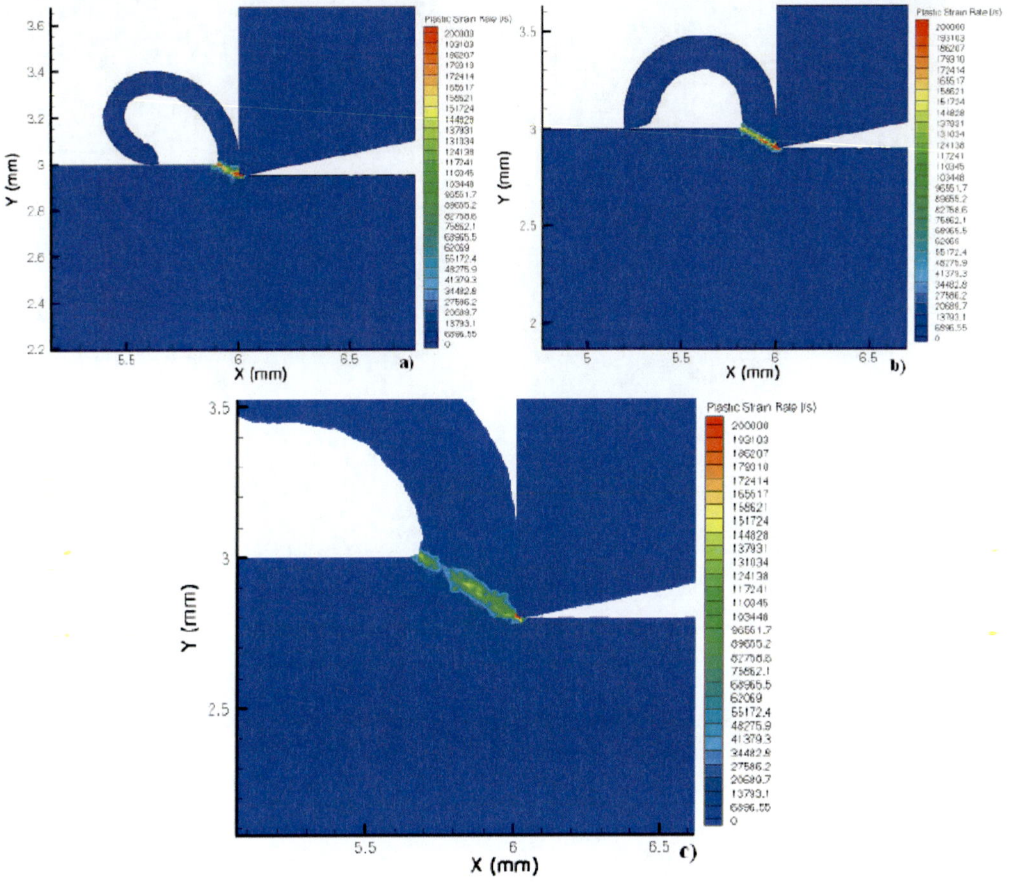

Figure 15. Detailed comparison of plastic strain rate in the tool, workpiece, and chip at the end of cut length, in machining of aluminium alloy with cutting speed of 500 m/min, depth of cut of 2 mm, and the following variable feed rates: a) 0.05 mm/rev b) 0.1 mm/rev c) 0.2 mm/rev.

As Figure 16 illustrates a detailed distribution of maximum shear stress in the tool, workpiece, and chip in at the end of length of cut, where steady state conditions exist. Note that as the feed rate increases, the amount of material removed by the cutting tool also increases and is reflected in the affected zone by the maximum shear stress. Note also that as the feed rate increases, the affected zone in the workpiece increases by a similar magnitude. Although maximum shear stress values the primary shear zone are similar in all cases, the affected zone increases with as the feed rate.

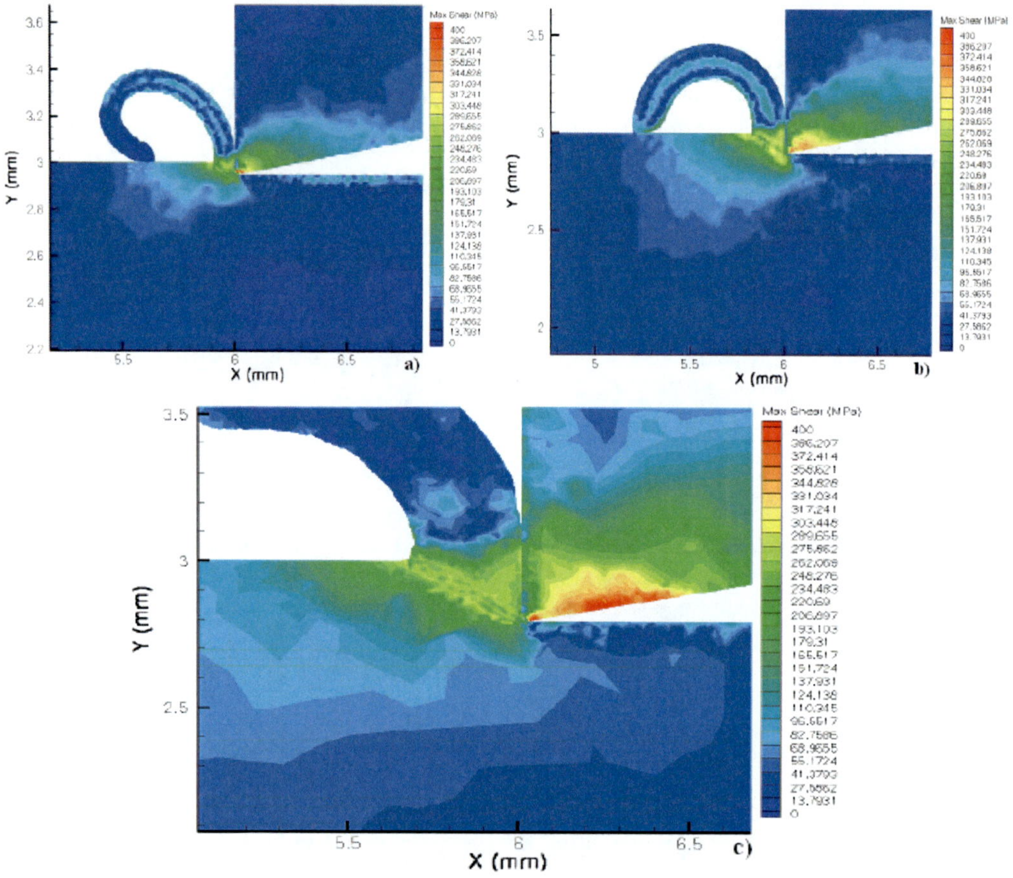

Figure 16. Detailed comparison of maximum shear stress in the tool, workpiece, and chip at the end of cut length in machining of aluminium alloy with a cutting speed of 500 m/min, a depth of cut of 2 mm, and the following variable feed rates. a) 0.05 mm/rev, b) 0.1 mm/rev, c) 0.2 mm/rev.

5. Conclusions

From all the simulations made, it is apparent that the friction coefficient greatly influences simulated run conditions. Unfortunately, friction coefficient cannot be measured with precision and must be iterated. In experimental tests, a Coulomb friction coefficient was found for each case was used the starting point to simulate machining operations. However, evaluation of the results of the simulations using the Coulomb friction coefficient indicate that the results differ from experimental values. Note that the software overestimated all variables in aluminium alloy and with enormous differences in some cases. Several iterations were conducted on the friction coefficient in order to decrease these differences.

The Coulomb model overestimates friction in turning of aluminium alloy. However, actual metal cutting produces dynamic conditions, along with deformation and thermal gradients, and show a decrease in the friction coefficient. Also, the experimental workpiece was a bit harder than the simulated workpiece, which suggests near identical flow stress curves in the experimental and simulated workpieces. This effect may be related to the reduction in the friction coefficient in the machining of aluminium alloy. The software uses a standard, default friction coefficient of (0.5), but Coulomb model gives a approximate friction coefficient that is closer to the "correct" friction coefficient. The Coulomb model, therefore, is helpful as a starting point for further iterations, but should not be used as a final input value. Nevertheless, friction coefficients for all cases are within the expected range of values (i.e., lower than 0.5).

It appears that using numerical simulation is a valid approach to determine parameters such as cutting forces, feed forces, cutting power, maximum cutting temperature, plastic strain and plastic strain rate, as well as maximum shear in the orthogonal cut of the Al 7075 provided that the correct friction coefficient is used. Otherwise, the differences in output results are unacceptable.

The completion of this work makes it possible to conclude the following:

- Friction coefficient greatly influences simulated results and cannot be measured with precision
- Coulomb model is helpful as a starting point for further iterations but should not be used as a final input value
- Coulomb model overestimates the friction for turning aluminium
- Numerical simulation is a valid approach to determine parameters like cutting forces, feed forces, cutting power, maximum cutting temperature, plastic strain, plastic strain rate, and maximum shear stresses

Acknowledgment

The authors wish to thank the Foundation for Science and Technology (FCT), Portugal, project POCTI/EME/61676/2004, for the use of the Advantedge® software licence.

References

[1] Bouzid W. (2005) An investigation of tool wear in high-speed turning of AISI 4340 steel, Int. J. Adv. Manuf. Technol. 26: 330–334

[2] Li L., He N., Wang M., Wang Z. (2002) High speed cutting of Inconel 718 with coated carbide and ceramic inserts, Journal of Materials Processing Technology. 129 127-130

[3] Bouzid W. (2005) Cutting parameter optimization to minimize production time in high speed turning, Journal of Materials Processing Technology. 161 388–395

[4] Kim J., Kang Y. (1997) High speed machining of aluminium using diamond endmills, International Journal Machine Tools and Manufacture. Vol. 37, No. 8, 1155-1165

[5] Ceretti E., Fallbohmer P., Wu W., Altan T. (1996) Application of 2D FEM to chip formation in orthogonal cutting, Journal of Materials Processing Technology. 59 169-180

[6] Davim J., Maranhão C., Jackson M., Cabral G., Grácio J. (2008) FEM analysis in high speed machining of aluminum alloy (Al7075-0) using polycrystalline diamond (PCD) and cemented carbide (K10) cutting tools, Int. J. Adv. Manuf. Technol. 39: 1093-1100.

[7] Ikawa N., Donaldson N., Komanduri R., Koenig W., Aachen T., Mckeown P., Moriwaki T., Stowers I. (1991) Ultra-precision metal cutting—the past, the present and the future, Ann. CIRP 40/2 587–594

[8] Mackerle J. (2003) Finite-element analysis and simulation of machining: an addendum bibliography (1996–2002), Journal of Materials Processing Technology. 43 103–114

[9] Kelly J., Cottrell M. (2002) Minimal lubrication machining of aluminium alloys, Journal of Materials Processing Technology. 120, 327-334

[10] Ezugwu E., Bonney J., Yamane Y. (2003) An overview of the machinability of aeroengine alloys, Journal of Materials Processing Technology. 134, 233-253

[11] Fukui H., Okida J., Omori N., Moriguchi H., Tsuda K. (2004) Cutting performance of DLC coated tools in dry machining aluminium alloys, Surface and Coatings Technology. 187 (1), 70-76

[12] Shen C. (1996) The importance of diamond coated tools for agile manufacturing and dry machining, Surface and Coatings Technology. 86-87 (part 2) 672-677

[13] Sartkulvanich P., Altan T., Gocmen A. (2005) Effects of flow stress and friction models in finite element simulation of orthogonal cutting – a sensitivity analysis, Machine Science and Technology. 9, 1-26

[14] Merchant E. (1945) Mechanics of the Metal Cutting Process. I. Plasticity Conditions in Orthogonal Cutting, J. Appl. Phys. 16, 318

[15] Boothroyd G., Knight W. (1989) Fundamentals of Machining and Machine Tools, Marcel Dekker, New York

In: Metal Cutting: Research Advances
Editor: J. Paulo Davim pp. 135-164
ISBN: 978-1-60876-207-1

Chapter 5

DIAMOND-COATED CUTTING TOOLS FOR MACHINING APPLICATIONS

Kevin Chou*

Mechanical Engineering Department; The University of Alabama
Tuscaloosa, AL 35487, USA

ABSTRACT

Cutting tools coated with thin diamond films are a novel engineering application and have the potential to replace costly polycrystalline diamond (PCD) tools in machining nonferrous materials, which is of special significance in the manufacture of lightweight, high-strength components for vehicle fuel efficiency. Despite the unsurpassed tribological characteristics of diamond coated cutting tools, their use is not wide spread.

This chapter provides tooling users and the machining research community with an overview of diamond-coated tools, including substrate materials and geometry, surface treatments, diamond film deposition, coating characterizations, and in particular, machining applications. In addition, recent developments from the author's group are reported to address additional aspects of diamond tool investigations. Research needed to achieve further advances in diamond-coated tools technology are also discussed.

1. INTRODUCTION

Modern machining often encounters advanced materials such as composites,\ and requires cutting tools with wear resistance sufficient to achieve high throughput, top-quality parts, and minimum environmental hazard in a cost-effective manner. Diamond coating is an ideal material for machining nonferrous materials because of its many unique tribological properties, including superior hardness and low friction. Synthetic polycrystalline diamond (PCD), made by high-pressure, high-temperature sintering, is commonly used in manufacturing. The increased usage diamond tools is driven in part by the use of lightweight, high-strength materials and demands of dry machining . However, the processing and

* Corresponding author, E-mail: kchou@eng.ua.edu

fabrication of PCD tools are costly. On the other hand, advanced surface-engineering technologies such as chemical vapor deposition (CVD) also have been widely explored for wear resistant applications, including cutting tools. Produced via CVD processes, diamond coating, developed about three decades ago, is recognized as one of the most distinguished engineering achievements of the 20th century. CVD-grown diamond films also were explored for various tooling applications. In addition, this coating has definite advantages in fabrication of tools with complex geometry such as drills, end-mills, grooved tools, and others. In addition, various CVD diamond-deposition processes and other related technologies, are constantly evolving to improve the diamond-coated tool fabrication processes and performance. However, despite the superior characteristics of diamond coatings, some of which are even better than PCD, application of diamond-coated tools is still limited because of a few crucial remaining obstacles.

This chapter offers an overview of different aspects of diamond-coated cutting tools, including substrate materials, surface treatments, diamond film deposition, coating characteristics, and in particular, machining performance. In addition, recent studies conducted by the author's group, in collaboration with other researchers, are reported in detail. Additional research needs to further the use of advance diamond-coated tools and optimize machining performance are also discussed.

2. Diamond-Coated Tools - State of the Knowledge

Chemical vapor deposition (CVD) involves gas-phase chemical reactions that occur above a solid surface and lead to substance depositions onto the target surface. Polycrystalline diamond films can be produced by CVD using a hydrocarbon gas (typically methane) with a large quantity of hydrogen gas [1]. All CVD techniques generally involve different ways to activate gas-phase carbon-containing precursor molecules [1]. As important as the deposition process itself, substrate materials and surface conditions are also critical to the functionality of diamond coated tooling systems.

2.1. Substrate Materials and Geometry

For diamond coated tool applications, most substrate materials are cobalt-cemented tungsten carbides (WC-Co), though silicon nitride materials (typically, 89.3%Si_3N_4, 7.0% Y_2O_3, 3.7% Al_2O_3) also have been studied to some extent by a few research groups [2-4]. WC-Co tools are widely used in the machining industry; however, one of the major difficulties in depositing diamond on WC-Co tools arise from chemical incompatibility, namely, the Co binder, which induces graphitization during deposition. Thus, a common procedure performed prior to diamond deposition is the removal of Co at the tool surface in a particular way. WC-Co tools vary in composition (i.e., have varying Co percentages) and in microstructure (i.e., varying grain sizes differ). A large amount of Co is unfavorable to diamond film depositions. Typically, no more than 6% Co is preferred. However, it is also known that Co increases toughness, which is also important in interrupted cutting processes. On the other hand, ultrafine-grained microstructures may not necessarily be favorable for adhesion by mechanical interlocking.

Various types of tooling, such as indexable inserts for turning and milling, have been successfully surface coated with diamond. Rotary tools like drills [5,6], end mills, and dental burrs, and others, also have been diamond coated, but are not as well documented. Diamond coating of rotary tools with complex shapes is especially advantageous, but challenging, and compete favorably with their PCD counterparts because of the difficulty in applying brazed-on techniques to tools with complex geometry. Cutting inserts with chip-breaker features also have been demonstrated to be diamond coatable [7]. Moreover, diamond coating of micro-scaled endmills (300 μm diameter) also has been attempted [8,9], but the authors indicate a potential size issue.

The micro-level geometry of tooling has not received much attention although it was recently proved that micro-scale geometry, such as edge radius, is critical [10]. While the edge geometry affects deposition stress distributions around the cutting edge, it also complicates thermal and mechanical loads imposed during machining. Literature on the effects of tool geometry on coating tool performances is rare. Almeida et al. experimentally investigated edge preparations of diamond coated tools in machining hard metal (WC) [11]. The authors tested three types of edge conditions, i.e., up-sharp, chamfered, and honed. These edge conditions were found to be significant to machining forces, wear pattern, and tool life, and other characteristics. The authors further report that coating delamination occurred first in the honed tool.

2.2. Substrate Surface Treatments

As discussed, Co in carbide tools has an adverse effect on diamond film depositions because the solubility of C in Co leads to graphite formations [12]. Thus, it is necessary to remove surface Co prior to diamond deposition. The traditional application of chemical solution etching to remove the surface Co is still widely used, although other, different solutions and procedures are proposed [13-15]. Results of surface treatment in general show good improvement in adhesion and tool performance, compared to untreated tools. However, some treatments—chemical etching, for example, may not be well controlled, and under- or over-etching may result in poor interface quality, which impairs film adhesion. Moreover, chemical etching is probably one of the major sources of inconsistent quality and performance of diamond-coated tools. In addition, due to high temperatures required for the application, Co in the tool body may diffuse to the tool surface during deposition, and the surface cobalt problem will remain. Other different approaches have also been suggested to remedy surface Co issues including Ar-H_2 etching decarburization [16], heat treatment [17], and addition of interlayers as a Co barrier [18]. For example, Polini pointed out that heat treatment as-ground WC-Co will modify the substrate surface morphology and reduce surface Co concentration, resulting in improved adhesion level of diamond coatings [17].

Among the different techniques studied, interlayers seem to hold the most promise to enhance diamond coating adhesion and prolong tool life. Different interlayer systems such as TiN-TiCN [19], SiC [20-21], Cr/CrN/CrC [18], Co-B [22], and BN [23] have been investigated. In general, interlayers seem to bring qualitative improvement to adhesion enhancements. For example, Cabral indicates that thin SiC interlayers are effective in providing adhesion to diamond films on WC–Co for dry machining of highly abrasive commercial graphite rods; and that the results show lower wear than PCD machined tools [20].

2.1.3. CVD Processes

Typical chemical vapor deposition (CVD) diamond deposition has gas chemistry of H_2 and CH_4. The CVD process requires near vacuum and very high temperatures (700 to 900 °C).In addition, energy sources—hot filament, microwave power, or arc discharge, are applied to induce the breakdown of gas molecules, which results in subsequent complex chemical reactions and eventually diamond formation. The diamond films produced are very sensitive to process parameters and can result in different structures, composition, and diamond purity. Although traditional CVD diamond growth techniques are 30 years old, diamond films produced by these methods— known as microcrystalline diamond (MCD) coatings—are not ideal for tooling applications because of their polycrystalline nature and crystal sizes on the order of a few microns.. Gas concentration in CVD plays an important role in the machining performance of CVD diamond tools. Almeida et al. discovered that smoother nanocrystalline diamond (NCD) coatings produce good workpiece surface roughness deposited by hot filament CVD (HFCVD) [24]. The authors also show that good surface finishing can be achieved by smaller diamond crystallite sizes when hard metal is dry turned with diamond coated Si_3N_4 inserts by HFCVD [2]. A volume ratio of 0.02 (CH_4/H_2) is used to deposit microcrystalline diamond (MCD) diamond, while a volume ratio of 0.03 results in nanocrystalline diamond (NCD).

In addition, a very low ratio gas chemistry is used to modify diamond film structures. For example, nitrogen gas may obtain nanostructured coatings that consist of nanocrystalline diamond grains embedded into a hard amorphous tetrahedral carbon (ta-C) matrix [25]. Liang et al. tested the nanostructured diamond-coated WC-Co inserts by microwave plasma CVD (MPCVD) and their results indicate that the inserts demonstrate excellent durability against the highly abrasive A390 aluminum-silicon alloys in high-speed dry turning [26]. The inserts coated with nanostructured diamond films have wear resistance comparable to that of PCD in the continuous turning, but fall short in interrupted turning.

On the other hand, adding Ar gas may achieve ultrafine nanocrystalline diamonds [27]. Ma et al. showed that when with an Ar concentration of 90%, diamond film exhibits rounded fine grains with an average grain size of approximately 60–80 nm [28] showing greater crack resistance along with a better surface finish. Lu et al. evaluated the performance of the nano-micro composite diamond film tools by High Current DC Extended Arc plasma CVD, and found the inserts with nanocrystalline diamond are superior to the microcrystalline diamond film tools, and further increase cutting life by 20–30% over that resulting from milling Al–12 wt.% Si [29].

2.4. Coating Characterizations

Diamond coatings produced by various CVD processes can have different properties. Diamond purity and bond characteristics are generally evaluated using Raman spectroscopy, which relates molecular vibration frequencies to the inter-atomic bonds [30]. Diamond crystalline sizes and qualities have been evaluated by x-ray diffractometry (XRD) [31]. Mechanical properties of diamond coatings are of great interests and nanoindentation has been used particularly to measure elasticity and hardness [30,32]. Diamond coatings have a wide range of elasticity and harness, from 500 to 1200 GPa and from 50 to 100 GPa, respectively [30]. In addition, during CVD processes, the coating-substrate system returns to room temperature from the deposition temperature. The largely mismatched thermal strains between diamond coating and substrates generate high stresses in the coated tool and stress discontinuity at the interface. Diamond coating, with a smaller thermal expansion coefficient, endures a compressive residual stress and the carbide substrate receives a stress in tension. In addition, intrinsic stresses occur during deposition, mainly from the effect of the non-diamond carbon content in the diamond crystals. In general, the intrinsic stresses reverse the thermal stresses [33]. Residual stresses in diamond coatings have been widely studied using various techniques, such as substrate curvature measurements [34], XRD [35,36], Raman spectroscopy [37],and others. The residual stresses in diamond coatings depend upon substrate materials, surface treatments, and deposition temperatures, and have been reported as high as 6 GPa in compression [38]. Such a high level of residual stress will compound the impact to coating performance [39]. Further, the deposition stress as reported in the literature assumes a nominal biaxial stress condition. Around any geometric changes, the local stress fields may be drastically altered. For machining applications, such high stresses around the cutting edge are obviously crucial as the cutting edge further bears localized thermal and mechanical loads during machining.

Although diamond coatings are ideal for tribological applications, coating delamination is, nevertheless, the major mode of failure in CVD diamond tools. The imperfect adhesion at the interface, combined with interfacial stresses lead to premature failure during machining by debonding. For diamond coating and WC substrate, adhesion depends on the substrate cobalt content, surface roughness, microstructures, and the like. A primary bonding mechanism is mechanical interlocking, which relates to surface textures [14]. Diamond coating adhesion has been qualitatively evaluated using a Rockwell hardness tester [26,40-42]. Polini et al. [42] and Liang et al.[26] employ bulk indentation (Rockwell indenter) to obtain the delamination size at different loads and report that the slope of the delamination vs. load curve implies the toughness of the interface. Grogler et al. also estimate the critical energy release rate of the coating interface using indentation testing [38]. On the other hand, Buijnsters et al. applied a sliding/scratch test to estimate the critical load for diamond coating failure [43]. However, quantitative modeling of interface strength and delamination behavior of diamond coatings has not been achieved.

2.5. Machining Performance

Applications of CVD diamond coating tools are widely reported. A wide range of work materials has been tested using diamond-coated tools. These materials include Al-Si alloys, Al-Mg alloys, electrode graphite, Al matrix composites, wood products, polymer matrix composite, and less common materials such as carbon fiber reinforced carbon disc [44], a-SiAlONs based ceramics [44], hard metals [2,45-46], dental alloys [47-48],and others. Al alloys were also used in micro-milling applications [8,9]. In most cases, diamond-coated tool are largely suitable in machining Al-Si alloys and graphite. On the other hand, composite work materials seem to be the most challenging due to severe tool wear and part quality issues. The machining parameters studied range from 15 m/min cutting speeds for hard metals to over 1200 m/min for graphite and Al-Si alloys. Feed rates are in the range of 0.1 mm/rev to

1.0 mm/rev, and cut depths used were between 0.1 mm and 2 mm. In most studies, no coolant was used in dry machining attempts. Without a doubt, CVD diamond tools are as much as than 10 times better than uncoated tools. In 1991, Leyendecker et al. published results showing that tool life of diamond coated drills is more than 25 times longer in than that of their uncoated counterparts [49]. The authors also caution that application-driven development of diamond-coated carbide tools must be based on analysis of how the tool, coating, and machining conditions affect quality.

Another common observation is that conventional MCD diamond tools generate poor surface finish compared to PCD tools due to multi-faceted crystals at the surface and an enlarged edge radius. Machining forces of diamond coated tools also have been analyzed by force transducers. In general, slightly higher forces are expected because of increased edge rounding due to added coating materials, also possibly due to rough coating surfaces. However, the effects from machining parameters, primarily the feed, are the dominant factor, rather than machining forces. Davim and Mata report that CVD diamond tools result in lower cutting forces compared to those in carbide and PCD tools in machining polyetheretherketone (PEEK) composites, both with and without carbon fibers [50]. Torres et al. report that NCD micro-endmills require much lower machining forces than their uncoated carbide counterparts in micro-milling of Al alloys [8]; and they attributed diminished adhesion of diamond to Al chips to significant reduction in required force.

However, there are mixed results regarding estimates of tool life in CVD diamond tool performance; most results claim that wear resistance of CVD diamond tools is much inferior to that of the PCD counterparts [51-57], though some argue that CVD diamond tools also have potential economic benefits because of their multiple edges [54].

In a few applications, CVD diamond shows tool life comparable to that of PCD tools in machining high-silicon (Si) aluminum (Al) alloys [13,58,59]. Some studies also demonstrate that CVD diamond coated tools outperform PCD tools in graphite machining [20,60]. For example, Cabral et al. tested time-modulated CVD (TMCVD) tools with varied cutting speed, and found the TMCVD tools exhibit a better wear resistance than inserts coated with conventional diamond coatings, PCD, and bare WC-Co inserts when turning graphite [60]. Another successful example was given by Karner et al., who investigated CVD diamond tools in turning copper armature, [61], and whose results showed that the CVD diamond tools double the tool life of PCD tools.

Several other studies further observe coating delamination as the dominant wear mode, which can be of catastrophic nature [62]. High stresses and/or degraded adhesion during machining may result in coating failure, and the exposed substrate then suffers from massive deformation and rapid wear.

3. Recent Developments

In collaboration with other researchers, this author's group has extensively studied diamond coated tools in machining. The cutting tool substrates used were 6% Co fine-grained tungsten carbides. The geometry used was mostly an insert-type, square shape (SPG422), 12.7 mm wide, 3.2 mm thick, with a 0.8 mm nose radius.

3.1. Diamond Deposition

The carbide inserts were first chemically etched to remove surface cobalt to facilitate diamond coating. In addition, the tools were ultrasonically vibrated in a slurry of diamond/water to increase the nucleation density The diamond film was produced by a microwave plasma-assisted CVD process using an in-house built reactor (30 kW, 915 MHz) at Vista Engineering [10]. A gas mixture of methane in hydrogen, 750-1000 standard cubic centimeters per minute (sccm), with 4.4-7.3 % of methane/hydrogen ratio, was used as the feedstock gas. Nitrogen gas, 2.75-5.5 sccm, was inserted to obtain nanostructures by preventing columnar growth. The pressure was about 30 to 55 Torr, and the substrate temperature was about 685 to 830 °C.

(a) MCD

(b) NCD

Figure 1. SEM view of microstructures at the rake face of different diamond tools: (a) MCD-coated tool, and (b) NCD-coated tool [63].

3.2. Coating Characterizations

Together with microcrystalline diamond (MCD) coated tools (commercial grade) and polycrystalline diamond (PCD) tools, the fabricated diamond tools, coated with nanocrystalline diamond (NCD) particles, were examined in detail, including surface topography, microstructures, grain sizes, and cutting edges, by scanning electron microscopy (SEM),. Figure 1 shows SEM images comparing the microstructures of MCD- and NCD-coated tools [63]. The MCD tool exhibits clear multi-facet diamond crystals and the grain sizes are estimated around 3 to 5 μm in average. On the other hand, the NCD tool has ultrafine grains, on the order of 100 nm, under a high-power SEM. In addition, it was observed that the MCD tool has the roughest surface because of the diamond crystal size. On the other hand, the NCD tool also shows a fairly smooth surface which replicates the texture of the substrate surface.

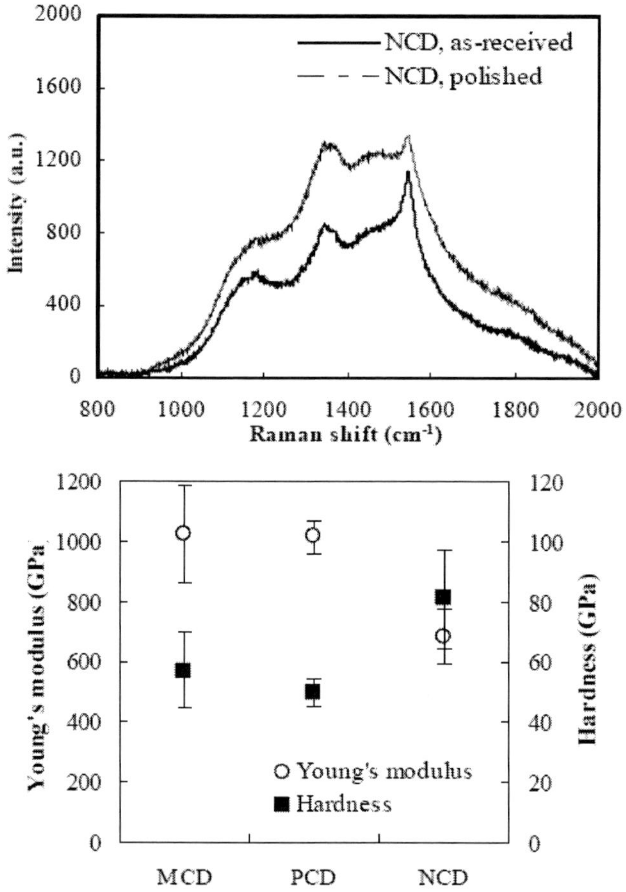

Figure 2. (a) Raman spectra of diamond-coated tools: as-received and polished NCD tools, and (b) mechanical properties of different diamond tools [63].

Analysis using Raman spectroscopy was performed on the tools to examine their carbon bonding characteristics [63]. The Raman scans were conducted using a Jobin-Yvon HR800 UV confocal microscope. The Raman spectra, figure 2a, show that the NCD tools have the clear peak associated with natural diamond (sp^3, 1332.5 cm^{-1}), as well as amorphous carbon (1540 cm^{-1}), graphite (1585 cm^{-1}), and disorder sp^3 (1150 cm^{-1}) and sp^2 (1350 cm^{-1}),and other characteristics. However, no noticeable peak-location shifting was observed from the Raman spectrum of the diamond coating tools. The Raman spectrum image also show that NCD tools have much smaller crystal sizes. The polished NCD specimen shows notable decrease of the graphite peak (1585 cm^{-1}), also figure 2a.

Nanoindentation was performed using a Hysitron TriboIndenter with a Berkovich pyramidal indenter [63]. The results, shown in figure 2b, indicate that the NCD tool has a much greater hardness (about 81 GPa) than the MCD and PCD tools 57 and 50 GPa, respectively. On the other hand, among the three diamond tools, the NCD tool shows the lowest elasticity—684 GPa, vs. 1027 and 1019 GPa for the MCD and PCD tools, respectively. Since the deposition-induced thermal stress is proportional to the elasticity, the low elasticity of the NCD tool implies a lower deposition-induced residual stress generated in the NCD tool.

In addition, residual stress measurements in diamond-coated tools were attempted by x-ray diffratometry (XRD), as well. Due to the difficulty in using low angle peaks such as (111) and (220), the (311) peak was selected and the $\sin^2\psi$-method with omega-tilting mode was applied [64]. The preliminary results, figure 3, show about 1.0 GPa compressive stresses in the NCD tools.

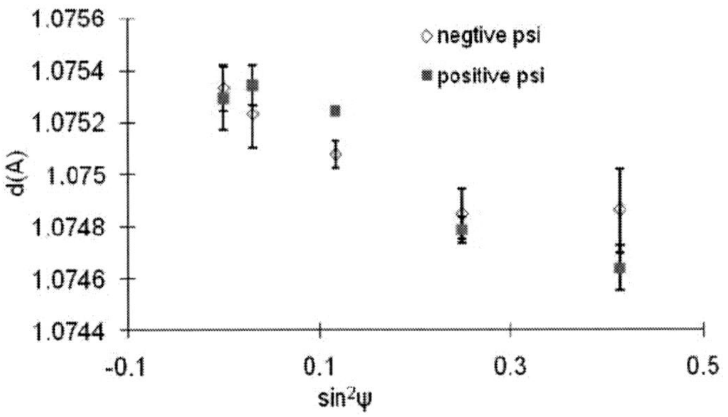

Figure 3. Sample residual stress measurement results for diamond-coated tools [64].

3.3. Machining Performance

3.3.1. NCD vs. MCD and PCD Tools

A series of tests on NCD-coated tools during machining were conducted and compared to commercial CVD and PCD tools. Two different, high-strength aluminum-based materials were used in testing: Al matrix composite (A359/SiC-20p) and Al-Si alloy (A390, 18 wt.% Si). A computer numerical control (CNC) lathe, Hardinge Cobra 42, was used to perform machining experiments, outside diameter turning, to evaluate the wear behavior of the two types of diamond tools. The diamond cutting inserts formed a 0° rake angle, 11° relief angle, and 75° lead angle when mounted on the tool holder used (CSRNL-164D). All machining was done without coolant. In addition, cutting forces were monitored during machining using a

Kistler dynamometer (9257B). In some studies, an acoustic emission (AE) sensor was also employed to acquire both AE-RAW (raw data) and AE-RMS (root-mean-square value) data during the machining operation. The sensor used was 8152B Piezotron sensor from Kistler. The AE-RAW and AE-RMS were collected at a 500 kHz sampling rate for each.

For Al matrix composite, machining conditions were 4 m/s, 0.05 mm/rev, and 1 mm depth of cut [65]. During machining testing, the cutting inserts were periodically inspected by optical microscopy to measure the flank wear-land width (VB). Worn tools after testing also were examined by SEM. Figure 4 plots tool wear developments of the diamond tools. CVD diamond experienced coating failure after a brief cutting time. The PCD tool reached about 0.11 mm wear-land width after about 10 min. cutting time. For the NCD tools, small wear, comparable to that of the PCD tool, with some deposits around the tool tip were observed. After the deposit was cleaned (at around 8 min. cutting), the tool continued cutting with little noticeable wear increase. Note however, that when the deposit was not removed, catastrophic delamination occurred quickly.

For A390 machining, 10 m/s of cutting speed, 0.8 mm/rev of feed, and 1 mm of depth of cut were used [66]. The flank wear development of the three different diamond tools is shown in figure 5. Note that as in Al matrix composite machining, a metal deposit sometimes appeared around the cutting zone, including the wear land. Aggressive machining conditions, 10 m/s and 0.8 mm/rev, caused rapid tool wear. For the MCD tool, premature coating failure occurred and led to a sharp tool wear growth, reaching about 1.0 mm VB at just 1.2 minutes of cutting time. On the other hand, the NCD tool had greater adhesion and the tool wear was similar to that of the PCD tool. However, coating failures eventually occurred and dictated the tool life, at about 2.6 min with a flank wear-land of 0.6 mm.

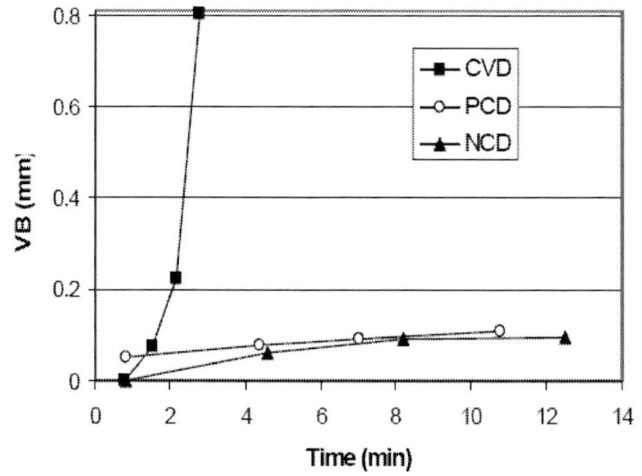

Figure 4. Tool flank wear vs. time in machining Al composite [65].

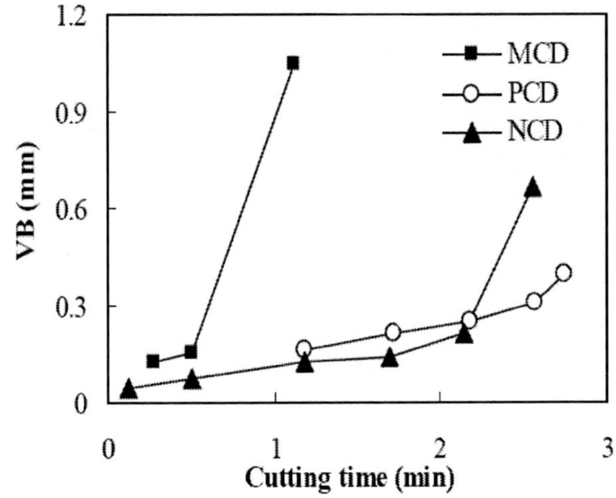

Figure 5. Flank wear of different diamond tools in machining A390 alloy [66].

3.3.2. Machining Parameter Effects

In machining of A390 using NCD tools, different machining parameters also were evalauted, including four combinations of two cutting speeds: 3 m/s and 10 m/s, two feeds: 0.2 mm/rev and 0.8 mm/rev, and fixed 1 mm depth of cut [66]. Figure 6 compares flank wear vs. cutting time under four different machining conditions. For all except the low-speed and low-feed condition, abrupt wear growth occurred during the last cutting pass and consistently showed catastrophic coating failure as the tool life limit. It is also observed that the cutting speed is the dominant factor in the tool life. For the most aggressive machining conditions (10 m/s and 0.8 mm/rev), the tool life was about 2.6 min. with a flank wear-land of 0.6 mm. On the other hand, for the 3 m/s and 0.2 mm/rev condition, coating failures were not noted after about 16 min. of cutting time with a VB less than 0.1 mm.

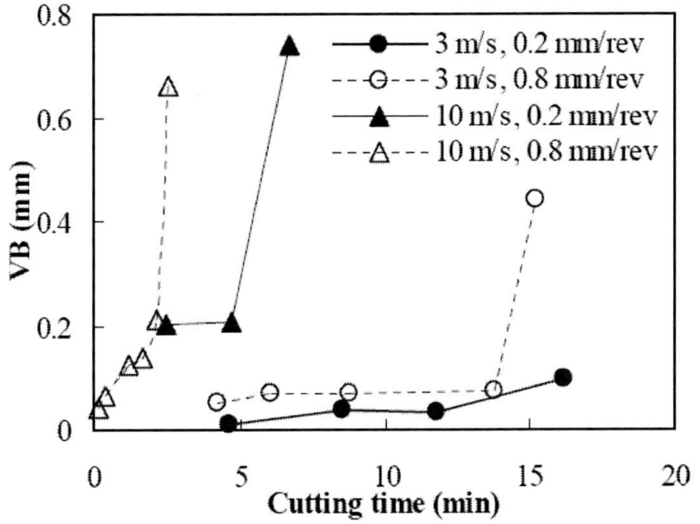

Figure 6. Flank wear of NCD tools at different conditions in dry turning of A390 [66].

In machining of A359/SiC-20p composite using NCD tools, machining parameters consisted of two levels of cutting speed (4 m/s and 8 m/s) and two levels of cutting feed (0.15 mm/rev and 0.3 mm/rev) [67]. The depth of cut was kept constant at 1 mm. Figure 7 shows tool wear growth results of two replicates, along cutting time under different machining conditions. In general, the tools show a slow increase in tool wear followed by an abrupt increase of wear-land in one single pass. It is believed that during that specific pass, coating delamination occurred and resulted in rapid wear of the exposed substrate material, i.e., carbide. Tool wear and onset of coating delamination (abrupt wear increase) are strongly dependent on machining conditions. In contrast to A390 machining, with A359/SiC-20p composite the feed rate, instead of the cutting speed, is the dominant cause of coating delamination,. Thus, it is hypothesized that under these conditions delamination is due to excessive mechanical loads. Further, as a baseline comparison, a plain carbide tool, with the same substrate as the coated tools, also was tested at the gentle machining condition, namely, 4 m/s and 0.15 mm/rev. Flank wear-land in the plain carbide tool reached 0.6 mm in just about 15 s, in contrast to 0.3 mm at between 130 s and 150 s cutting time for the NCD coated tool. Thus, it is concluded that the NCD-coated tool is far more effective, in fact an order of magnitude better, than widely used carbide tools.

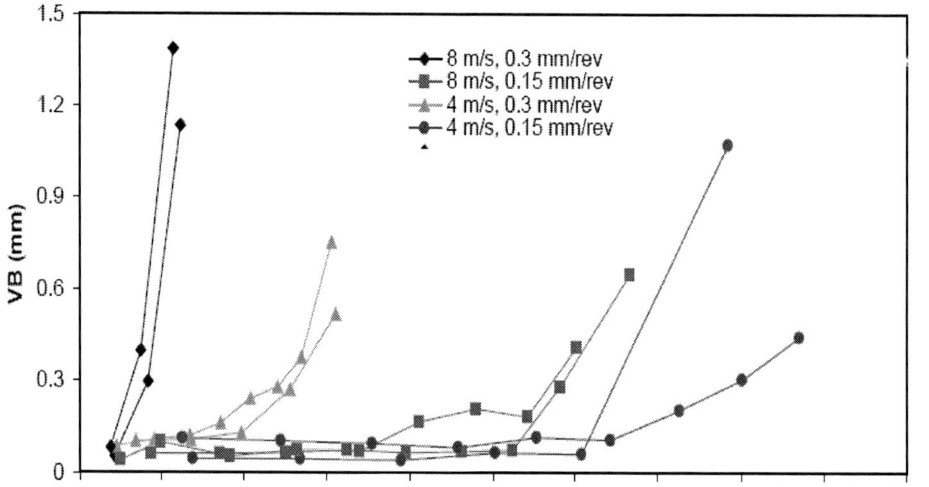

Figure 7. Flank wear-land (VB) vs. cutting time under various cutting conditions in Al composite machining [67].

Figure 8 shows typical wear features of a diamond-coated tool after machining. The machining conditions were at 4 m/s and 0.15 mm/rev condition and was cleaned, after machining, by 10% hydrochloric acid solution to remove the deposit [67]. As noted, coating delamination is the major wear mode.

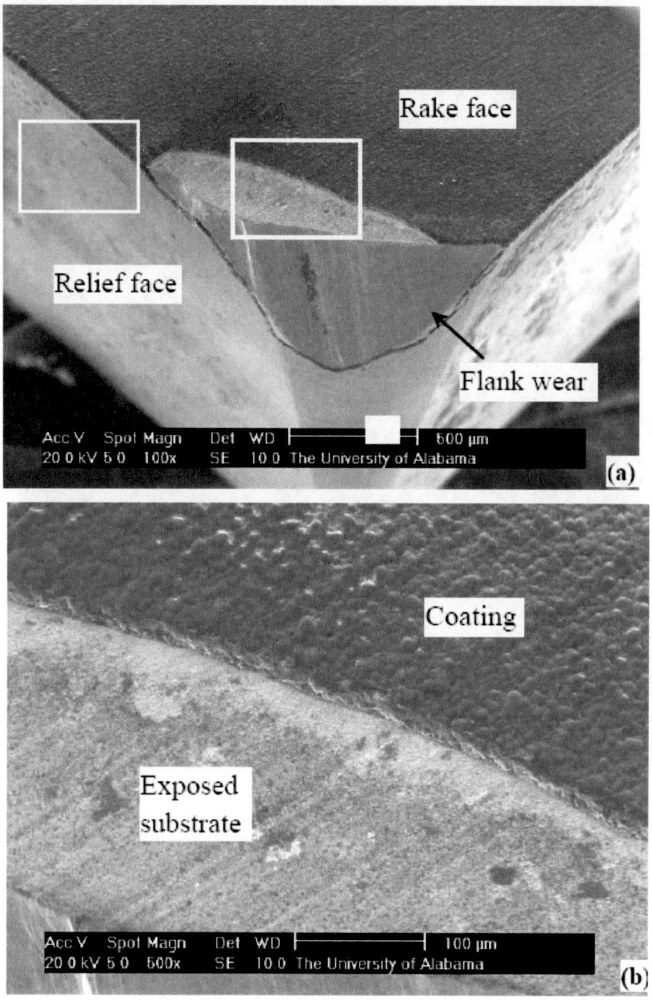

Figure 8. Worn diamond-coated tool after testing showing: (a) overall wear feature and (b) detail around coating failure boundary [67].

3.3.3. Cutting Edge Radius Effects

As mentioned above, cutting geometry may also greatly influence diamond-coated tool functions. In this study, the four levels of edge radii evaluated were 5 μm, 15 μm, 30 μm and 65 μm [10]. The edge radius of cutting inserts (SPG422) was measured prior to coating by a white-light interferometer, NT1100 from Veeco Metrology. Measurement results indicate that the edge radii were: on average, 3.79 μm, 13.7 μm, 29.8 μm, and 66.4 μm. Figure 9 shows examples of cutting edge images, uncoated tools, from the interferometer. Surface textures of the inserts also were assessed by the interferometer to analyze surface roughness. The surface roughness of the inserts was in a similar range, that is, between 0.29 μm and 0.32 μm of R_a.

The inserts after diamond coating also were inspected using the interferometer to measure the edge radius and to estimate the coating thickness. Figure 10 shows cutting edge images of coated tools. The coating thickness is estimated between 5 μm and 8 μm. Surface roughness, R_a, was about 0.5 μm for coated tools.

The workpieces used were A359/SiC-20p composite bars. Two machining conditions were used: the first one at 4 m/s and 0.05 mm/rev, and the other at 1.3 m/s and 0.15 mm/rev. The depth of cut for both was set at 1 mm. For each machining condition, two tests were done.

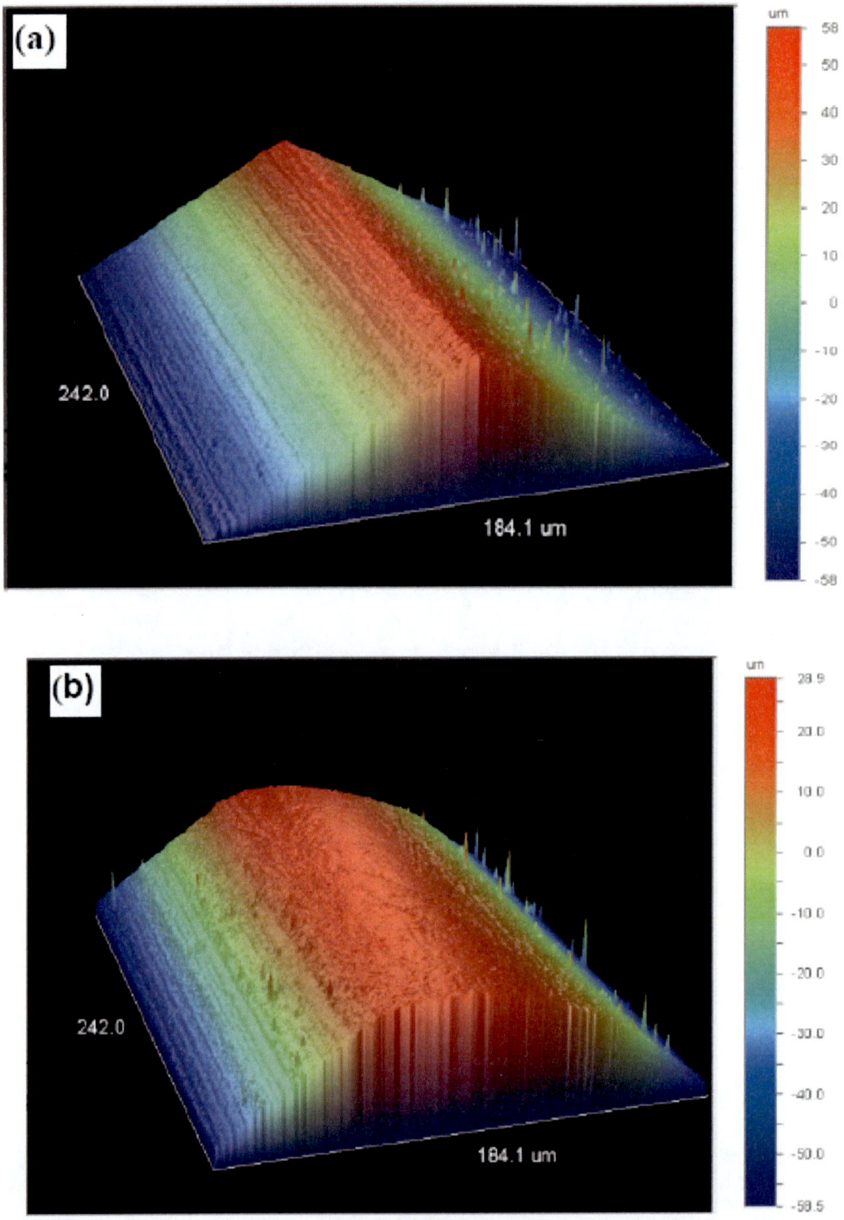

Figure 9. Cutting edge images from the substrates: (a) 5 μm and (b) 65 μm edge radius [10].

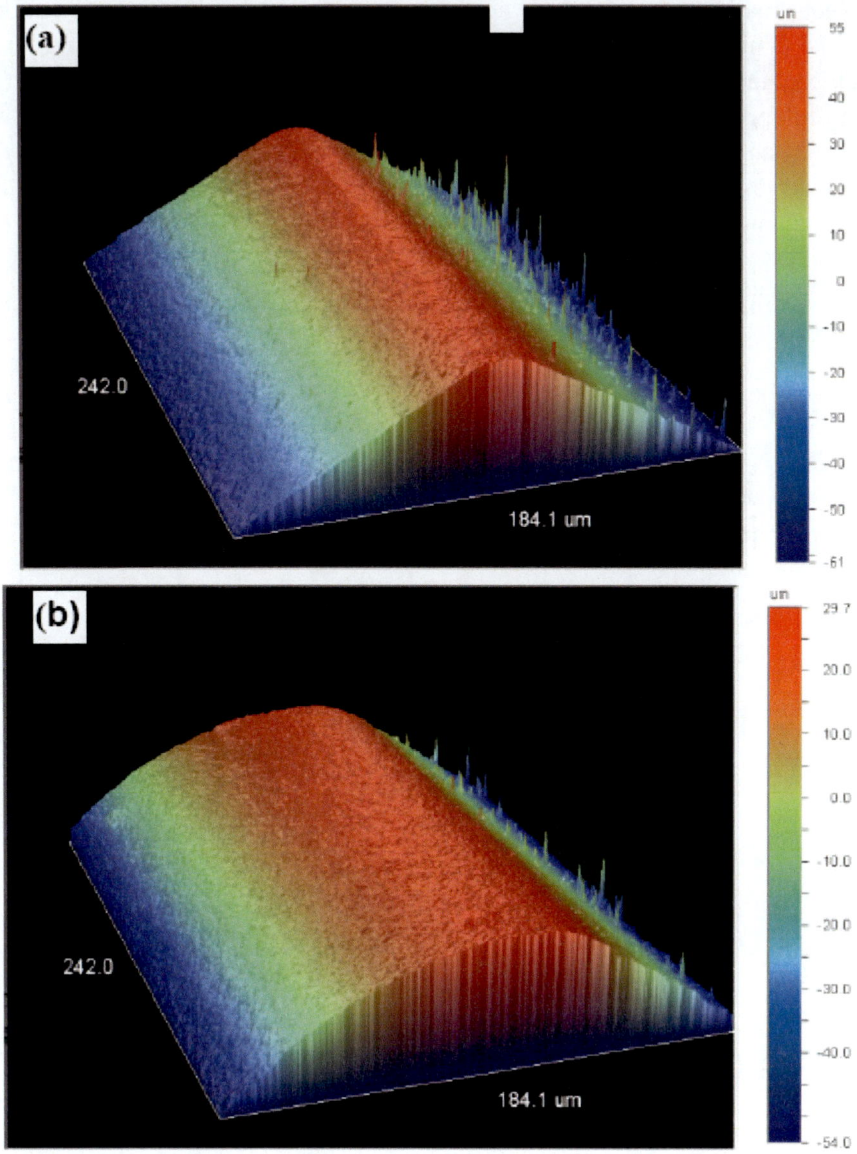

Figure 10. Cutting edge images of coated tools: (a) 5 μm and (b) 65 μm [10].

Figure 11a shows tool wear, flank wear-land width (VB), along cutting time under the 4 m/s and 0.05 mm/rev condition for different edge radii. Results of two replicate tests are shown. In general, the tools show a gradual increase of tool wear followed by an abrupt increase of wear-land in one or two passes. Tool wear and the onset of coating delamination (abrupt wear increase) are dependent on the edge radius. The tools with 5 μm and 15 μm substrate edge radii have similar wear growth curves. The 65 μm radius tools show slightly greater wear resistance and delay of delamination-induced catastrophic wear. Surprisingly, the 30 μm radius tools show the poorest tool life.

On the other hand, figure 11b shows flank wear-land width (VB) vs. cutting time at the 1.3 m/s and 0.15 mm/rev condition for different edge radii. The tools with 5 μm and 15 μm substrate edge radii have a rather rapid linear wear growth and the shortest tool life. The 5-μm tool also shows high initial wear, which might be caused by high deposition stress and result in a significant increase in cutting forces. The 30 μm tools show a somewhat better tool life, but most strikingly, 65 μm radius tools demonstrate significant delay in abrupt wear. A general trend can be noted: the larger the edge radius, the better the wear resistance among the edge radii tested. Using 0.5 mm VB as the life criterion, 65 μm radius tools have an average of ~20 min tool life vs. ~3 min for 5 μm radius tools.

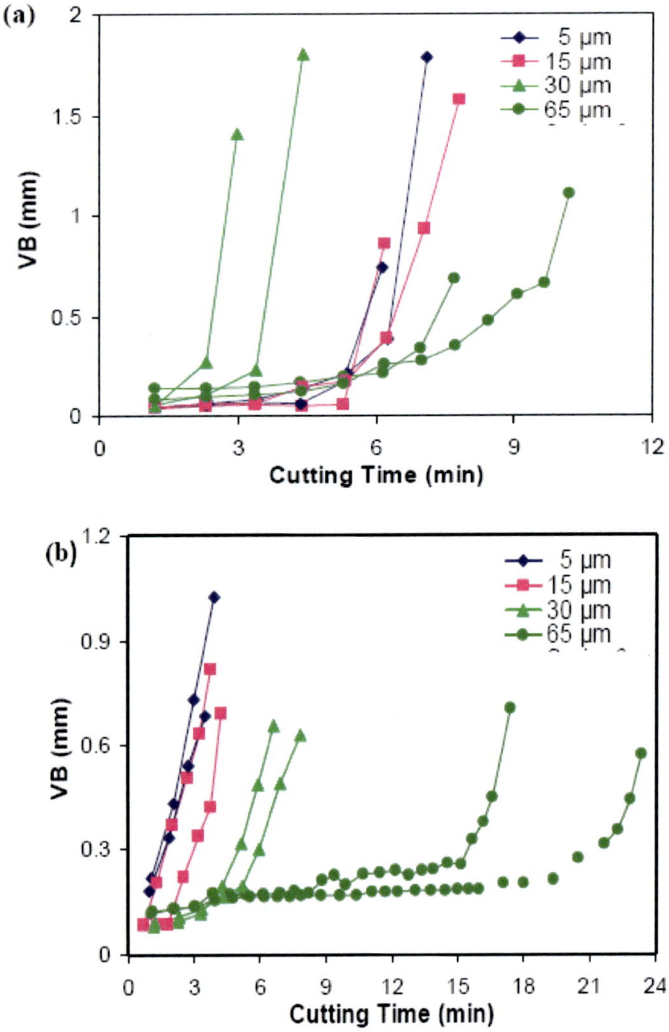

Figure 11. NCD-coated tool wear development at (a) 4 m/s and 0.05 mm/rev, and (b) 1.3 m/s and 0.15 mm/rev [10].

It is also noted that for the sharp tools (5 μm and 15 μm), the tool life at 1.3 m/s and 0.15 mm/rev is shorter than at 4 m/s and 0.05 mm/rev, which is consistent with observations from the previous studies, because the mechanical effect seems to be the more dominant factor in the onset of delamination. However, for the large honed tools (30 μm and 65 μm), the tool life at 1.3 m/s and 0.15 mm/rev is longer than that at 4 m/s and 0.05 mm/rev, contradictory to the sharp tool cases, and implies that the alleviation of deposition stresses by the rounded edge may outweigh the added machining loads due to the enlarged edge radius.

3.3.4. Coating Thickness Effects

Different levels of coating thickness also were obtained [68]. The coated inserts were further inspected by the interferometer to measure the edge radius, estimate the coating thickness, and evaluate the surface morphology. Measurement results indicate that the edge radius of the uncoated substrates averaged about 14 μm, with less than 1 μm variation. For coated tools, edge radius measurements showed 3 distinct ranges: (a) 17 to 19 μm, (b) 29 to 32 μm, and (c) 40 to 45 μm, with average coating thicknesses for coated tools estimated at (a) 4 μm, (b) 17 μm, and (c) 29 μm. Figure 12 shows tool wear and flank wear-land width (VB), along cutting time for the 4 m/s and 0.15 mm/rev condition at different coating thicknesses. Both tool wear and the onset of coating delamination (abrupt wear increase) are shown to be dependent on coating thickness. The coating thickness ranges tested indicate that a thicker coating gives a longer tool life. It is also noted that for the 17 μm- and 29 μm-thick coated tools, the range of the wear-land sizes prior to delamination showed a similar range, indicating a comparable abrasive wear resistance. In addition, the surface roughness of machined parts was similar in tools with differing coating-thickness.

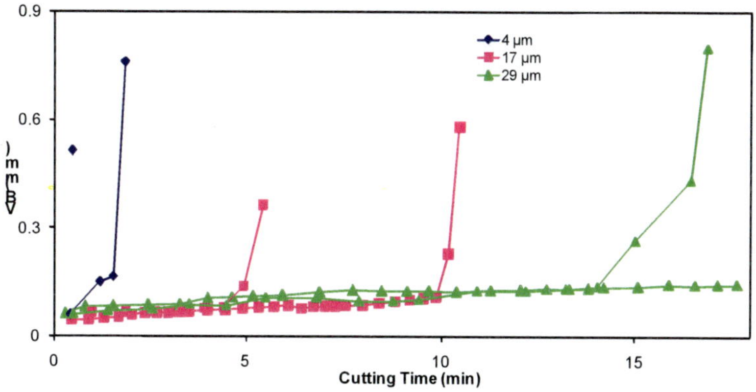

Figure 12. Coating thickness effects on tool wear development at 4 m/s and 0.15 mm/rev [68].

3.4. Tool Condition Monitoring

As described above, coating delamination is the major wear mechanism of diamond-coated cutting tools, followed by catastrophic tool failure/breakage that may damage the part or machine tool. The occurrence of delamination is still very difficult to predict. Thus, the monitoring of coating delamination events can prevent production loss and assist process planning. Figure 13a compares AE-RMS evolutions in Al composite machining using NCD tools at different cutting conditions, during two cutting passes: the initial cutting pass and the cutting pass at tool failure [69]. Compared to initial cutting, the RMS intensity during the tool failure pass is lower and shows a noticeable, continuous decrease and a reduction in intensity of over 20% in less than 10 s of cutting. The intensity decrease is due to the change of contact pairs (diamond-composite vs. WC-composite), which strongly affect acoustic emission (AE) generation. The frequency responses of AE signals also shows distinct features between before and after coating delaminations. Figure 13b compares Fast Fourier Transform (FFT) spectra of AE signals. For the tool failure pass, the amplitude of low-frequency peaks (~20 to 30 kHz), which are correlated to chip formations, was substantially reduced. On the other hand, the intensity of high-frequency AE components (~120 to 130 kHz) becomes relatively greater due to intense tribological contact with the increased wear-land area; and the ratio of high- to low-frequency components is greater than 1.0 prior to delamination vs. less than 0.5 at initial cutting. Thus, the significance of the AE signal is that it may provide tool users with a means of condition monitoring of diamond coated tools because of the AE response sensitivity to coating delamination: that is, AE-RMS values decrease noticeably the point of contact of the tool material with the workpiece changes from diamond to carbide.

For further analysis of the AE signals in the coating failure pass, the AE-RAW data was divided into 4 equal cutting periods and FFT was performed for the AE subset data. Figure 14 compares AE-FFT spectra at different cutting periods during the failure pass. The intensity reduction is very clear from period 3 to period 4, where the ratio of low to high frequency peaks decreases significantly. Thus, the AE FFT spectra into subdivided time periods during one cutting pass may clearly indicate the coating failure transition.

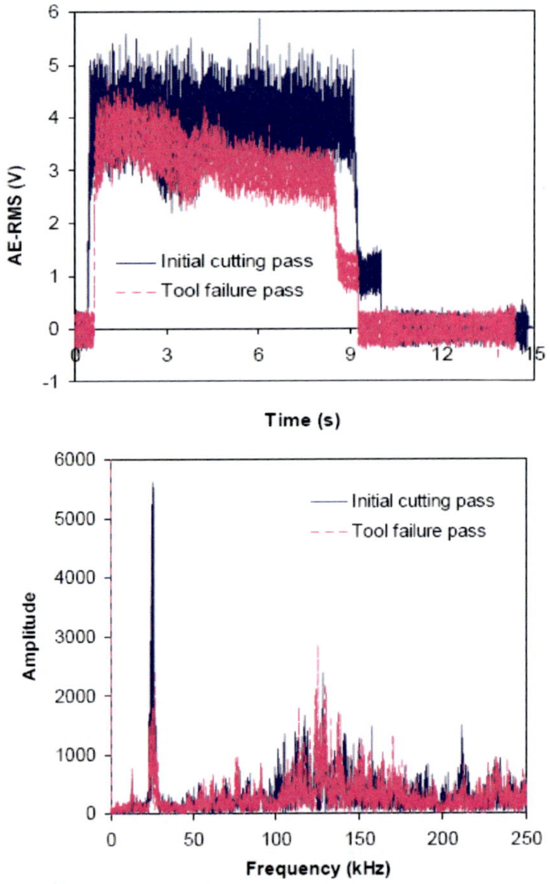

Figure 13. AE signal comparison between initial cutting pass and tool failure pass showing: (a) RMS signal and (b) FFT spectra (8 m/s, 0.3 mm/rev) [69].

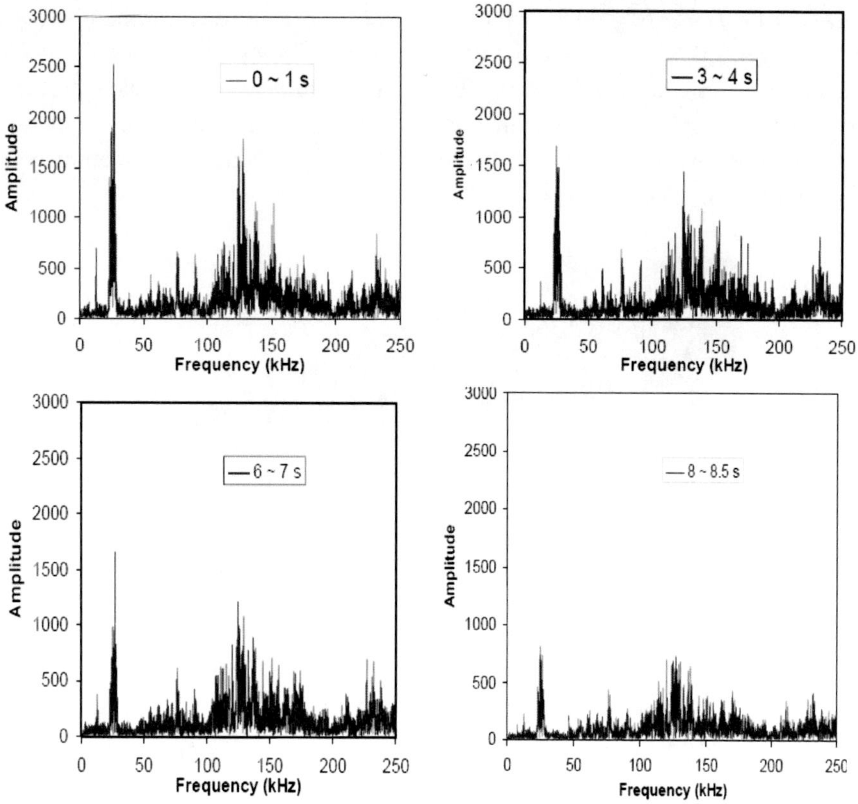

Figure 14. Evolution of FFT spectrum of AE signals during the failing pass [69].

3.5. Deposition Residual Stress Simulations

Deposition residual stresses in diamond-coated cutting tools, due to thermal mismatch, significantly impact the coating-substrate adhesion, and also affect the stress states of diamond coated tools during machining. In this study, diamond coated cutting tools were modeled using CAD software (Pro/Engineer) according to their actual geometry: square-shaped inserts that are 12.7 mm wide and 3.2 mm thick with an 11° relief angle and a 0.8 mm corner radius. The CAD models of the tool (both substrate and coating) were then imported into commercial finite-element (FE) software, ANSYS, for thermal stress simulation. Because the tools are symmetrical, only a quarter model is needed for analysis. The element used for structural analysis was Solid45. First meshing was generated in the coating using the default setting, and the edge area was refined. Then the substrate was meshed using the default setting.

Static structural analysis with thermal strains considered was conducted. A uniform deposition temperature of 800 °C was set as the initial condition, and a room temperature of 25 °C was set as the final temperature. Linear-elastic material models independent of temperature were used for both diamond and WC tools. The elasticity, Poisson's ratio, and thermal expansion coefficient of diamond and WC used were 1200 GPa, 0.07, 2.5 μm/(m·K), and 620 GPa, 0.24, 5.5 μm/(m·K), respectively. The boundary conditions used in the mechanical analysis included two symmetric planes and a fully constrained point at the bottom corner of the quarter model. After the model setup, a structural analysis was executed to obtain displacement, strain, and stress data. Further, the stress data along the interface was extracted and then transformed into the local polar coordinate to evaluate interface stresses along the cutting edge, including three components: the radial normal stress (σ_r), the circumferential normal stress (σ_θ), and the shear stress ($\tau_{r\theta}$), which are functions of location. The procedures are detailed in Renaud et al. [70].

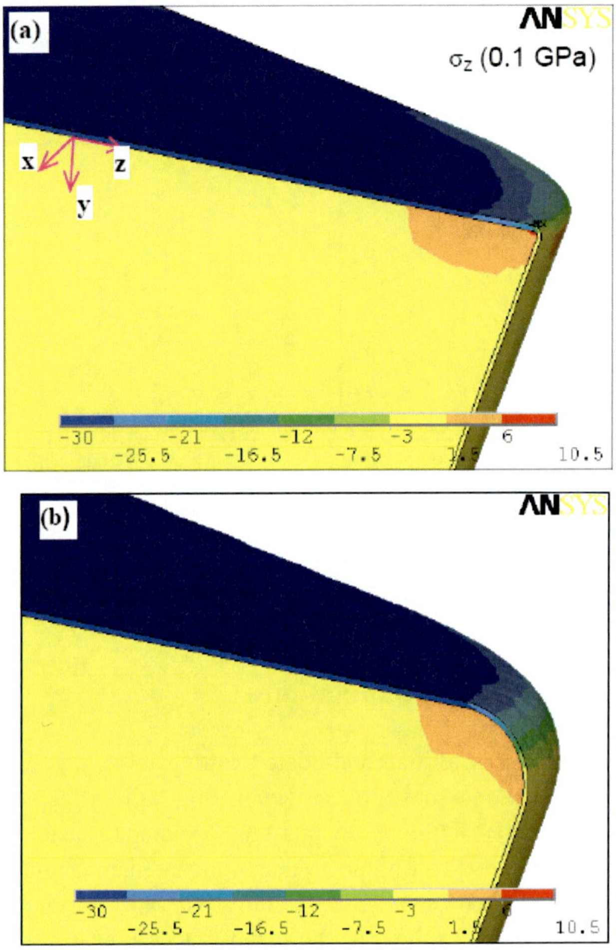

Figure 15. Examples of stress contours (σ_z) around the following cutting edge radii: (a) 5 μm edge radius and (b) 65 μm edge radius [10].

3.5.1. Cutting Edge Radius Effects

For observation of the effects of differing cutting edge radii in cutting tools, different edge radii ranging from 5 μm to 65 μm were incorporated into CAD insert substrate models. The diamond coatings on the substrates have a uniform thickness, 5 μm, at the rake and the relief faces, extending to at about 1.6 mm from the substrate bottom. Figure 15 shows the cutting tool stress contours (for σ_z component, with one major axis at the tool rake surface) in diamond-coated inserts with differing edge radii [10]. Note that the stress in the coating, away from the edge, is about 3.0 GPa in compression, which is consistent with the biaxial stress analysis. Moreover, the stress distribution around the tool edge shows considerable stress concentration for the 5 μm radius tool.

Figure 16 compares the interface stress points around cutting tool edge radius for various edge radii. The abscissa in figure 15 is the normalized distance (by the arc length of the rounded edge) from 0 to 1, where 0 is the beginning of edge rounding at rake and 1 is the end of rounding at the relief surface [70]. High stress concentrations due to the edge sharpness can be noted. For the radial normal stress (σ_r), the maximum decreases from 0.93 GPa for 5 μm to 0.2 GPa for 65 μm edge radius. The large edge radius also gives smooth stress gradients along the edge. However, the circumferential normal stress (σ_θ) shows a close range, though the sharp edge tends to have a larger gradient. Also noted is that the minimum stress increases and then decreases with an increase in the edge radius. Moreover, the maximum shear stress ($\tau_{r\theta}$) decreases from about 0.5 GPa for 5 μm to about 0.1 GPa for 65 μm edge radius.

3.5.2. Coating Thickness Effects

Coating thickness effects on deposition residual stresses also were modeled. An edge radius of 15 μm was incorporated into the models. The coating thickness varied from 5 μm to 30 μm. The approach described above also was applied to evaluate the effects of edge radius on the radial normal stress (σ_r), the circumferential normal stress (σ_θ), and the shear stress ($\tau_{r\theta}$) as a function of the location. Figure 17 shows the stress contours (σ_z component) in diamond coated inserts with two different coating thicknesses [68]. Note that the stress in the coating, away from the edge, is about 3.0 GPa in compression. Moreover, the stress distribution around the tool edge area shows considerable stress concentration for the 30 μm coating thickness with a greater tensile stress in the substrate.

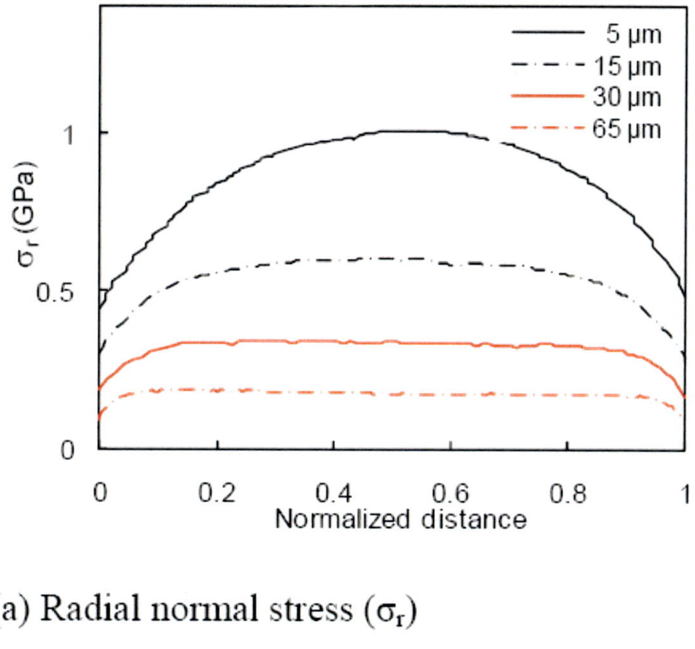

Figure 16. Cutting edge radius effects on interface stresses: (a) σ_r.

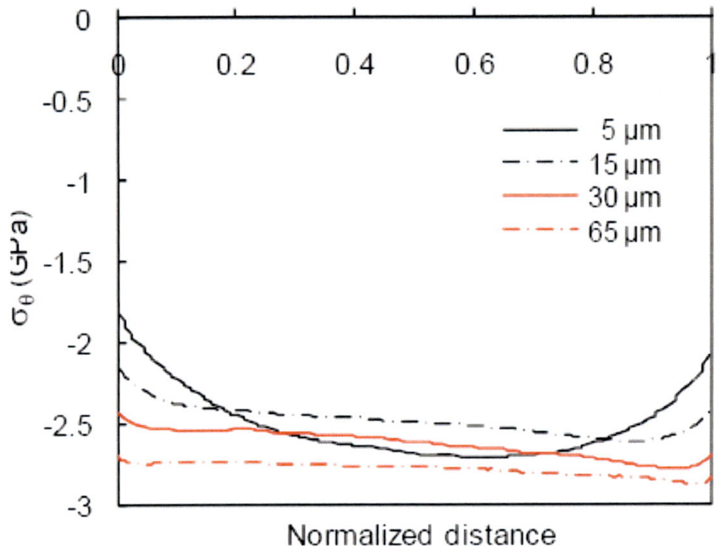

(b) Circumferential normal stress (σ_θ)

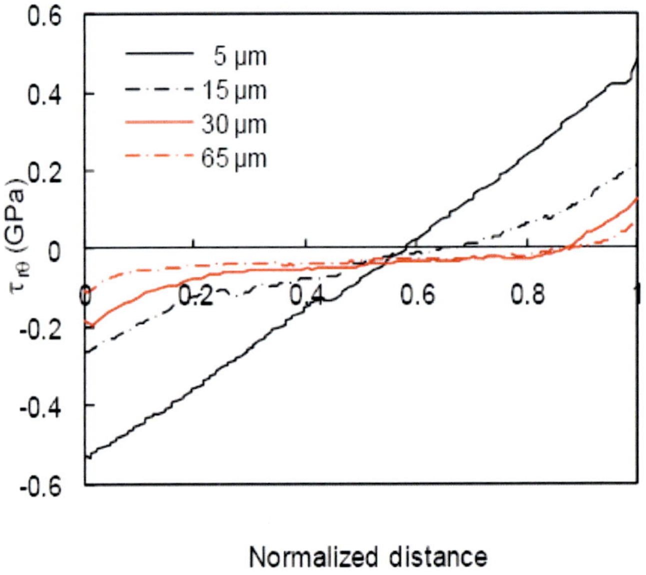

(c) Shear stress ($\tau_{r\theta}$)

Figure 16. Cutting edge radius effects on interface stresses: (b) σ_θ., and (c) $\tau_{r\theta}$ [10].

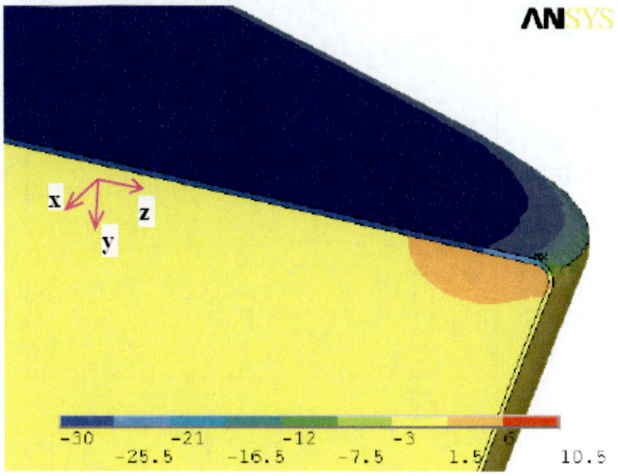

(a) 5 μm coating thickness

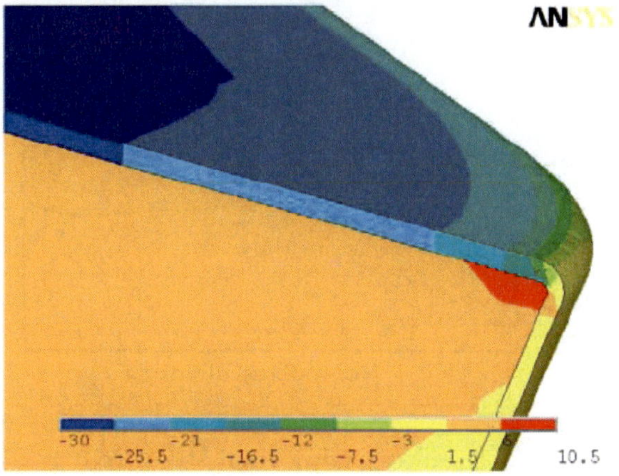

(b) 30 μm coating thickness

Figure 17. Examples of stress contours (σ_z) around the cutting tip with an edge radius of 15 μm [68].

Figure 18 compares interface stresses around the edge for different coating thicknesses. High stress concentrations resulting from thicker coatings are observable. For the radial normal stress (σ_r), the maximum stress area increases from 0.6 GPa for 5 μm to 1.2 GPa for 30 μm coating thickness. Such high tensile stresses may be detrimental in brittle fractures due to crack propagation and may require greater adhesion strengths. The thin tool coatings also give smooth stress profiles along the edge. However, the maximum magnitude of circumferential normal stress (σ_θ) increases from 2.5 GPa for 5 μm coating thickness to 3.1 GPa for 30 μm coating thickness. Moreover, the maximum shear stress ($\tau_{r\theta}$) increases from about 0.25 GPa for 5 μm to about 0.7 GPa for 30 μm coating thickness.

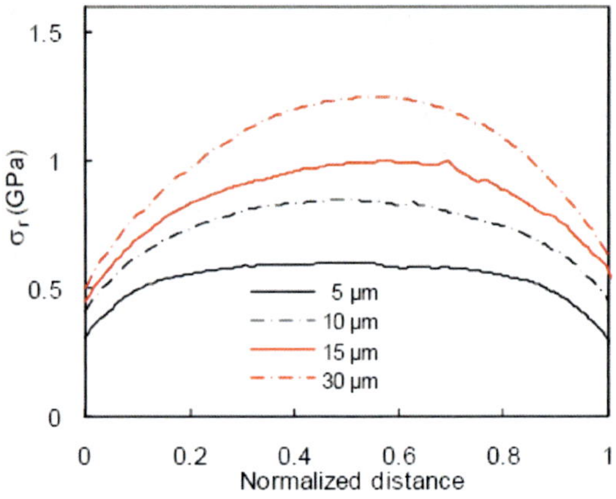

(a) Radial normal stress (σ_r)

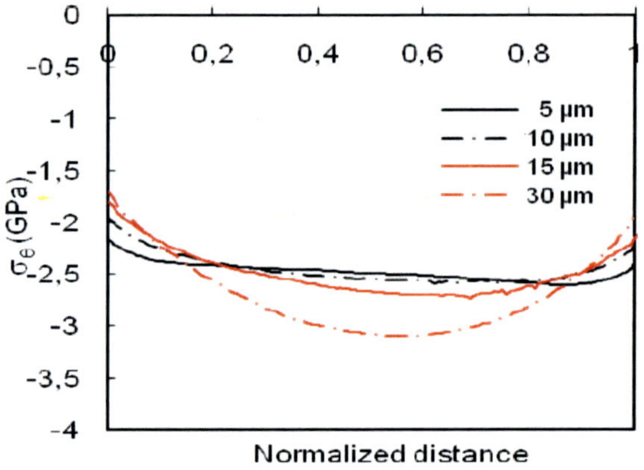

(b) Circumferential normal stress (σ_θ)

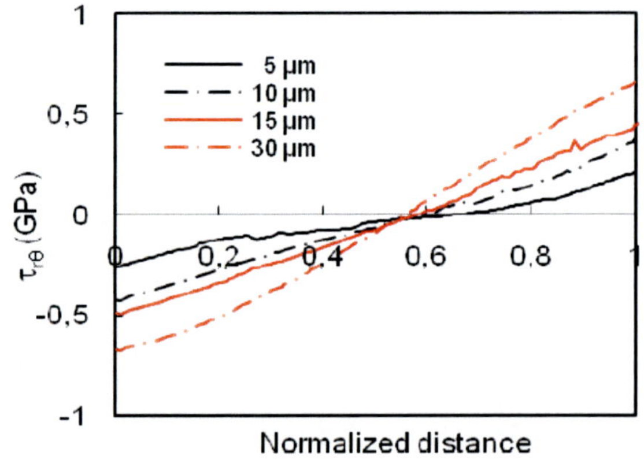

(c) Shear stress ($\tau_{r\theta}$)

Figure 18. Coating thickness effects on interface stresses for (a) radial normal stress (σ_r), (b) circumferential normal stress (σ_θ), and (c) shear stress ($\tau_{r\theta}$) [68].

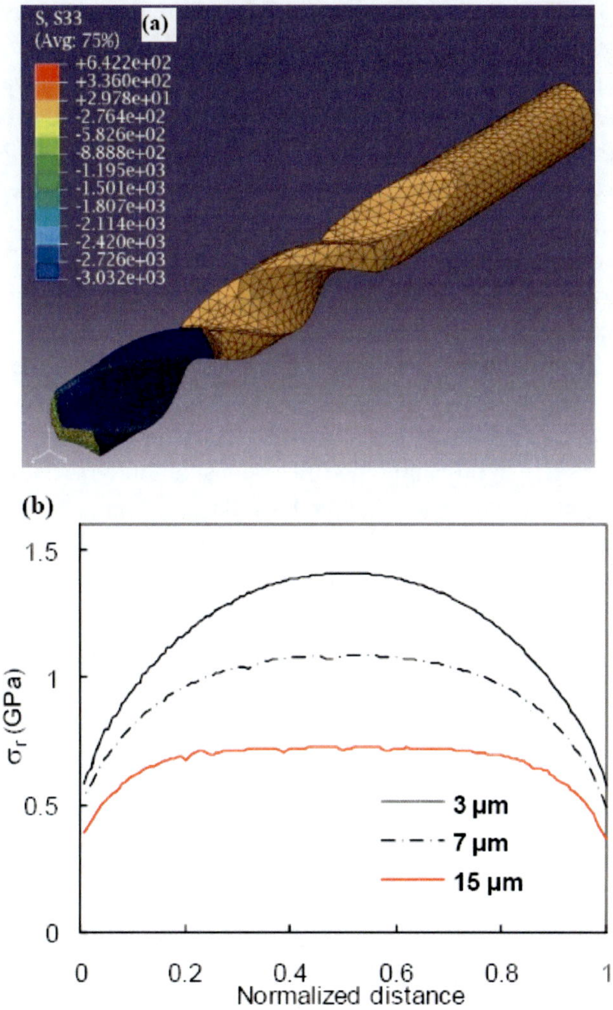

Figure 19. (a) An example of stress distributions (longitudinal normal stress) in a diamond-coated drill. (b) Edge radius effects on interface stress profiles of diamond-coated drills (radial normal stress) [71].

3.5.2. Diamond Coated Twist Drills

The approach above was expanded to examine residual stresses in diamond coated twist drills. Commercial CAD software, Solid works, was used to develop diamond-coated, two-flute twist drills [71]. FEA software, ABAQUS, was further applied to simulate 3D deposition residual stresses in a diamond-coated drill. Further, 2D FEA stress simulations of drill cross-sections at the cutting edge, at different locations, were approximated to obtain stress distributions around the cutting edge which were then extracted and transformed to quantitatively evaluate local interface stresses of various components.

Figure 19a shows the stress contours (longitudinal normal stress) of a diamond-coated two-flute twist drill [71]. The diamond-coated drill has 6.34 mm diameter, 34.5 mm flute length, 30° helix angle, 135° point angle, 1.33 mm web-thickness, and 3 µm edge radius. The length of the coating area is 5 µm thick, with a longitudinal direction length of 12.7 mm. Note that on the coating surface with moderate curving, the longitudinal stress is in the neighborhood of ~3.0 GPa in compression. This measurement is consistent with results obtained from the biaxial stress analysis. However, the substrate has tensile stresses on the order of 100 MPa, and the tensile stress is generally small in the area away from the coating area. Note also that large stress gradients are prominent around the drill tip.

This modeling methodology also was employed to develop drill models with varying geometry parameters—such as helix angle, point angle, and web thickness; and the models were used for 2D residual stress simulations. Thus, drill geometry effects, including effects of the cutting edge radius on the deposition stresses, were quantified. Detailed simulation procedures are described in [71]. The helix angle, the point angle, and the web thickness are the major drill geometry parameters investigated. Micro-level of drill geometry, namely, cutting edge radius, also was investigated. Three levels of edge radii were simulated: 3 μm, 7 μm, and 15 μm.

Figure 19b plots the normal stress components affected by the drill cutting edge radius. Clearly, the edge radius (r_e) significantly modifies deposition interface stresses. For the radial normal stress (σ_r), the maximum reduces from 1.4 GPa for 3 μm r_e to 0.7 GPa for 15 μm r_e. The large edge radius also shows smooth stress gradients along the edge. From simulation results, it is concluded that the effects of micro-level geometry such as the edge radius are the dominant effects on the interface stresses. In particular, the radial normal stresses can become largely tensile, over 1.0 GPa, which may affect the adhesion integrity. On the other hand, the macro-level geometry, such as the helix angle, has only marginal effects on deposition residual stresses. However, macro-level geometry is expected to impact the drilling capability as well.

4. Conclusion

Diamond-coated cutting tools have been developed and advanced for about 20 years. Significant progress has been made in different aspects of their performance, for example, improved CVD processes produce nanostructured, high-purity diamond crystals with extreme hardness and smooth surface. In addition, interface adhesion problems also have been studied and alleviated using various surface modification techniques. More recently, tool geometry issues also have been examined for their complex effects on diamond deposition and subsequent use in cutting operations.

Through a wide literature survey (in addition to those cited herein), it appears that diamond coated tools have gained acceptance in machining of graphite and Al-Si alloys. However, machining of composite materials still present great challenges to diamond-coated tooling because of their nonhomogeneous structures and hard phases that promote fatigue in the coating and interface of diamond coated tools during machining. Despite many advances, diamond-coated tools are not widely used in the machining industry. Discussions in the literature and other recent developments in diamond-coated cutting tools indicate that the following subjects must be successfully resolved in order to further extend the potential of diamond-coated tooling and to optimize their performance in machining.

1. Interface Engineering. The consensus is that coating-substrate adhesion is one of the biggest challenges to drastically improving diamond-coated tool performance.

Interface adhesion, degraded by Co in carbide tools, is closely related to delamination resistance. Precision methods to control Co etching (rate and uniformity) also are required to improve interface conditions for adhesion integrity. The literature also indicates that interlayers have the potential to enhance the diamond coating adhesion. A primary goal is to minimize cobalt-induced graphitization. Additionally, different types of interlayers are suggested; however, no systematic investigation exists that furthers understanding of the process, characteristics, and functionality. The basic mechanisms and limitations of interlayers are not completely known, and interface mechanics of diamond coating systems have not been widely studied. Moreover, quantitative characterizations of coating adhesion and interface behaviors are not yet established.

2. Diamond Film Optimization. The literature indicates that NCD films are far more effective than MCD for machining applications. However, a variety of CVD processes and many different types of gas chemistry are employed to attain different NCD films. Moreover, CVD parameters, such as deposition temperature, feed-gas chemistry, chamber pressure, and power all affect the quality of the diamond film and need to be optimized to achieve overall best wear resistance and adhesion of the NCD film. For example, the relative concentration of CH_4 and N_2 (in a balance of H_2) is shown to strongly influence coating structure, hardness, residual stresses, and adhesion. In addition, quality control and tooling consistency must be maintained in order for NCD-coated tools to be accepted in the industry.
3. Tool Geometry Design. The literature also demonstrates that tool geometry and in particular, micro-level geometry, is critical to deposition-induced residual stresses that will affect the ability of the tool to withstand machining resistance. Therefore, the tool geometry design consider both deposition and cutting in an integrated fashion. Current diamond-coated cutting tools on the market use off-the-shelf products and may not have drill geometry optimized for high performance. It is suggested that cutting simulations that incorporate coating attributes such as interface characteristics and residual stresses may provide a methodology to model/analyze the thermo-mechanical evolution of diamond-coated tools in machining, and may correlate the tool states with coating failures. The simulations may be applied to study the effects of both coating characteristics and machining parameters on diamond-coated tool performance.

ACKNOWLEDGMENTS

This research is supported by National Science Foundation: CMMI 0728228. Research work reported was conducted by several graduate students: J. Hu, F. Qin, X. Liu, P. Lu, and C. Miao, as well as undergraduate students: A. Renaud, D. Chen, C. Ibrahim, R. Nelms, G. Sthur, A. Klasterka, S. Hamilton, and A. Ogilvie. In addition, this project has been collaborated with other organizations and researchers: R. Thompson and D. Noland (Vista Engineering), W. Ni (Kennametal), E. Novak (Veeco Metrology) and R. Kountanya (Diamond Innovations). Many helpful discussions with A. Kumar (University of South Florida) and X. Xiao (General Motor) are greatly appreciated.

REFERENCES

[1] May, P.W. *Endeavour.*1995, vol. 19, 101.

[2] Almeida, F.A.; Amaral, M.; Oliveira, F.J.; Silva, R.F. *Diam. Relat. Mater.* 2006, *vol, 15, 2029.*

[3] Martinho, R.P.; Silva, F.J.G.; Baptista, A.P.M. *Wear* 2007, *vol, 263, 1417.*

[4] Itoh, H.; Shimura. S.; Sugiyama K.; Iwahara, H.; Sakamoto H. J. Am. *Ceram. Soc.* 2008, *vol, 80, 189.*

[5] Chen, M.; Jian, X.G.; Sun, F.H.; Hu, B.; Liu, X.S. *J. Mater. Process. Technol.* 2002, *vol, 129, 81.*

[6] Hanyu, H.; Kamiya, S.; Odagi, H.; Murakami, Y.; Saka, M. *Thin Solid Films,* 2002,*vol,413, 139.*

[7] Thompson, R.G. Private communication, 2006.

[8] Torres, C.D. ; Heaney,P.J. ; Sumant,A.V.; Hamilton,M.A.; Carpick, R.W.; Pfefferkorn, F.E. *Int. J. Mach. Tools Manuf.* 2009, *vol, 49, 599.*

[9] Heaney, P. J.; Sumant, A. V.; Torres, C. D.; Carpick, R. W.; Pfefferkorn, F. E. *Diam. Relat. Mater.* 2008, *vol, 17, 223.*

[10] Qin, F.; Chou, K.; Noland, D.; Thompson, R.G.; Ni, W. *Transactions of NAMRI/SME, 2009, vol, 37, in press.*

[11] Almeida, F.A.; Oliveira, F.J.; Sousa, M.; Fernandes, A.J.S.; Sacramento, J.; Silva, R.F. *Diam. Relat. Mater.* 2005, *vol, 14, 651.*

[12] Kumar, A. Private communication, 2007.

[13] Oles, E.J.; Inspektor, A.; Bauer, C.E. *Diam. Relat. Mater.* 1996, *vol, 5, 617.*

[14] Mallika, K.; Komanduri, R. *Wear,* 1999, *vol, 224, 245.*

[15] Vohra, V.; Catledge, S.; Vohra, Y.K. Mater. Res. Soc. Symp. Proc., 2004, vol, 791, P. Q 8.22.1

[16] Sun, F. H.; Zhang, Z. M.; Chen, M.; Shen, H. S. *Diam. Relat. Mater.* 2003, *vol, 12, 711.*

[17] Polini, R. Thin Solid Films, 2006, vol, 515, 4.

[18] Xu, Z.Q.; Lev, L.; Lukitsch, M.; Kumar, A. *Diam. Relat. Mater.*2007, *vol, 16, 461.*

[19] Li, Y.S.; Tang, Y.; Yang, Q.; Shimada, S.; Wei,R.; Lee, K.Y.; Hirose, A. *Int. J. Refract. Met. Hard Mater, 2008, vol, 26, 465.*

[20] Cabral, G; Gabler, J; Lindner, J; Gracio, J; Polini, R. *Diam. Relat. Mater.* 2008, *vol, 17, 1008.*

[21] Wang, Q.; Zhang, Q.; Wang, S.G.; Yoon, S.F.; Ahn, J.; Zhao, B; Tang, W.Z.; Miao, J.Q.; Lu, F.X. J. *Vac. Sci. Technol. A.* 2003, *vol, 21, 1939.*

[22] Tang, W.; Wang, S.; Lu, F. *Diam. Relat. Mater.* 2000, vol, 9, 1744.

[23] Cappelli, E.; Orlando, S.; Mattei, G.; Armigliato, A. *Surf. Coat. Technol.* 2003, *vol, 180-181, 184.*

[24] Almeida, F.A.; Sacramento, J.; Oliveira, F.J.; Silva, R.F. *Surf. Coat. Technol.* 2008, *vol, 203, 271.*

[25] Catledge, S.A.; Vohra, Y.K. *J. Appl. Phys.* 1995, *vol, 78, 7053.*

[26] Liang,Qi; Vohra, Y. K.; Thompson, R. *Diam. Relat. Mater.* 2008, *vol, 17. 2041.*

[27] Naguib, N.N.; Elam, J.W.; Birrell, J.; Wang, J.; Grierson, D.S.; Kabius, B.; Hiller, J.M.; Sumant, A.V.; Carpick, R.W.; Auciello, O.; Carlisle, J.A. *Chem. Phys. Lett.* 2006, *vol,* 430, 345.

[28] Ma, Y.P.; Sun, F.H.; Xue, H.G.; Zhang, Z.M.; Chen, M.*Diam. Relat. Mater*. 2007, *vol, 16, 481.*
[29] Lu,F.X.; Tang, W.Z.; Tong,Y.M.; Miao, J.Q.; He, L.F.; Li, C.M.; Chen, G.C. *Diam. Relat. Mater.* 2006, *vol, 15, 2039.*
[30] Chowdhury, S.; de Barra, E., Laugier, M.T. *Diam. Relat. Mater*. 2004, *vol, 13, 1625.*
[31] Rats, D.; Bimbault, L.; Vandenbulcke, L.; Herbin, R.; Badawi, K.F. J. Appl. Phys. 1995, vol, 78, 4494.
[32] Catledge, S.A.; Borham, J.; Vohra, Y.K.; Lacefield, W.R.; Lemons, J.E. *J. Appl. Phys.* 2002, *vol, 91, 5347.*
[33] Moulin, D.; Raymond, O.; Chevrier, P.; Lipinski, P.; Barre, T. *Mater. Sci. Forum.* 2006, *vol, 256, 55.*
[34] Windischmann, H.; Epps, G.F.; Cong, Y.; Collins, R.W. *J. Appl. Phys.* 1991, *vol, 69, 2231.*
[35] Ferreira, N.G.; Abramof, E.; Leite, N.F.; Corat, E.J.;Trava-Airoldi, V.J. *J. Appl. Phys.* 2002, *vol, 91, 2466.*
[36] Liu, E.; Li, L.; Blanpain, B.; Celis, J.P. *J. Appl. Phys*. 2005, vol, 98, *073515.*
[37] Ager, III, J.W.; Drory, M.*D. Phys. Rev. B.* 1993, *vol, 48, 2601.*
[38] Grogler, T.; Zeiler, E.; Horner, A.; Singer, R.F.; Rosiwal, S.M. *Surf. Coat. Technol.* 1998, *vol, 98, 1079.*
[39] Kitamura, T.; Hirakata, H.; Itsuji, T. *Eng. Fract. Mech*. 2003, *vol, 70, 2089.*
[40] Fan, Q.H.; Gracio, J.; Pereira, E. *J. Appl. Phys*. 1999, *vol, 86, 5509.*
[41] Ikeda, R.; Hayashi, M.; Yonezu, A.; Ogawa, T.; Takemoto, M. *Diamond Relat. Mater.* 2004, *vol, 13, 2024.*
[42] Polini, R.; D'Antonio, P.; Casto, S. L.; Ruisi, V. F.*; Surf. Coat. Technol*. 2000, *vol, 123, 78.*
[43] Buijnsters, J.G.; Shankar, P.; van Enckevort, W.J.P.; Schermer, J.J.; ter Meulen, J.J. *Thin Solid Films*, 2005, *vol, 474, 186.*
[44] Lanna, M.A.; Abro, A.M.; Neto, F. L.; Santos, C.; Da S., C.R.M. *Mater. Sci. Forum*, 2008, *v 591-593, 537.*
[45] Almeida, F.A.; Fernandes, A.J.S.; Silva, R.F.; Oliveira, F.J. *Surf. Coat. Technol*. 2006. *vol, 201, 1776.*
[46] Almeida, F.A.; Oliveira, F.J.; Sousa, M.;A.J.S. Fernandes,; Sacramento, J.; Silva, R.F. *Diam. Relat. Mater.* 2005, *vol, 14, 651.*
[47] Polini, R.; Allegri, A.; Guarino, S.; Quadrini, F.; Sein,H. ; Ahmed, W. *Thin Solid Films*, 2004, *vol,469-470, 161.*
[48] Ahmed, W.; Sein, H.; Jackson, M.; Polini, R. *Tribology. Int*. 2004, *vol, 37, 957.*
[49] Leyendecker, T.; Lemmer, O.; Jiirgens, A.; Esser, S.; Ebberink J. *Surf. Coat. Technol.* 1991. *vol,48, 253.*
[50] Davim J. P.; Mata, F. Materials & Design. 2008, vol, 29, 1568.
[51] Arumugam, P.U.; Malshe, A.P.; Batzer, S.A.; Bhat, D.G. *J. Manuf. Process. 2003, vol, 5, 163.*
[52] Schafer, L.; Fryda, M.; Stolley, T.; Xiang, L.; Klages, C.P. *Surf. Coat. Technol.* 1999, *vol, 116-119, 447.*
[53] Andrewes, C.J.E.; Feng, H.Y.; Lau, W.M. J. *Mater. Process. Technol.* 2000, *vol, 102, 25.*
[54] D'Errico, G.E.; Calzavarini, R. J. Mater. Process. Technol. 2001, vol, 119, 257.

[55] Davim, J.P. J. Mater. Process. Technol. 2002, vol, 128, 100.
[56] Belmonte, M.; Ferro, P.; Fernandes, A.J.S.; Costa, F.M.; Sacramento, J.; Silva, R.F. *Diamond Relat. Mater.* 2003, *vol, 12, 738.*
[57] Polini, R.; Casadei, F.; D'Antonio, P.; Traversa, E. *Surf. Coat. Technol.* 2003, *vol, 166, 127.*
[58] Shen, C.H. Transactions of NAMRI/SME, 1994, vol, 22, 201.
[59] Shen, C.H. *Surface and Coatings Technology,* 1996, vol, 86-87, 672.
[60] Cabral, G; Reis, P.; Polini, R.; Titus, E.; Ali, N.; Davim, J.P.; Gracio, J. *Diam. Relat. Mater.* 2006, *vol, 15, 1753.*
[61] Karner, J.; Pedrazzini, M.; Reineck, I.; Sjostrand, M.E.; Bergmann, E. *Mat. Sci. Eng., A.* 1996, *vol, 209, 405.*
[62] Chou, Y.K.; Liu, J. Surf. Coat. Technol. 2005, vol, 200, 1872.
[63] Hu, J.; Chou, Y. K.; Thompson, R.G.; Burgess, J.; Street, S. *Surf.* Coat. Technol. 2007, *vol. 202, 1113.*
[64] Liu, X.; Schad, R.; Chou, K. *Proceed. Mat. & Sci. Tech.* 2009, submitted.
[65] Hu, J.; Chou, Y. K.; Thompson, R.G. Transactions of NAMRI/SME, vol, 35, 177.
[66] Hu, J.; Chou, Y. K.; Thompson, R.G., *Int. J. Refract. Met. Hard Mater.*2008, *vol. 26, 135.*
[67] Qin, F.; Hu, J.; Chou, Y. K.; Thompson, R.G. *Wea*r, *in press*, 2009.
[68] Qin, F.; Chou, K.; Noland, D.; Thompson, R.G. *Surf. Coat. Technol.* 2009, submitted.
[69] Hu, J.; Qin, F.; Chou, Y. K.; Thompson, R. G. Proceed. 2008 Int. Manuf. Sci. Eng. Conf., MSEC2008-72507.
[70] Renaud, A.; Hu, J.; Qin, F.; Chou, Y. K. *IJMMS*, 2009, *vol.* 2, *in press.*
[71] Miao, C.; Qin, F.; Stuhr, G.; Chou, K.; Thompson, R.G. *CAD and Applications*, in press, 2009.

In: Metal Cutting: Research Advances
Editor: J. Paulo Davim pp. 163-185

ISBN: 978-1-60876-207-1

Chapter 6

GENERATION AND MODELLING OF SURFACE ROUGHNESS IN MACHINING USING GEOMETRICALLY DEFINED CUTTING TOOLS

W. Grzesik* and S. Brol
Opole University of Technology, Department of Manufacturing Engineering and Production Automation, 45-271 Opole, 5th Mikolajczyka St., Poland

ABSTRACT

This chapter discusses possible sources of discrepancies between real-surface profiles and ideal reproduction of the tool nose represented by regular feed marks. Of the numerous factors influencing surface roughness, such factors as the minimum undeformed chip thickness and the corresponding real tool rake geometry, the spring back effect, the plastic side flow of removed material, defects of the cutting edge, and deformation of the tool nose traces were implemented into surface roughness models. In particular, apart from several surface roughness models available in the literature, a new decomposition method of generating the surface roughness profile with recognizable tool nose traces is presented. In the procedure applied, deterioration of individual tool nose traces in the vertical direction is quantified by means of the coefficient of profile deformation K_{pd}. It is documented that this specific measure of profile deterioration is sensitive to changes in cutting parameters, friction induced by tool coatings, and tool wear. Moreover, Continuous Wavelet Transform (CWT) is used to visualize the process-induced disturbances of surface roughness profiles and as a result to validate qualitatively the proposed models.

1. INTRODUCTION

The surface finish of machined parts, including surface roughness, is one of the most important factors that characterize product quality. As a result, the generation of surface roughness in turning and milling processes, which generally represent machining processes

* Corresponding author, E-mail:w.grzesik@po.opole.pl

performed with geometrically defined cutting edges (roughly about 60-70% of all machining operations), has been modelled with varying degrees of success for several decades. In many cases the procedures available to predict surface roughness produced by single-point turning and milling tools do not correlate satisfactorily with real machined surfaces. However, it is desirable for both designers and manufacturers of precision and ultraprecision parts to have tools that incorporate process disturbances and enable the accurate prediction of surface roughness.

For instance, many metal-cutting researchers [1] report that turning and facing operations at a defined feed rate *f* with a single-point tool of nose radius r_ε, usually generate peak-to-valley (P-V) surface roughness, termed also the SL maximum roughness height Rz, equal to or exceeding the ideal stereometric-kimematic value calculated by a simple formula $R_{zt}=f^2/(8r_\varepsilon)$. Figure 1 shows that below some limiting value of feed rate *f*, Rz increases or sometimes becomes independent of *f* with further reduction of feed rate. This effect was reported and modelled for the first time by Brammertz [2]. It should be noted that these unique effects exist in machining both ferrous and non-ferrous work materials with cemented carbide and using both single crystal diamond (SCD) and polycrystalline diamond (PCD) cutting tools, as well as for polycrystalline cubic boron nitride (PCBN) tools [1].

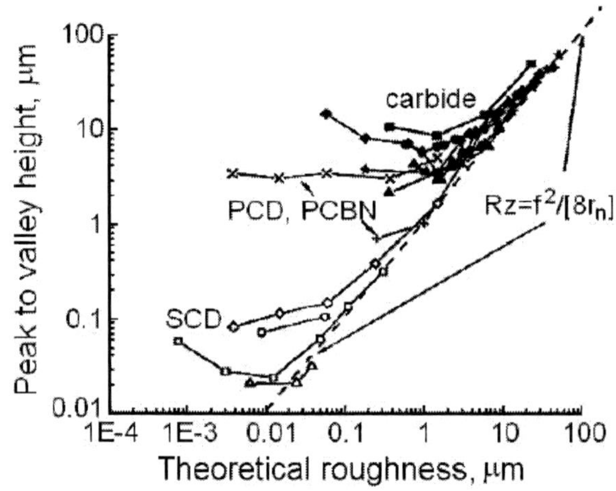

Figure 1. Surface roughness (P-V) variations with feed and tool nose radius in log-log scale for carbide, PCD, SCD and PCBN tools [1].

The author's literature survey shows three main causes of natural surface roughness that exceeds ideal (theoretical) roughness. The causes include machine tool limitations (dynamic stiffness, feed control, and precision of bearing system), the nature of the work material, the machining process itself (e.g., mechanical properties and microstructure, work hardening effect, chip separation mechanism and material plastic flow, and post-machining elastic recovery), as well as cutting tool defects and tribological behaviour (e.g., cutting edge preparation--mainly, its sharpness and radius in the normal plane, micro-chipping, friction intensity, built-up-edge phenomenon, etc.) [1,3-5]. It can be concluded based on this analysis that improving surface roughness predictability also involves the dynamic behaviour of chip removal and elastic-plastic deformation, which penetrates to the subsurface layer during the cutting process.

Schematically, the influence of some of the important process-related factors mentioned above can be expressed by adding the appropriate components to the ideal roughness height R_{zt}.

$$R_z = R_{zt} + \Delta R_{zp} + \Delta R_{ze} + \Delta R_{zf}$$

where: ΔR_{zp} is Rz increase due to material plastic flow (termed side-flow effect), ΔR_{ze} is the component resulting from elastic recovery of the subsurface layer (spring-back effect), and ΔR_{zf} is P-V increase caused by friction between the workpiece and tool flank. It should be noted that in precision machining and especially in ultraprecision machining, tool defects and edge roughness also may be important contributors to surface roughness[1,3].

This chapter presents, discusses, and experimentally validates a range of surface roughness predictive models that incorporate important geometrical and mechanical factors such as the sharpness of the cutting edge, the minimum undeformed chip thickness h_{min}, elastic recovery of the machined surface, plastic side flow, and relative displacements of the tool nose traces distorting the real surface profile.

2. Survey of Surface Roughness Models

2.1. Geometrical/Kinematical Models

According to the kinematical reproduction of the tool corner and its distortion, the final surface roughness obtained during a practical machining operation may be expressed as a sum of the theoretical (ideal) and natural components [7]. The first component results from the geometry of the tool and the feed or cutting speed, whereas natural component is affected by a number of negative effects, such as built-up edge phenomenon, tool wear, chatter, inaccuracies in machine-tool movement, and so on. Moreover, the resulting dynamic and static stability of the total process-system influences the quality of consequence to the quality of the surface texture achieved.

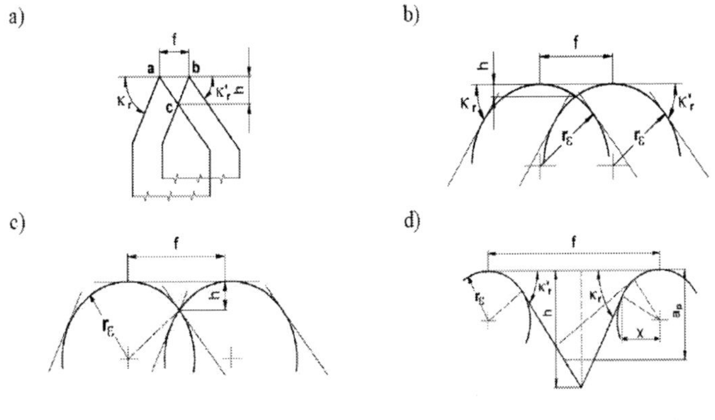

Figure 2. Types of feed marks produced by cutting tools with different kinematical-geometrical features [1].

The theoretical surface roughness can be calculated for turning and milling operations. In case of longitudinal turning operations with sharp-cornered and round-cornered tools, the theoretical maximum achievable value of surface roughness can be calculated for the four characteristic models illustrated in figure 2.

They are as follows:

1. Linear reproduction (figure 2a)

$$R_{zt} = f\frac{\sin\kappa_r \sin'\kappa'_r}{\sin(\kappa_r + \kappa'_r)} \tag{1}$$

Theoretical values of maximum and arithmetic average roughness are equal to $R_{maxt}=R_{zt}$ and $R_{at}=R_{maxt}/4$.

2. Arc reproduction for feed $f \leq 2\, r_\varepsilon \sin\kappa_r'$ (figure 2b)

$$R_{zt} = r_\varepsilon - \sqrt{r_\varepsilon^2 - \frac{f^2}{4}} \approx \frac{f^2}{8\, r_\varepsilon} \qquad \text{or} \tag{2}$$

$$R_{at} = \frac{f^2}{18\sqrt{3}\, r_\varepsilon} \approx \frac{0.0321 f^2}{r_\varepsilon} \tag{3}$$

3. Arc-linear reproduction of the 1st type for $f \geq 2r_\varepsilon \sin\kappa_r'$ (figure 2c)

$$R_{zt} = r_\varepsilon\left(1-\cos\kappa'_r\right) + f\sin\kappa'_r\cos\kappa'_r - \sin\kappa'_r\sqrt{f\sin\kappa'_r\left(2r_\varepsilon - f\sin\kappa'_r\right)} \tag{4}$$

4. Arc-linear reproduction of the 2nd type for $x > r_\varepsilon \sin\kappa_r$ (figure 2d)

$$R_{zt} = r_\varepsilon\left(1-\cos\kappa'_r\right) + \frac{f - r_\varepsilon\left(\sin\kappa_r + \sin\kappa'_r\right) + r_\varepsilon\, ctg\kappa_r\left(ctg\kappa_r - ctg\kappa'_r\right)}{ctg\,\kappa_r + ctg\,\kappa'_r} \tag{5}$$

In practice, the well-known surface roughness model (2) cannot be directly applied to the prediction of surface roughness in a milling operation due to insert face runouts in the milling cutter body. Among the axial and radial runouts existing in milling operations, the radial runout, which changes the feed rate per tooth during spindle rotation, also influences the surface roughness, which is mainly determined by the feed rate.

Figure 3 illustrates the formation of roughness in *face* and *peripheral* (slab) milling operations using multi-toothed.

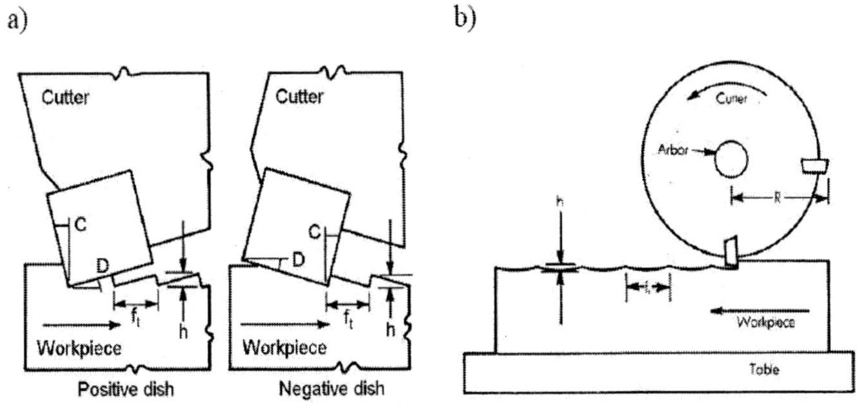

Figure 3. Schemes to calculate theoretical roughness in face (a) and peripheral (b) milling operations [5].

The surface roughness, or waviness, in *face milling* operations is determined by insert nose geometry, feed per tooth (insert), spindle and cutter runout, and stability of the workpiece and fixturing interaction. When a radiused insert is used, the theoretical roughness R_{zt} can be calculated as

$$R_{zt} = r - \sqrt{r^2 - \left(\frac{f_t}{2}\right)^2} \times 10^3, \mu m \tag{6}$$

where: r is nose radius and f_t is feed per tooth (insert).

The P-V roughness height generated by up and down peripheral milling (figure 3b) can be obtained by means of the following formula:

$$R_{zt} = \frac{f_t^2}{8 \times \left(R_c \pm \frac{f_t \times N}{\pi}\right)} \tag{7}$$

where: R_c is cutter radius, N is number of teeth in cutter, and f_t is feed per tooth (insert). Sign (+) is used for up milling and (-) for down milling.

It should be noted that Equation (7) is exact for a cutter without spindle runout. The surface roughness produced in slab milling with cutter of diameter D_c, including the runout effect, *e*, can easily be determined by modification of Eqn. (2), hence

$$R_{zt} = \frac{f_t^2}{4(D_c + 2e)} \tag{8}$$

In practice, the insert is fixed in the cutter body with negative and positive dishes, as depicted in figure 3a. In such a case, an estimate of peak-to-valley (P-V) roughness height is based on the following formula:

$$R_{zt} = \frac{f_t}{\tan C + \cot D} \tag{9}$$

where: f_t is feed per tooth (insert), D is face clearance angle (dish), C is lead angle (if dish is positive) or trailing angle (if dish is negative).

2.2. Models Considering Minimum Undeformed Chip Thickness

In order to improve the fitting of the theoretical surface roughness to the measured results (figure 4), some corrections of the ideal kinematical model for a rounded tool nose are needed.

The first model, which considers the minimum value h_{min} of the undeformed chip thickness, was proposed by Brammertz [2,5] in the form of a non-linear three-term equation.

$$R'_{th} = \frac{f^2}{8r_\varepsilon} + \frac{h_{min}}{2}\left(1 + \frac{r_\varepsilon h_{min}}{2}\right) \tag{10}$$

where: r_ε is the radius of tool nose, f is the feed rate, and h_{min} , which was originally termed „Spanzipfel," is the un-removed part of the cut due to geometrical limitations and not caused by real plastic deformation.

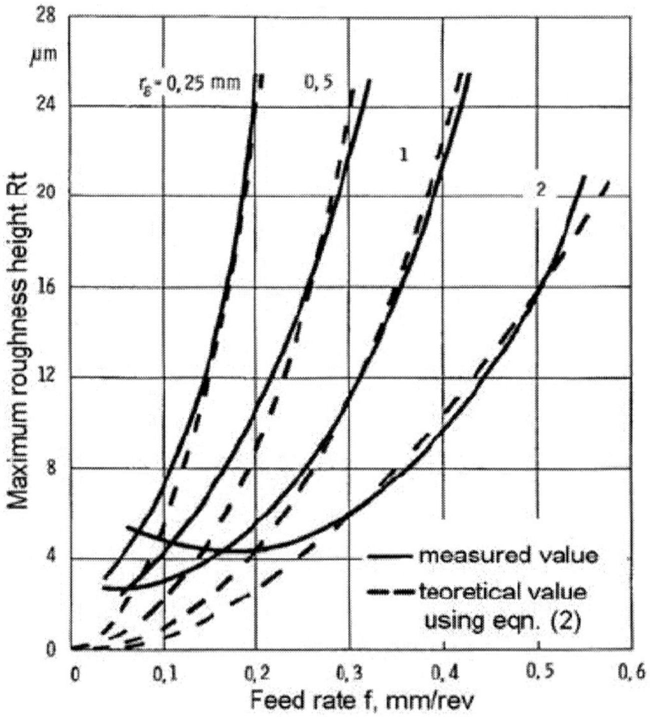

Figure 4. Influence of feed rate and tool nose radius on the maximum roughness height Rt [8].

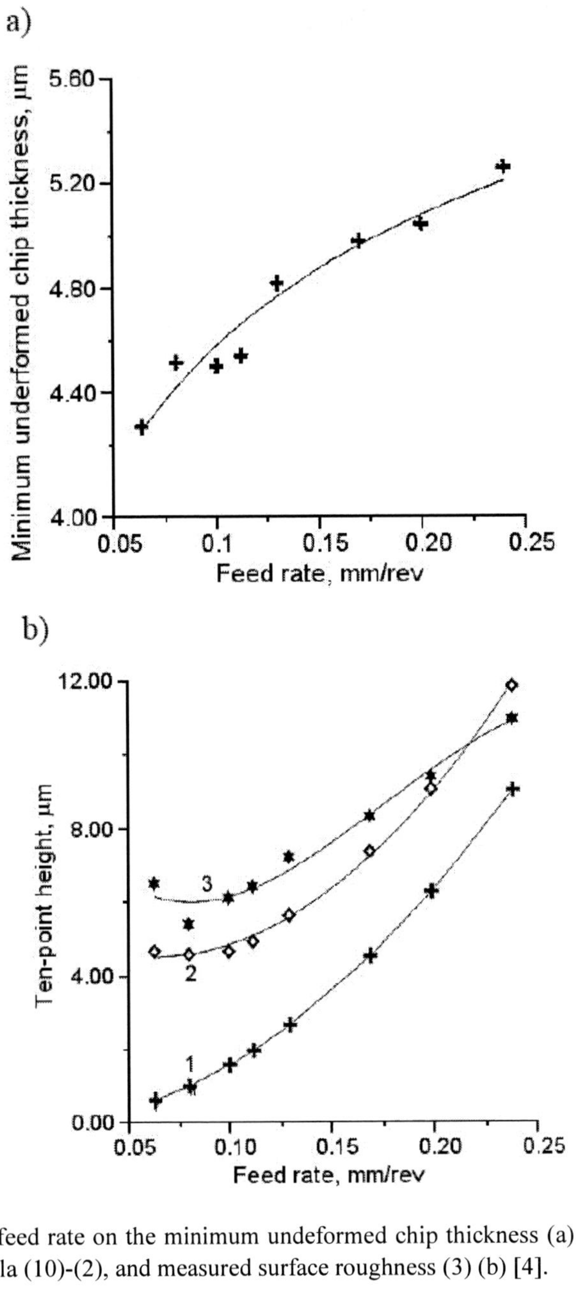

Figure 5. Influence of feed rate on the minimum undeformed chip thickness (a) and its effect on ideal (1), predicted by formula (10)-(2), and measured surface roughness (3) (b) [4].

As a result, in the Brammertz's model the Rz-f curve is shifted vertically by a correction value and achieves a relatively good agreement with measured roughness values for the feeds lesser than 0.1 mm/rev. For finishing conditions with small feeds, Brammertz [2,5] implicitly provides a material partition that remains on the surface due to the effect of the minimum undeformed chip thickness (h_{min}).

The magnitude of the minimum chip thickness depends on the cutting edge radius, and typically a ratio of $h_{min} \approx 0.293 r_n$ or even $0.1 r_n$ is used [3,4]. More recently, Grzesik [4,9] computed this critical value based on the condition for the transition from ploughing to micro-cutting known in the molecular-mechanical theory of friction as the Kragelski-Drujanov equations, expressed as follows:

$$h_r = 0.5r_n\left(1-\frac{2\tau_a}{Y_{sh}}\right) \tag{11}$$

where: h_r is the limiting depth of penetration of an indenter, r_n is the radius of indenter or rounded cutting edge radius, τ_a is the shear strength of the adhesive junction, and Y_{sh} is the yield stress of strain-hardened bulk material.

It is documented that for C45 steel and P10 carbide tools, h_{min} increases visibly with the increase of feed rate (figure 5a); and based on the adhesive interaction between tool and workpiece materials, the ratio of h_{min} to r_n is about 0.1 (4 μm versus 40 μm). Figure 5b shows that the predicted value of Rz (P-V) height is closer to the measured values of Rz and differ distinctly from the ideal surface roughness determined by the kinematical model (Eqn. 2).

2.3. Models Considering Real Tool Rake Configurations

Further improvement in determining the real chip formation can be achieved by considering the real rake angle (γ_{nr}) instead of the effective rake angle (γ_{ne}), as shown in figure 6. This approach is based on the assumption of an equivalent local plane rake face of the length l in the vicinity of the round cutting edge, which is obtained by joining the boundary points of the tool-chip contact [7].

The value of the real rake angle can be calculated using the following formula:

$$|\gamma_{nr}| = \arcsin\left(1-\frac{h_r}{r_n}\right) \tag{12}$$

where: h_r is the limited (minimum) undeformed chip thickness and r_n is the radius of cutting edge.

In figures 6 a and b, models 1 and 2 correspond to the undeformed chip thickness $h \geq r_n(1-\sin|\gamma_{nr}|)$ and undeformed chip thickness $h<r_n(1-\sin|\gamma_{nr}|)$, respectively. As a result, the values of real rake angles for these two cases of chip formation, considering the spring back effect δ_s, can be equal to

$$\gamma_{nr1} = \arcsin\left[\cos^2\gamma_{ne}\left(A+\tan\gamma_{ne}\sqrt{\sec^2\gamma_{ne}-A^2}\right)\right] \tag{13a}$$

where $A = \dfrac{r_n}{l\cos\gamma_{ne}} - \dfrac{(r_n-\delta_s)}{l}\tan\gamma_{ne} - \dfrac{\sqrt{(2r_n-\delta_s)\delta_s}}{l}$

$$\gamma_{nr2} = \arcsin \frac{l\sqrt{(2r_n - \delta_s)\delta_s} - (r_n - \delta_s)\sqrt{4r_n^2 - l^2}}{2r_n^2} \tag{13b}$$

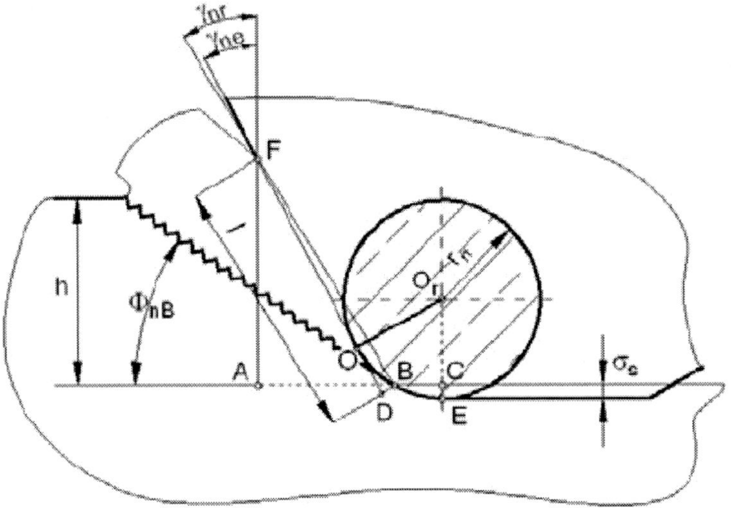

(a)

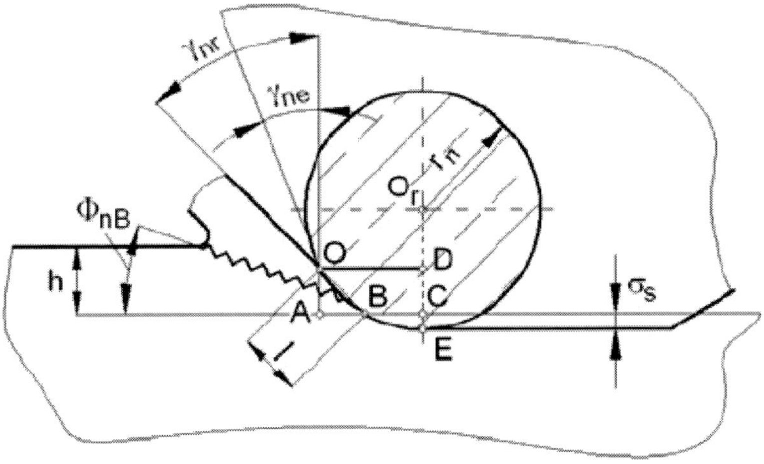

(b)

Figure 6. Models of chip formation with small undeformed chip thickness: (a) model given by Eqn. (13a), (b) model given by Eqn. (13b).

The value of the elastic recovery (spring back effect) of the machined surface δ_s is [15]

$$\delta_s = r_n \left(1 - \frac{1 + \mu_n}{\sqrt{2(1 + \mu_n^2)}} \right) \tag{13c}$$

where μ_n is the friction coefficient in the area of the cutting edge.

2.3. Models Considering Material Side Flow Effect

Another approach to the prediction of surface roughness in micro-turning considers the following three components [10]: kinematic surface roughness R_{th}, roughness component due to plastic side flow R_p causing material pile-up at the secondary cutting (trailing) edge, and roughness of the cutting edge R_{edge}, which obviously has little effect on the P-V height at conventional feeds. Hence,

$$R_{total} = R_{th} + R_p + \left|R_{edge}\right| \tag{14}$$

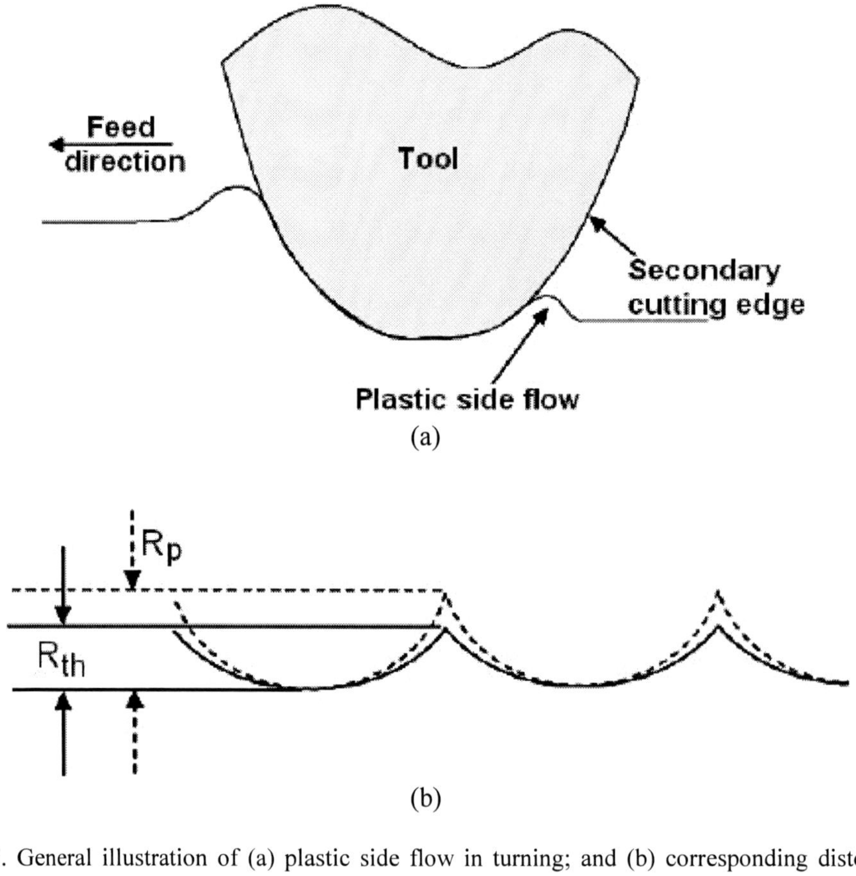

Figure 7. General illustration of (a) plastic side flow in turning; and (b) corresponding distortion of surface profile.

Machining tests with aluminium alloy Al5083 H-16 (v_c=200 m/min, f=30-150 μm/rev, a_p=100 μm) using PCD tools revealed that the roughness increase due to plastic side flow is a function of the material rheological factor *x* via the following empirical equation [10]:

$$R_p = 4.3408\ln(x) - 23.814(\mu m) \tag{15}$$

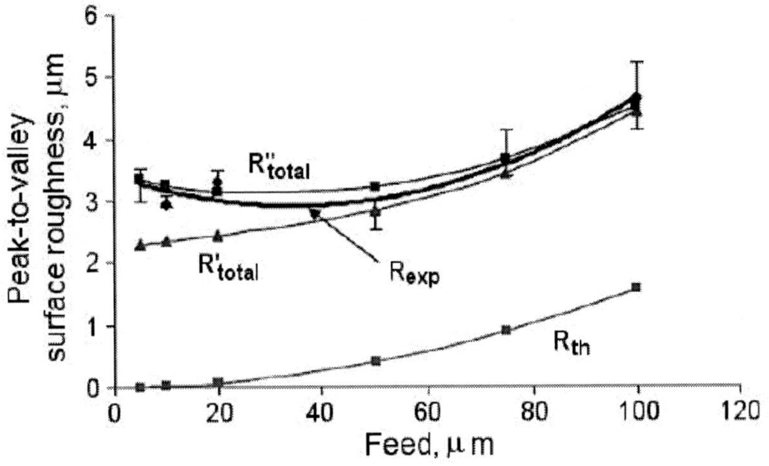

Figure 8. Surface roughness prediction using the model (15) [10].

Figure 8 shows the results of predictions using Eqn. 15, in which the symbols R'_{total} and R''_{total} denote the total surface roughness calculated without and with strain gradient strengthening of the material, respectively. The information illustrated in figure 8 shows that the consideration of strain gradient straightening of the workpiece material in the front of the tool results in better capture of the measured surface roughness R_{exp}.

Shaw [11] and Kishawy et al. [12] also indicate that plastic side flow is the most significant factor in and could partly contribute to the rise in surface roughness at very small feeds, less than certain minimum value (according to Shaw about 50 μm/rev for AISI 1045 steel).

2.4. Models Considering Quality/Preparation of the Cutting Edge=

It is documented that in hard turning, the surface roughness model should also incorporate other process disturbances such as cutting tool defects and machine loop vibration amplitude [3,13].

For this purpose, Knuefermann & McKeown [13] propose a numerical procedure based on the material partition equation, as follows:

$$p_{chip}(h)+p_{plastic}(h)+p_{elastic}(h)=1 \tag{16}$$

where: p_{chip}, $p_{plastic}$ and $p_{elastic}$ (relative chip removal, relative plastic deformation, relative elastic deformation, respectively) denote the contributions that each factor makes at a given position, expressed as functions of the chip thickness (h), not as functions of location (x).

The results of simulations are presented in figure 9a, in which the modeled roughness generated for feeds lower than 10 μm is visibly unstable. These authors concludes that at feed rates higher than 100 μm plastic deformation has very little impact on the generated surface. On the other hand, this effect rises to well over 90% for f=20 μm.

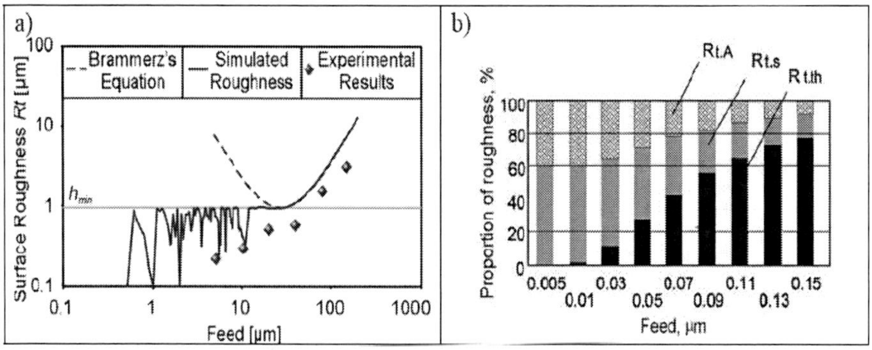

Figure 9. Influence of feed rate on the maximum roughness height Rt (b) and their partitions in the entire surface roughness height [3,13].

Figure 9b indicates that it is possible to consider several factors influencing the surface roughness Rt in micro-scale related to feed rate, tool geometry (Rt.th-roughness component related to the tool profile), tool defects (Rt.s-roughness component resulting from tool defect size), and machine vibration (Rt.A-asynchronous machine tool error). In special cases of machining such as hard turning, the achievable surface roughness is mainly influenced by the size and shape of defects on the cutting edge for feeds lower than 50-80 µm, and by the macroscopic tool geometry for feeds higher than 100 µm [3]. It is very important to note that for microcutting operations at feeds lower than 50 µm, machine vibration becomes a significant contributor to roughness height, and that for such machining cases, the ideal component of the surface roughness predicted by Eqn. (2)) can be neglected.

2.5. Models Considering Deterioration/Relative Displacement of Tool Nose Traces

2.5.1. Fundamentals of Profile Decomposition Method

This part of the chapter discusses novel results obtained by means of a new decomposition method of the recorded surface profile, which consists of a series of tool nose traces. The shapes of tool nose traces generated for different feed rates are shown in figure 10a. Figure 11 illustrates the recorded tool traces, which are visibly higher than the height of the idealized surface profile obtained by means of the artificial kinematical model. Moreover, their shapes seem to be elongated in the vertical direction and, as a result, can be well fitted by a parabolic curve. This trend showing the increase of the surface roughness was also noticed by Liu & Melkote [10] but they relate it only to plastic side flow generated by the secondary cutting edge (see figure 7b).

The application of profile decomposition procedures starts with definition of the boundary coordinates of each tool trace in the surface profile. This analysis is performed by means of a technique similar to Motif analysis [14], in which maxima occurring within the selected part of the profile are detected in a moving window (its width A is equal to feed rate), as shown in figure 12.

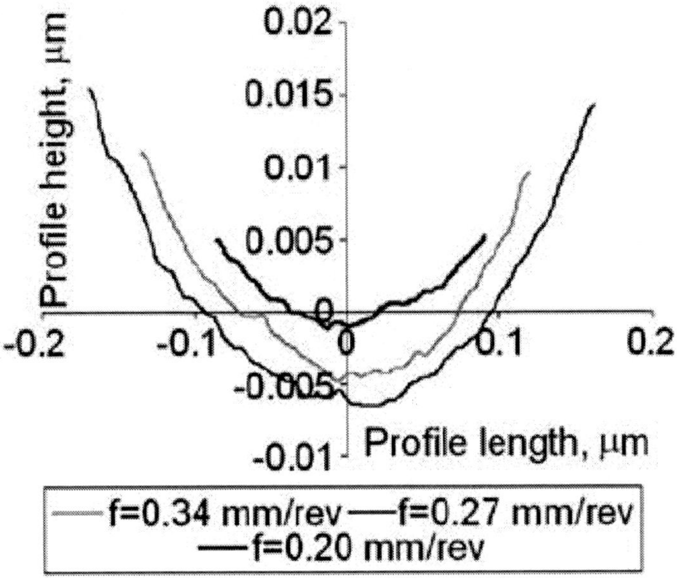

(a)

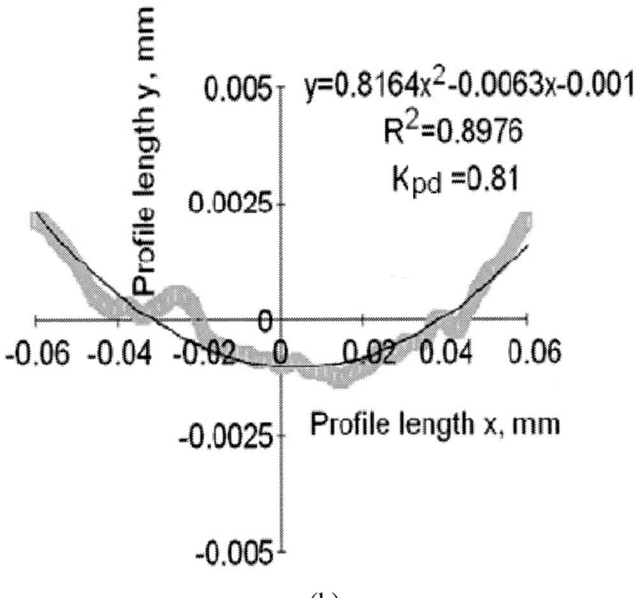

(b)

Figure 10. Different shapes of TNT's for: (a) turning C45 steel at varying feed rate (v_c=250m/min, a_p=0.85 mm), and (b) polynomial fitting of one cutting edge trace for f=0.20 mm/rev.

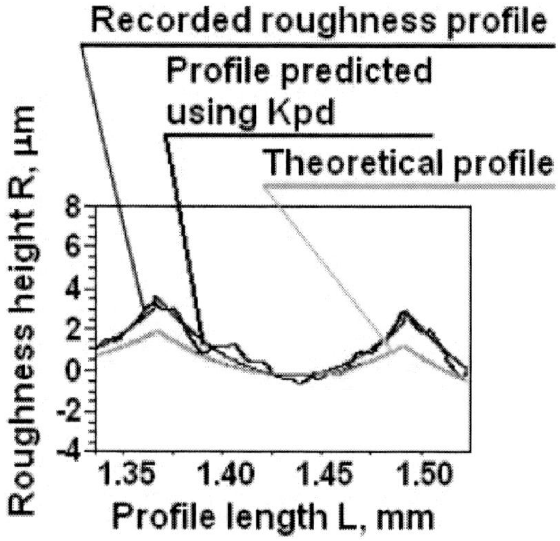

Figure 11. Comparison of the fragments of recorded and predicted; and theoretical roughness profiles.

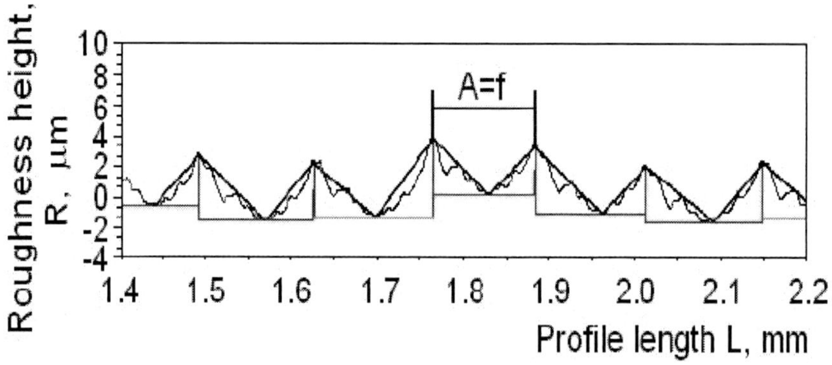

Figure 12. Local maxima and minima in the recorded roughness profile.

Each individual tool nose trace is defined as a profile part between two local maxima inside the selected window. Subsequently, the extraction of this trace and non-linear regression is performed using the Levenberg-Marquardt's algorithm [15,16]. Shapes of all tool nose traces distributed along the profile length are fitted to the second-degree polynomial $y=ax^2+bx+c$, as shown in figure 10b. Therefore, three constants a, b and c in this model define the translation of a parabola in the vertical and horizontal directions.

In general, the value of constant a in the polynomial equation defines the slenderness of the parabola; and in relation to the local tool nose trace, the constant a was termed the profile deformation coefficient K_{pd}. The value of the K_{pd} coefficient for an ideal profile was calculated to be equal to 0.624. However, when its is value greater than 0.642, it implies elongations of feed marks resulting from elastic-plastic deformations (spring-back effect of the machined surface) in the cutting zone and the relative displacement of the tool nose traces. These computations result in a number of sets of a, b and c constants for all tool traces considered.

Finally, a new roughness model, which utilizes the values of K_{pd} coefficient calculated for each feed rate, is proposed as:

$$R_{th} = K_{pd}\left(\frac{f}{2}\right)^2 \qquad (16)$$

The values of K_{pd} can be used for the prediction of total deformation of the surface profile h_{pd}. The deformation height h_{pd} corresponds to the difference between tool trace heights calculated with use of K_{pd} actual value and that of 0.624, determined for the ideal stereometric-kinematic model. Hence, the new expression for the profile deformation h_{pd} is

$$h_{pd} = \left(K_{pd} - 0.624\right)\left(\frac{f}{2}\right)^2 \qquad (17)$$

2.5.2. Machining Tests and Measurements of Surface Roughness

Machining tests were performed on C45 steel bars 60 mm in diameter, with a few segments of 20 mm width. Machining parameters used were as follows: cutting speed of 250 m/min (also 95 m/min for P20 carbide tools); four feed rates—0.1, 0.13, 0.27, and 0.34 mm/rev; and constant depth of cut of 0.85 mm. Several flat-rake inserts were used, including uncoated P20, TiC, and $TiC/Al_2O_3/TiN$ (Seco symbol NT15) coated carbide grades, with nose radius of r_ε=0.8mm were used. Tool angles were: κ_r=90°, α_o=5°, γ_o=-5°, λ_s=-6°.

Surface roughness was measured on each specimen segment using a 3D profilometer TOPO–01P. It is important that only selected profiles without surface defects were analyzed with SCILAB scripts [15,16]. Finally, average values of K_{pd} coefficients were determined and used to carry out data profile decompositions. All experiments and computations were performed in machining and surface metrology labs by W. Grzesik and S. Brol.

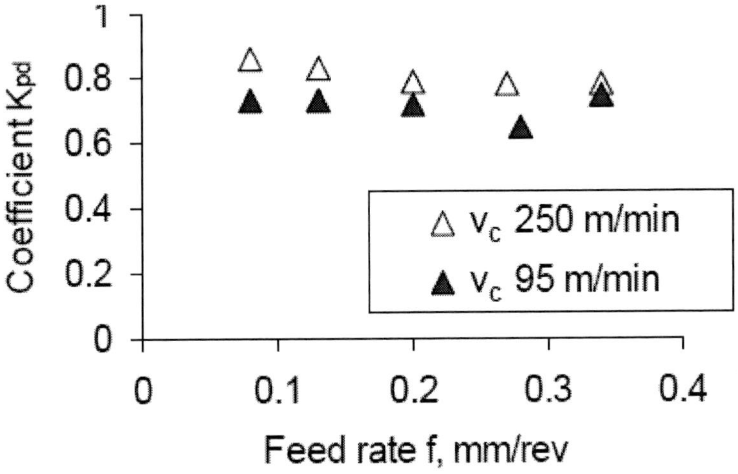

Figure 13. K_{pd} values determined for turning with P20 carbide insert (v_c=95 and 250 min, a_p=0.85mm.

2.5.3. Coefficient of Profile Deformations Versus Process Variables

Figure 13 shows that for varying cutting speed, the values of K_{pd} coefficient change from 0.65 do 0.82, depending on the feed value. The general observation is that for the cutting speed of v_c=250 m/min, the profiles of tool nose traces are sharper than for lower cutting speed of v_c=95 m/min (corresponding to the K_{pd} difference of 0.08).

The influence of single TiC and multilayer TiC/Al_2O_3/TiN tool coatings on the value of the K_{pd} coefficient is illustrated in figure 14. In particular, the multilayer coating causes the higher value of K_{pd}, especially for the smallest feeds. This observation implies that the relevant tool traces are higher than in case of uncoated P10 carbide tools. However, a TiC coated carbide tool does not influence the mechanism of surface profile generation in comparison to uncoated carbide tool.

This specific effect observed for TiC/Al_2O_3/TiN coated tools contributes to visible increase of the surface profile height at minimum feed rate, as previously reported by Shaw [11] and Grzesik [4]. The problem is to specify some physical reasons for such process behaviour: probably, partly a result of the roughness of the insert faces, or side flow effects, or some combination of these two influences.

The measured values of Ra roughness of the flat rake face were equal to Ra=1.10 μm, Ra=0.51 μm, and Ra=0.57 μm, for TiC/Al_2O_3/TiN, TIC, and uncoated carbide tools, respectively. The measured cutting edge radii for the these same cutting tools were equal to 64 μm, 58 μm, and 48 μm. Consequently, it is difficult to conclude that the influence of the cutting edge quality is a decisive factor. But it is evident that an additional correction is necessary with respect to the tribological characteristic of the cutting tool and workpiece material pair. Also, based on figure 13, the influence of the cutting speed seems to be taken into account.

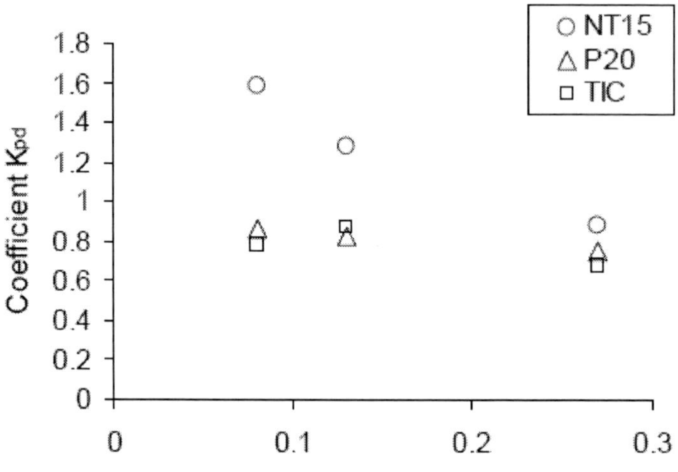

Figure 14. K_{pd} values obtained after turning with different inserts (v_c=250 m/min, a_p=0.85mm).

In a real machining process, dynamic interactions cause the vertical relative displacements (VRD's) of tool nose traces occurring in the surface profile. In order to quantify these dynamic tool-work interactions, for surface profiles with such distinct displacements of tool nose traces, corresponding variations of Kpd coefficients were recorded along the profile length. As an example, figure 15 illustrates the relation between the relative displacement of tool nose traces obtained as the difference between the vertical displacements of subsequent parabolas and the corresponding values of the Rzp roughness height predicted by means of Kpd coefficient.

Figure 15 illustrates that experimental points lie mostly on the straight regression line, but a relatively low value of regression coefficient R^2 suggests that vertical displacements of tool nose traces can be a primary source of variations of Kpd coefficients. Other influences can include deterioration of tool geometry due to elastic deformation of the tool body and probably in homogeneity of the workpiece material.

In general, the roughness models presented in Section 2.1 are applicable for ideal sharp tools and do not consider tool wear. Hence, additional investigations were carried out to determine the values of K_{pd} coefficient when machining with controlled wear VB_C of the tools, as shown in figure 16.

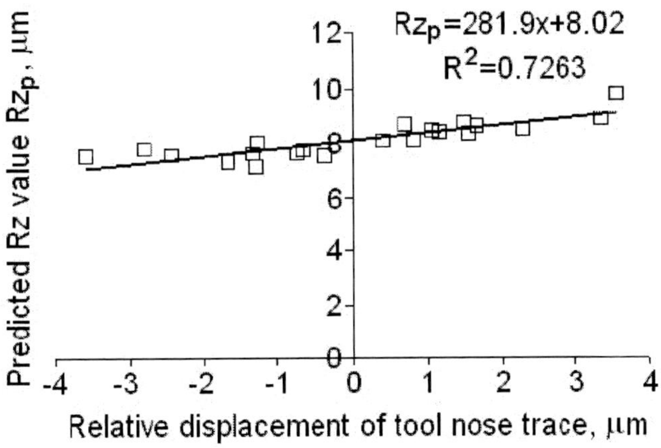

Figure 15. Predicted roughness Rzp versus relative displacements of tool nose traces. Cutting conditions: P20 insert, v_c=250 m/min, a_p=0.85mm..

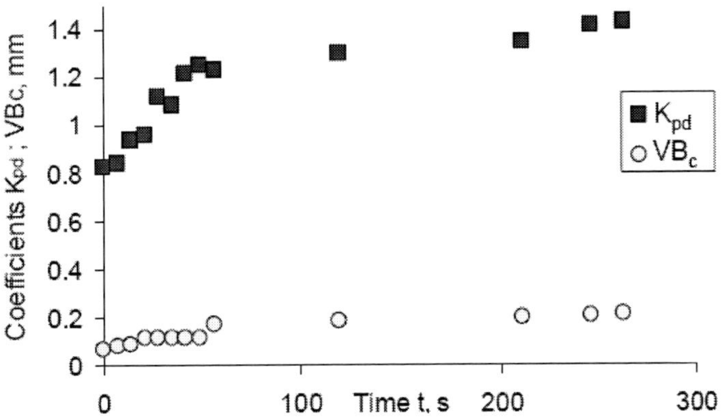

Figure 16. K_{pd} and corner wear VB_C versus cutting time (P20 carbide insert, v_c=215 m/min, f=0.13mm/rev, a_p=0.85mm).

These machining wear tests were performed for uncoated P20 and C45 steel, keeping cutting speed at v_c=215 m/min, feed rate at f=0.13 mm/rev, and depth of cut at a_p=0.85mm. As shown in figure 16, the changes of K_{pd} coefficient are well correlated with VB_C flank wear indicator measured within the tool nose area. For this case, the regression coefficient is about R=0.826 (α=0.05).

2.5.4. Predictions of Profile Deformations and Surface Roughness Parameters

As noted in Section 2.5.1, the proposed profile decomposition method allows local disturbance of feed marks to be recognized. As a consequence, two components of the recorded surface profile (i.e., distorted tool nose traces and the process-induced disturbances within each individual trace) can be distinguished, as shown in figures 17a and b. In this investigation, some characteristic modes of disturbances were identified regardless of the surface profiles considered. One of these characteristic features is the appearance of local cavity (pit) and elevation of the part of the feed mark produced by the secondary cutting (trailing). The other is that the profile stepped down in the vicinity of the corner and piled-up again in the area of the primary cutting edge, as shown in figure 17b. It can be concluded that characteristic elevation (material swelling) of the tool nose traces close to the trailing edge can result from the spring-back effect of the machined surface, as well as from plastic side flow effects.

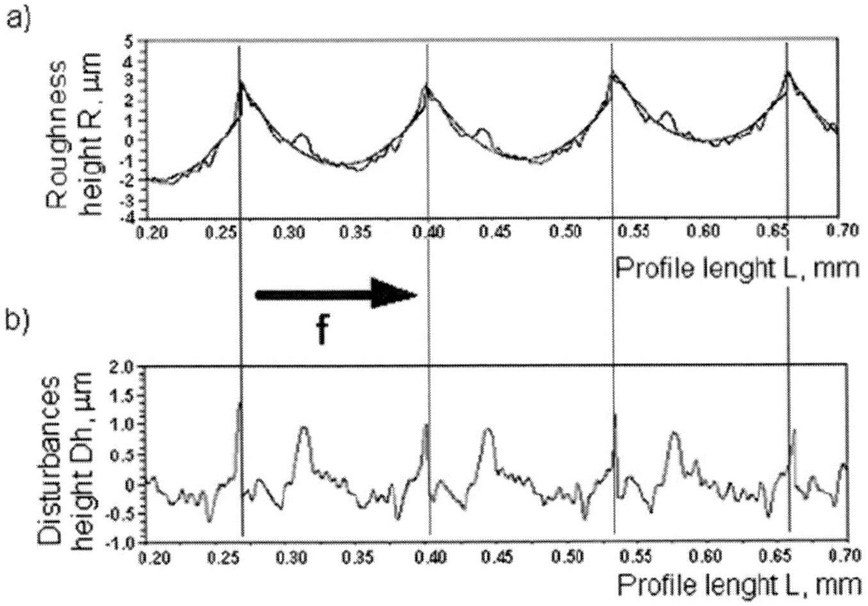

Figure 17. Comparison of recorded and modeled roughness profiles using (a) K_{pd}-based procedure, and (b) disturbances of a series of tool nose traces. Cutting conditions: P20 insert, v_c=250 m/min, f=0.13 mm/rev, a_p=0.85 mm.

In this analysis, in addition to the Rz parameter, the average motif depth R defined within selected window of width A/2 (figure 12) was also considered. The changes in motif R, theoretical roughness Rth, and Rz roughness parameters, are presented in figure 18.

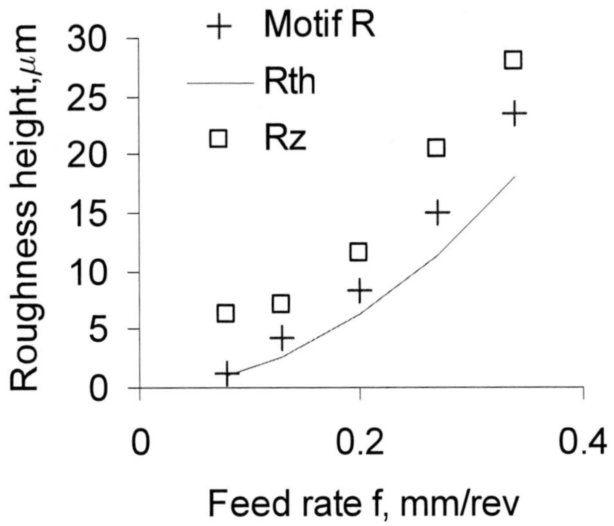

Figure 18. Roughness parameters Rz, Rzt, and Motif R after cutting with P20 insert (v_c=90 m/min, a_p=0.85mm).

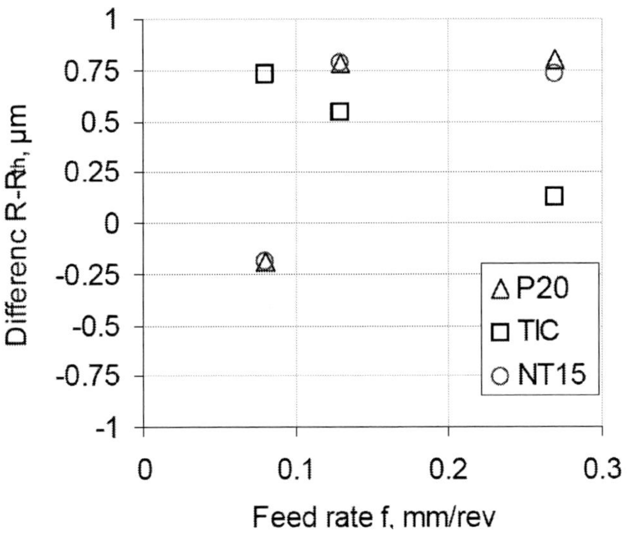

Figure 19. Difference R-R_{th} *vs.* feed rate (v_c=250 m/min, a_p=0.85mm).

As shown in figure 18, the parameters Rz and R change in a similar way and their values are comparable; and on average the difference between them is not higher than 3 μm. In general, Rz and R are greater than the R_{th} value predicted from the kinematic model (the same trend is seen in both Brammertz's and Grzesik's models). In addition, these roughness parameters are greater than motif R. In terms of the process behaviour, this discrepancy means that in comparison to the Rz parameter, motif R is not sensitive to dynamic profile disturbances.

The results of predictions with the use of formulae (16) were compared with Motif R parameter. Figure 19 shows that the biggest difference of 0.8μm was obtained for the feeds of f=0.13mm/rev and f=0.27 mm/rev when machining with uncoated and NT-15 coated cutting tool, respectively.

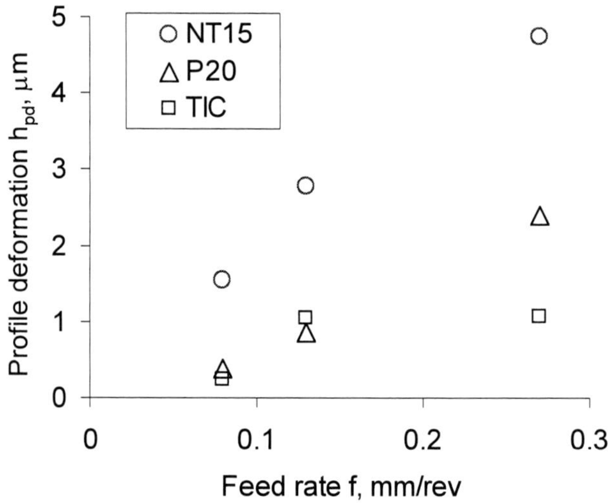

Figure 20. Values of h_{pd} vs. feed rate for different inserts (v_c=250 m/min, a_p=0.85mm).

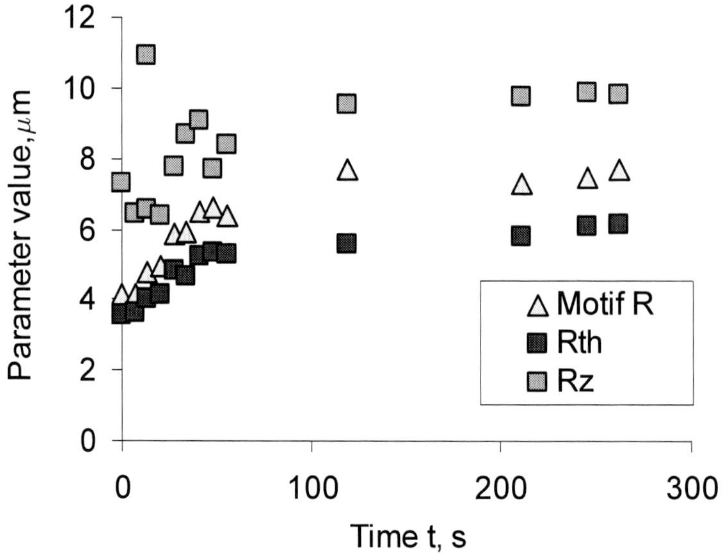

Figure 21. Motif R, R_{th}, and Rz values *vs.* cutting time (P20 inserts, v_c=215 m/min, f=0.13mm/rev, a_p=0.85mm).

Figure 20 presents the results of computations of h_{pd} using Eqn. (17) for the three different cutting tool materials. It should be noted that the higher values of profile deformation are computed for three-layer coated tools.

The use of h_{pd} values, which were predicted by means of appropriate K_{pd} coefficients, highlights the deterioration of the surface roughness deteriorates as tool wear progresses.

Figure 21 indicates that tool wear contributes to the increase of the surface roughness parameters considered. At the same time, the K_{pd} coefficient increases (equivalent to the height of tool nose trace) as tool wear progresses. This fact is confirmed by the increase of the Motif R parameter in time.

This confirmation is very important because, for the period of a uniform tool wear (middle part of the Lorentz's wear curve), the maximum difference between Motif R and the theoretical roughness R_{th} varies between 1.5μm and 2.0 μm. In contrast, the minimum difference for a fresh tool is about 0.3 μm.

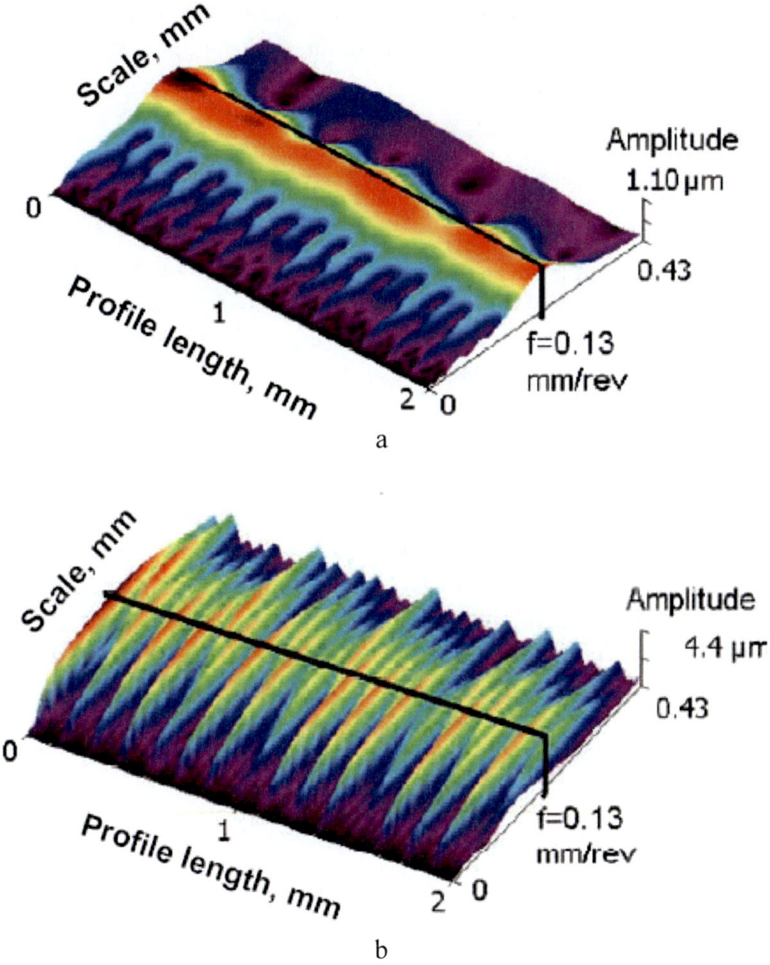

Figure 22. Continuous Wavelet Transforms using (a) Morlet's wavelets and (b) 'mexican hat" wavelets (P20 carbide insert, v_c=215 m/min, f=0.13mm/rev, a_p=0.85mm).

2.5.5. Wavelet Image of Surface Profile Distortions

Continuous wavelet transform (CWT) is one of the recently developed, advanced techniques for 3D visualization of the surface profile and its deformation, if it appears. This technique was previously applied by Grzesik & Brol [17] to distinguish characteristic geometrical features of the turned profiles obtained for different workpiece materials [16]. In this investigation, the applicability of Morlet and "mexican hat" wavelets are tested for assessing the regularity of the recorded surface profile.

As shown in figure 22a, the CWT representation of the machined surface includes many local ridges corresponding to local maxima when the scale is equal to the feed rate. The ridge is waved in both axes, which confirms earlier conclusions that the individual tool nose traces have different heights and they are not regularly distributed along the profile. The differences in height of feed marks are depicted by CWT representation using the "mexican hat" wavelet, as shown in figure 22b. In this case, local maxima corresponding to local irregularities can be observed for the scale equals the feed rate.

It should be noted that for both matrixes (figures 22a and b) and for scales lower than feed rate, the local maxima, which denote the amplitudes/distances of the disturbances of tool traces, are clearly visible.

3. Summary

Some latest developments in surface roughness modelling include such important factors as the minimum undeformed chip thickness, plastic side flow, the quality of the cutting edge, and displacements of tool nose traces resulting from process behaviour and tool wear.

A new decomposition method of the recorded surface profile is proposed. This new method is based on the fact that the dynamic interactions between the tool and the workpiece material result in deformations of the tool nose traces, in addition to those from plastic side flow..

The total deformation of each individual tool nose trace is described by means of the corresponding value of the coefficient of profile deformation. It is verified that the K_{pd} coefficient allows modelling the natural surface roughness for variable cutting conditions, different cutting tool materials, and for worn tools.

Continuous Wavelet Transform (CWT) was applied to visualize the distortion of the real surface profile due to process disturbances and to quantify its scale depending on the process instability.

Acknowledgments

My sincere thanks to Dr. K. Żak for his assistance in carrying out the surface roughness measurements and preparing graphics for Section 2 of this chapter.

References

[1] Childs, T. H. C., Sekiya, K., Tezuka, R., Yamane, Y., Dornfeld, D., Lee, D-E., Min, S.& Wright P. K. (2008). Surface finishes from turning and facing with round nosed tools. CIRP Annals-Manufacturing Technology, 57/1, 89-92.

[2] Brammertz, P. H. (1961). Die Entstehung der Oberflächenrauheit beim Feindrehen. Industrie Anzeiger, 2, 25-32.

[3] Klocke, F., Brinksmeier, E. & Weinert K. (2005). Capability profile of hard cutting and grinding processes. Annals of the CIRP, 54/2, 557-580.

[4] Grzesik, W. (1996). A revised model for predicting surface roughness in turning. Wear, 194, 143-148.

[5] Grzesik, W. (2008). Advanced Machining Processes of Metallic Materials. Amsterdam: Elsevier.

[6] Boothroyd, G. & Knight W. A. (2006). Fundamentals of Machining and Machine Tools. NY: Marcel Dekker.

[7] Grzesik, W. (1987). Generalized model of free oblique cutting process. DSc thesis, Monograph No. 15, Opole: Publishing House of High Engineering School.

[8] König, W. & Klocke, F. (1997). Fertigungsverfahren 1. Drehen, Fräsen, Bohren. Berlin: Springer.

[9] Grzesik, W. (1999). Experimental investigation of the influence of adhesion on the frictional conditions in the cutting process. Tribology International, 32, 15-23.

[10] Liu, K., Melkote, S. N. (2006). Effect of plastic side flow on surface roughness in micro-turning process. International Journal of Machine Tools and Manufacture, 46, 1778-1785.

[11] Shaw, M. C. (1989). Metal Cutting Principles. Oxford: Clarendon Press,.

[12] Kishawy, H. A., Haglund, A. & Balazinski M. (2006). Modelling of material side flow in hard turning. Annals of the CIRP, 55/1,85-88.

[13] Knuefermann, M.M.W. & McKeown, P.A. (2004). A model for surface roughness in ultraprecision hard turning. Annals of the CIRP, 53/1, 99-102.

[14] Griffiths, B. (2001). Manufacturing Surface Technology. Surface Integrity and Functional Performance. London: Penton Press.

[15] (2008), Programming with SCILAB, www.scilab.org.

[16] Brol, S., Żak, K. & Grzesik, W. (2007). Analysis of the surface profile generated in the turning of different materials. In: Proceedings of 6th Int. DMC (Development of Metal Cutting) Conference, Slovakia, Kosice, pp.129-132.

[17] Grzesik, W. & Brol, S. (2009). Wavelet and fractal approach to surface roughness characterization after finish turning of different workpiece materials. *Journal of Materials Processing Technology, 209*, 2522-2531.

In: Metal Cutting: Research Advances
Editor: J. Paulo Davim pp. 187-211

ISBN: 978-1-60876-207-1

Chapter 7

CHATTER MODELING IN DRILLING AND MICRODRILLING

Mark J. Jackson* and Tamara Novakov
Center for Advanced Manufacturing, College of Technology, Purdue University, 401 North Grant Street, West Lafayette, IN 47907, USA

ABSTRACT

Drilling chatter (vibration) is one of the main problems in degradation of the quality of machined surfaces and it limits tool life and productivity and impairs tolerances. Chatter in milling and turning operations has been extensively analyzed, but the role of chatter in drilling operations has been largely neglected due to the complexity of drilling tools and problems that develop in the modeling of the tool. In this chapter, an overview of chatter vibrations and chatter suppression in drilling is presented. In addition, the problem of chatter analysis and chatter suppression in micromachining processes is also addressed. Differences in chatter at the macro and micro scales are explained and provide an understanding of the direction of research on microdrilling operations.

1. INTRODUCTION

Chatter is considered to be a self-excited vibration between the machine tool and the workpiece [1, 2]. Regenerative chatter is the most important type of chatter and it is a consequence of forming a previously cut surface and current chip formation. As the tool is creating a new cut surface by removing material in a form of a chip, it leaves waviness in the surface profile, which consequently affects the next cut. The new wavy surface induces vibrations in the tool that then affect the surface. Chatter is a great problem in cutting because it causes a decrease in tool life, limited productivity, poor surface finish and tolerances, as well as producing unacceptable noise. Chatter problems in turning and milling have been studied extensively throughout the years and results can easily be found in literature [1-4].

* Corresponding author, E-mail: jacksomj@purdue.edu

There are many papers dealing with the analytical models for prediction of maximum stable depths of cut as a function of spindle speeds. In this way, optimal operating regimes are defined for both turning and milling operations, resulting in the use of large depth of cut and large spindle speeds simultaneously. Some researchers go even further and develop computer programs for chatter prediction [5,6].

Drilling has not been researched to the same extent as turning and milling processes as drilling tools have a more complex geometry than milling and turning tools. For example, chisels create a ploughing action, cutting lips have varying rake angles, chip thickness changes along the edge, and so on. This complexity creates a number of problems in the modeling process. Chatter in drilling often involves large amplitude torsional vibrations, which are not present in of milling or turning. Through analysis of chatter in drilling, several models investigated in the literature are proposed: torsional–axial model, bending model, and a combination of axial and bending models. Apart from these different models, several other influential factors in chatter analysis in drilling must be considered, such as frictional interactions, chisel edge, margin engagement, grinding errors, misalignment of the tool, drill flank, pilot hole size, length of the drills, and other factors.

Reference [7] suggests two different approaches for detecting chatter that used throughout the literature. The first approach is the frequency domain method (FDM), which is based on the appearance of increase in frequency or equivalent signals, such as acceleration or sound, with the onset of chatter. The second approach is the time domain method (TDM), which is based on on-line determination of a time series. Both of these techniques still encounter certain problems in determination of the proper chatter onset threshold, especially if the cutting conditions show variability. The analytical part of this research is covered in a number of papers that deal with analysis of stability and bifurcations of non-linear differential equations [8]. The use of stability curves obtained by these analyses significantly increase the efficiency of cutting operations, although the boundary conditions of chatter onset still remain extremely important. Determination of whether the bifurcation is subcritical or supercritical is considered to be crucial.

2. Investigation of Chatter In Drilling

Ema and his colleagues initiated analysis of chatter problems in drilling in 1986 and 1988 [9-11], by keeping the tool stationary and feeding it into the rotating workpiece. These authors also investigated the effect of the drill geometry, chisel edge, drill flank, and pilot holes and showed that whirling vibrations, which result from the regeneration effect at the major cutting edges, result in distorted polygonal holes with an odd number of sides. This phenomenon was later investigated by Bayly and his team, who showed that the chisel edge and flank surfaces cause damping in the system. However, because these investigations were conducted in an unconventional way, the results are not completely trustworthy: There is rarely a situation where one would rotate the workpiece instead of the drill. This method also neglects the effect of the spindle and bearings, as well as other conventional problems in drilling [12].

The state of the art in turning, milling, drilling, and grinding also has been described [13], in terms of present mathematical models and chatter suppression methods. This works discusses the dynamics of the orthogonal metal cutting and chatter stability lobes, defining the linear dependency of the cutting forces on the width of cut and feed per revolution (or static chip thickness):

$$F_t = K_t a h_0$$
$$F_r = K_r a h_0$$

Where: a is the width of cut, h_0 is the feed per revolution or static chip thickness, and Kt, Kr are tangential and radial cutting coefficients, respectively. The dynamic chip thickness and transfer function of the closed loop system are defined as:

$$h(t) = h_0 - \mu\left[y(t) - y(t-T)\right]$$

$$\frac{h(s)}{h_o(s)} = \frac{1}{1+(1-e^{-sT})K_r a \Phi(s)}$$

Where, T is the spindle rotation period, μ is the overlap factor, y(t) and y(t – T) are present and past vibration amplitudes in the radial direction, and Φ (s) is the frequency response function. Stability lobes are presented, showing that if the depth of cut (DOC) and spindle speed are selected below the stability lobe, the process is stable, and if they are above the lobe, the process becomes unstable with growing vibration. The issue of stability at high and low spindle speeds is discussed, stressing the difficulty of defining the stability region for the low speeds due to the complexity of modeling process damping (clearance angle becomes zero or negative if there is a lot of waviness in the surface) [14]. These authors also address the complexity of the drilling process and suggest that in drilling, the chisel has a plowing action and the cutting lips have a varying rake angle and chip thickness along the edge. They also explain that drilling forces and torque are a function of chip thickness, width of cut, material properties, drill geometry, and drill tip vibration both at the current time and one tooth earlier. It is stated that in order to predict the cutting load when there are vibrations in the drill, the chip thickness distribution along the cutting edge must be identified considering vibrations in every direction. These vibrations influence the cutting forces in drilling, therefore making it very challenging to solve the equation of motion. Thus,

$$[M]\begin{Bmatrix} \ddot{x}(t) \\ \ddot{y}(t) \\ \ddot{z}(t) \\ \ddot{\theta}(t) \end{Bmatrix} + [C]\begin{Bmatrix} \dot{x}(t) \\ \dot{y}(t) \\ \dot{z}(t) \\ \dot{\theta}(t) \end{Bmatrix} + [K]\begin{Bmatrix} x(t) \\ y(t) \\ z(t) \\ \theta(t) \end{Bmatrix} = \begin{Bmatrix} F_x \\ F_y \\ F_z \\ T_c \end{Bmatrix}$$

Where, x, y are the lateral distances, z is the axial distance, θ is the torsional deflection of the drill, matrices M, C, K are mass, damping, and stiffness characteristics at the drill tip, Fx, Fy are lateral forces, the thrust force is Fz, and Tc is the torque. It is suggested that one of the ways to start the analysis of a process as complex as drilling is to consider individual vibration mechanisms. Therefore, torsional-axial, lateral, and whirling vibrations are covered individually below [13].

2.1. Torsional-Axial Model

The coupling of torsional and axial vibrations happens when the drill starts to lengthen and shorten, respectively, due to the torsional vibration and creates a wavy surface at the bottom of the hole. As the previous material surface profile influences the next cut, a regenerative effect of thrust and torque acts on the tool. This action can lead to an increase in the vibrations of the system, leading to unstable vibrations during machining. For this reason, it is necessary to determine stability boundaries for the torsional-axial model. Bayly and his colleagues started research on the torsional axial model in 2001 [3], by developing a mathematical model of the torsional-axial coupling that causes chatter in drilling. In this work, the researchers took into consideration three translations and one rotation, leaving the bending moments out of the equation. Their equation of motion can be written as:

$$M\ddot{x} + C\dot{x} + Kx = F$$

where:

$$x = \left[u_1, u_2, .., u_N, v_1, v_2, .., v_N, w_1, w_2, .., w_N, \theta_1, \theta_2, .., \theta_N\right]^T$$

$$F = \left[F_{x1}, F_{x2}, .., F_{xN}, F_{y1}, F_{y2}, .., F_{yN}, F_{z1}, F_{z2}, .., F_{zN}, M_1, M_2, .., M_N\right]^T$$

u, v, w are deflections in x, y, z directions , and θ is the angle of twist about the axis. The time varying component of the modal equation can be written as:

$$m_p\ddot{\eta}_p + c_p\dot{\eta}_p + k_p\eta_p = -\alpha C_1 b\left[\eta_p - \eta_p(t-\tau)\right]$$

Where: m, k, c are modal mass, stiffness, and damping coefficient; $\alpha = C_2/C_1 + \theta_{Np}R_{av}$, θ_{Np} is the torsional axial coupling parameter; R_{av} is the average radius of the cutting force; C_1, C_2 are tangential and thrust forces per unit area of uncut chip; b is the radial depth of cut; η_p is the modal coefficient of the pth mode; and τ is the time delay between passage of cutting edges of the drill.

The difference between this equation and classical chatter equations is in the constant, α, which includes the effects of tangential and axial forces, as well as torsional axial coupling. Another difference lies in the fact that $-\alpha C_1$, which presents the effective cutting pressure, is not always positive as it is in milling and turning operations.

Predictions of stability regions have been performed by frequency domain analysis and confirmed by both experimental work and computer simulations. The drill dynamics were described by the frequency response function (FRF), which is defined as displacement at the given location in response to a unit force, as a function of excitation frequency.

Radial depth of cut is expressed as:

$$b = \frac{-1}{\alpha C_1 2\,\mathrm{Re}\left[Gp(\omega)\right]},$$

Where, Re[Gp(ω)] is the real part of the frequency response function for the pth mode.

The consequence of the derivations is that, unlike the onset of chatter in turning and milling, in drilling the chatter frequency must be slightly below the natural frequency of the corresponding mode, not slightly above, for the chatter episodes to occur. This phenomenon was experimentally confirmed by cutting tests. In their experiment, Bayly et al., have used a symmetric "winged" steel collar, attached 25mm from the tip of the drill. The collar was designed symmetrically to avoid inertial coupling between bending and torsion. However, the added inertia of the collar lowers the natural frequencies of the drill so that the phenomena usually apparent at high speeds is observed at lower speeds. This shift causes the natural frequencies in the test to be much lower than the frequencies of actual drills. Apart from the collar effect, the authors also point out that they neglect to include some effects that influence the model: burnishing, plowing, rubbing, and process damping due to the clearance face of the tool interfering with the cut surface. The process damping should be especially significant at lower speeds. Because this paper deals with the existing piloted hole, the phenomenon of damping is neglected. Also evident is the inevitable effect of the friction between the rake face and the chip, as well as the margin effect when the margin of the drill interacts with the sides of the hole. The effect of the axial loading of the drill during drilling is neglected even though it would cause some change in the torsional natural frequency. The fact that the workpiece deforms during the drilling process also is neglected, even though it will influence the surface and therefore the chip load. Other possible causes of torsional vibration in drilling that are covered are considered to be stick slip interaction between tool and wall of the hole, built-up edge, and the like.

Stone and Askari continued to study Bayly's work in 2002 in [15] developing a model that incorporated the friction effect into the axial-torsional vibrational model by studying the stick-slip interaction mode between the tool and workpiece material. The stick-slip mode provides non-linear process damping to the system. By interacting with the vibrational mode, stick-slip mode produces instability in the non-regenerative chatter case.

The idea that the drag of the chip on the rake phase of the cutter causes the drill to experience a stick-slip friction effect also is incorporated into the model, and the law has been mathematically defined. The friction law used to incorporate the effect of friction in the derivations is defined as:

$$\mu(u) = \mu_1 + (\mu_2 - \mu_1)\exp\left[\left({u}/{U_o}\right)^2\right]$$

Where, u is chip speed, Uo is friction cutting speed, and μ_1, μ_2 are threshold friction coefficients.

For obtaining the dynamic model of the drill reverse engineering has been used, where the drill has been laser scanned and the obtained 3-D model was imported into ABAQUS program for analysis. The motion of the tip during vibration has been obtained from ABAQUS. The equation of motion has been expressed as:

$$m\ddot{\eta} + c\dot{\eta} + k\eta = F_\eta$$

Where m, c, k are mass, damping coefficient, and stiffness; and F_η is the forcing function. The vibration mode with regenerative chatter that includes the chip thickness variation is expressed as:

$$\eta'' + \gamma\eta' + \eta + \beta(1 - (\eta - \eta_T)\cos\theta)(p_o + \tilde{p}_1\eta' + \tilde{p}_2\eta'^2) = 0$$

Where: η is the rescaled amplitude variable, η' is the derivative with respect to the rescaled time variable, γ is the damping coefficient, β = wτ/k and $\tilde{p}_1, \tilde{p}_2$ are the re-scaled p_1, p_2 values, η_T = η (t-T), and T is the time to complete one revolution of the workpiece. The dependence on the friction model is explained through the coefficient p_1 of G (force penetration rate function), which depends on the rake angle α and vibration angle θ and on the form of the friction relation μ=h(u):

$$p_1 = \frac{-2\cos(\theta + \phi_0)}{\sin^3\phi_0} \frac{\frac{-h_0'}{1+h_0^2}\frac{-\cos(\phi_0 - \theta)}{\cos(\phi_0 - \alpha)}}{2 + \frac{h_0'}{1+h_0^2}\frac{V\cos\alpha}{\cos^2(\alpha - \phi_0)}},$$

Where, θ is vibration angle, φ is shear plane angle, and α is the rake angle, and where

$$\frac{du}{d\eta} = \frac{-\cos(\phi_0 - \theta)}{\cos(\phi_0 - \alpha)}$$

$$\frac{du}{d\phi} = \frac{V\cos\alpha}{\cos^2(\alpha - \phi_0)}$$

The critical depth of cut is expressed as follows:

$$\omega_{crit} = \frac{k}{2\tau p_0 \cos\theta}(\Gamma \pm 2)\Gamma$$

Where + is used for $p_o\cos\theta>0$, and – for $p_o\cos\theta<0$, p_o is the constant coefficient of G, G is the force penetration function, Γ is the dimensionless damping coefficient, and τ is shear stress of the material. As the rake angle is increased from $\alpha<\alpha_{crit}$ to $\alpha>\alpha_{crit}$, the response frequency becomes less than the natural frequency and a shift in the stability threshold occurs. This mathematical phenomenon is seen on the shop floor as the tool starts digging into the material, as opposed to being forced away from it.

Stone and Campbell continued Stone and Askari's work [16] by studying torsional-axial chatter. This work confirms Bayly's finding that for the drilling case, the onset of chatter oscillation happens at a frequency less than the natural frequency of the vibration mode, which is in contrast to the traditional case where the onset chatter frequency is greater than the natural frequency. The authors point out that the previous research has an approximation that $\beta p_1 < \gamma$ (β - bifurcation parameter, p_1 – linear coefficient of G, γ – dimensionless damping coefficient), which causes the difference between the cases, resulting in the presence of closed loops in the traditional case. The criticality of the Hopf bifurcation has been analyzed considering the sign of the quantity:

$$a = \frac{1}{8}\left[\Psi_{12}(0)(3f_{111} + f_{122}) + \Psi_{22}(0)(f_{112} + 3f_{222}\right] - \frac{1}{8\omega}\left[(\Psi_{12}(0)^2 - \Psi_{22}(0)^2)f_{12}(f_{11} + f_{22}) + 2\Psi_{12}(0)\Psi_{22}(0)(f_{22}^2 - f_{11}^2)\right]$$

Where f_{jk} and f_{ijk} are functions of the parameters β, T, γ, p_0, p_1, p_2 (constant, linear and quadratic coefficients of G), θ is the vibration angle, the Hopf frequency is ω, and the centre manifold coefficients. Analysis of this equation suggests that the bifurcation in the traditional case is subcritical for high values of T and supercritical for intermediate and small values of T. In the axial-torsional case, it was determined that the bifurcation is always subcritical. An analysis for high-speed stability limits shows very interesting results, especially if one is interested in micro drilling, which would involve very high speeds. In high speed limits, T becomes zero (large cutting speeds) and the equations do not provide further delay, but can be analyzed by conventional techniques.

For the T→0 limit, the equation of motion is expressed as:

$$\eta'' + \gamma\eta' + \eta - \beta(p_o + p_1\eta' + p_2\eta'^2) = 0$$

If calculated, a=0 points out degeneracy of the system, suggesting that it would be necessary to determine higher order terms in order to define stability. These bifurcations are not well understood yet and are suggested as a good topic for future research. The information suggests that macro and micro drilling cannot be analyzed the same way.

Stone and Askari's work was followed by research [17] that illustrates the effect of varying thickness and speed with linear stability boundaries and considers the change with varying cutting width. This paper is based on the models developed in Stone and Askari's work and defines the stability regions with respect to depth of cut. Stone and Askari define the traditional case with a high cutting speed having a supercritical bifurcation, while the case with low cutting speed is followed by a subcritical bifurcation. In the axial-torsional vibration case, for all physically reasonable situations, the bifurcation is shown to be subcritical. The implications of switching between supercritical and subcritical Hopf bifurcation influence the stability of the cutting process. When approaching the supercritical Hopf bifurcation, the linear stability boundary determines when the cutting will become unstable. With the subcritical Hopf bifurcation, small perturbations can cause the instability.

The equation of motion used is derived in [15] and re-scaled in order to use depth of cut, t_1, as the bifurcation parameter instead of width of cut w and accounting for chip thickness variation:

$$\eta'' + \gamma\eta' + \eta = \frac{w\tau}{k}\{t_1 - [\eta(t) - \eta(t-T)]\cos(\theta)\}(p_0 + \tilde{p}_1\eta' + \tilde{p}_2\eta'^2)$$

Where: $\tilde{p}_1 = \omega_0 p_1$ and $\tilde{p}_2 = \omega_0^2 p_2$.

Further, Dilley et al., studied the effect of the chisel edge on the chatter frequency in drilling in [18]. Owing to the problem of complex boundary conditions of the margin and chisel edge, most researchers have been dissuaded from studying such problems. A bending motion was seen due to the movement of the chisel, and therefore it is proven to be wrong to use fixed end conditions that make the chisel stationary. Here, we consider a simplified model where a spring end condition, representing the chisel edge effect, is used. For proper frequency matching with the experimental data, a spring end condition is used to represent the contact between the chisel edge and workpiece. A transverse spring and axial thrust force are added to the simple fixed-free beam, approximating the equation of motion using the Rayleigh-Ritz procedure as follows:

$$[m]\ddot{q}(t) + [k]q(t) = 0$$

$$m_{ij} = \int_0^L \mu\phi_i(z)\phi_j(z)dz$$

$$k_{ij} = \int_0^L EI_y\phi_i''(z)\phi_j''(z)dz + K_{end}\phi_i(L)\phi_j(L) - \int_0^L F_z\phi_i'(z)\phi_j'(z)dz$$

Where, m is mass, k is stiffness, φ is the trial function, q represents generalized coordinates, and $(F_z L^2)/(EI_y)$ is a dimensionless quantity.

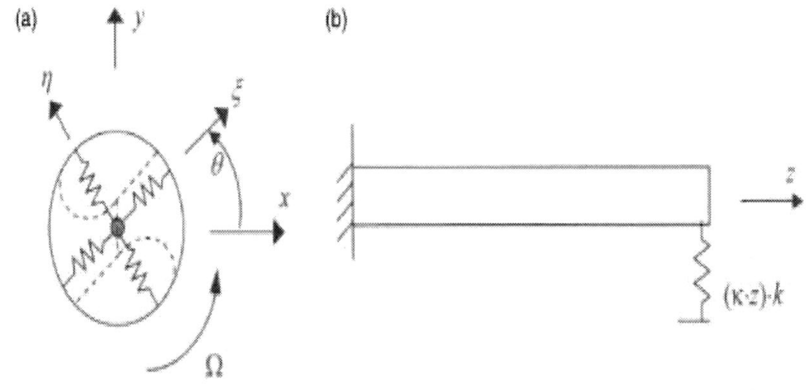

Figure 1. Drill model diagrams: (a) 2-DOF rotating single-mode system and (b) simplified drill model with spring-end condition representing effects of margin engagement. Tool fixed frame: ξ - η. Space fixed frame: x-y, where z is the depth into the hole.

After analyzing the effect of the chisel, Dilley at al., studied the problem of margin engagement and its effect on the frequency shift in drilling [19]. This paper shows that there is a significant frequency shift in the case of margin engagement. The drill model is presented in figure 1.

The effect of margin engagement is incorporated into the equation by adding a spring constant, making the model a function of the depth of penetration into the hole:

$$M_r \begin{bmatrix} \ddot{\xi} \\ \ddot{\eta} \end{bmatrix} + C_r \begin{bmatrix} \dot{\xi} \\ \dot{\eta} \end{bmatrix} + K_{rM} \begin{bmatrix} \xi \\ \eta \end{bmatrix} = \vec{F}_{ext}$$

$$M_r = \begin{pmatrix} m_{\xi\xi} & 0 \\ 0 & m_{\eta\eta} \end{pmatrix}, C_r = \begin{pmatrix} c_{\xi\xi} & -2m_{\xi\xi}\Omega \\ 2m_{\eta\eta}\Omega & c_{\eta\eta} \end{pmatrix}$$

$$K_{rm} = \begin{pmatrix} k_{\xi\xi}(1-k_{\xi\xi}z) - m_{\xi\xi}\Omega^2 & -c_{\xi\xi}\Omega \\ c_{\eta\eta}\Omega & k_{\eta\eta}(1-k_{\eta\eta}z) - m_{\eta\eta}\Omega^2 \end{pmatrix}$$

where: m, c, k are mass, damping, and stiffness, and $(\xi - \eta)$ is the tool's fixed frame.

It has been experimentally confirmed that the drill's natural frequency increases with the depth of the hole. By showing that this does not happen when machining a tube, it is proven that the margin effect is the sole reason for this frequency increase. Both of Dilley's papers are aimed at obtaining and constructing predictive models for engineers with the aim of drilling process optimization.

2.2. Bending Model

Apart from the axial and torsional vibrations, bending vibrations also have been found to be common in the drilling process. It is suggested that bending vibrations have a more detrimental effect on hole quality because the hole form changes drastically with the effect of bending.

The work of Arvajeh and Ismail focuses on the bending models and investigation of regenerative chatter in bending of both short and long drills. The first of their two papers [20] is based on the model delivered by Ulsoy [21]. However, unlike Ulsoy, Arvajeh and Ismail add the cutting forces with the regenerative effect.

The drilling equation of motion in this case is:

$$\begin{pmatrix} m & 0 \\ 0 & m \end{pmatrix}\begin{pmatrix} \ddot{x} \\ \ddot{y} \end{pmatrix} + \begin{pmatrix} c & -2m\Omega \\ 2m\Omega & c \end{pmatrix}\begin{pmatrix} \dot{x} \\ \dot{y} \end{pmatrix} + \begin{pmatrix} k - m\Omega^2 & -c\Omega \\ c\Omega & k - m\Omega^2 \end{pmatrix}\begin{pmatrix} x \\ y \end{pmatrix} = -b\begin{pmatrix} P & -Q \\ -P & Q \end{pmatrix}\begin{pmatrix} x(t) + x(t-\tau) \\ y(t) + y(t-\tau) \end{pmatrix}$$

Where, m, c, k are the bending dynamic properties of the drill, Ω is the rotational speed of the drill, τ is time between each cutting pass, and the matrix with P and Q is the matrix of bending cutting pressures.

Owing to the results presented in [22], which show that the chatter frequency is equal to the bending natural frequency when the tip of the drill is simply supported, the authors use an adjustment of the drill presented as a clamped-pinned beam with lumped mass at the middle . Further, the analytical model developed by Rincon [23] is used to obtain cutting pressure coefficients for bending stability analysis. The authors develop the relationships between the cutting angles and cutting forces, as well as rotational angles at the tip and deflections at the middle of the drill:

$$\begin{pmatrix} F_x' \\ F_y' \end{pmatrix} = b \begin{pmatrix} C_{11}' & C_{12}' \\ C_{21}' & C_{22}' \end{pmatrix} \begin{pmatrix} \Gamma_x \\ \Gamma_y \end{pmatrix}$$

$$\Gamma_x = kx$$
$$\Gamma_y = ky$$

Where, $C'_{11,12,21,22}$ are cutting coefficients, $\Gamma_{x,y}$ is the rotational angle at the tip, b is the radial width of cut, x, y are deflections at the middle of the drill, and k is a constant.

It is shown that in this case, the gyroscopic effect taken into consideration causes the natural frequencies of the system to vary according to the drill rotational speed. However, the drill rotational speed must be determined for the stability boundary. In this case, due to the dependency between the natural frequency and the speed at which the drill rotates, it is necessary to use the stationary drill natural frequency. Therefore, the relationship between the chatter frequency and natural frequency of the stationary drill was defined:

$$\omega = r\omega_n \pm \frac{2\pi r^2 \omega_n}{4\pi N + 2\theta \mp 2\pi r}$$

Where, r is the ratio, N is the integer number of full waves, and θ is the phase difference.

Three-dimensional laser scans in combination with Patran and Nastran models were used to obtain the drill surface and acquire mode shapes, natural frequencies and stiffness of drills. After the modeling and analysis were completed, stability lobes were developed, showing results for both the short and the long drill. It is shown that the higher frequencies can be ignored since they are less critical. It is also shown that the bottom of the lobes, which defines the minimum stability of the process, gets lower as the rotational speed increases due to the gyroscopic force, which causes the chatter frequency to change with the speed. Apart from lowering the stability minimum, the lobes get wider with the gyroscopic effect. Because the gyroscopic effect both widens and lowers the lobes, this effect must be considered when high drilling speeds are used, since this would greatly change the stability boundaries for those cases, reducing the stable area to a minimum.

2.3. Combination of the Bending and Torsional-Axial Models

Similarly, reference [24] presents a combination of the effects of torsion and bending on chatter in drilling using "short" and "long" drills. This analytical work regarding the torsional limits of stability in drilling, was developed using Bayly's work as its basis [3] and combines bending equations with two degrees of freedom with the equation of motion in the axial direction (z):

$$m\ddot{x}(t)+c\dot{x}(t)+(k-m\Omega^2)x(t)-2m\Omega\dot{y}(t)-c\Omega y(t)=-F_x^{'}(t)$$

$$m\ddot{y}(t)+c\dot{y}(t)+(k-m\Omega^2)y(t)-2m\Omega\dot{x}(t)-c\Omega x(t)=-F_y^{'}(t)$$

$$m_a\ddot{z}(t)+c_a\dot{z}(t)+k_a z(t)=-F_z(t)$$

Where $F_x^{'}(t), F_z(t)\ F_y^{'}(t)$ are variations of cutting forces; x(t), y(t), z(t) are drill deflections in the X, Y and Z directions; x(t), y(t) are deflections of the middle of the drills; z(t) is the axial elongation of the tip; m, c and k are dynamic properties of the bending modes; m_a, c_a , k_a are dynamic properties of the axial mode; and Ω is the drill's rotational speed.

A time domain simulation of chatter drilling shows strong agreement between the experiments and analytical derivations for the "long" drills, but not for the "short" drills due to inaccurate mathematical modeling. It was experimentally determined that two sets of chatter marks appear for the "long" drill, corresponding to the bending mode and the torsional mode. For the "short" drill, all chatter marks correspond to the torsional mode. This is explained by the fact that the support in the case of the "short" drill acts more like a spring than a rigid pin, and therefore makes the bending mode less important. Several parameters were not included in the study: (1) process damping due to the interaction between the flank faces of the drill and the just cut surface (i.e., the ''rubbing'' mechanism tends to dampen the vibration and it plays an important role in the stability of the tool), (2) the chisel edge in cases where there is no previous pilot hole, and (3) edge forces.

Dynamics, kinematics, and stability in drilling operations are reviewed in a two-part paper by Roukema and Altintas [6]. The first part of the study deals with the problem of combined lateral drill deflection, lateral chatter vibrations, and combined lateral and torsional-axial vibrations. The model discussed was used for prediction of lateral chatter and for prediction of the combination of lateral and torsional-axial vibrations. The general equation of motion for the dynamic drilling system in the stationary frame is:

$$[M]\begin{Bmatrix}\ddot{x}(t)\\ \ddot{y}(t)\\ \ddot{z}(t)\\ \ddot{\theta}(t)\end{Bmatrix}+[C]\begin{Bmatrix}\dot{x}(t)\\ \dot{y}(t)\\ \dot{z}(t)\\ \dot{\theta}(t)\end{Bmatrix}+[K]\begin{Bmatrix}x(t)\\ y(t)\\ z(t)\\ \theta(t)\end{Bmatrix}=\begin{Bmatrix}F_x\\ F_y\\ F_z\\ T_c\end{Bmatrix}$$

Where x, y are lateral displacements; z is the axial displacement; θ is the torsional deflection of the drill; matrices M, C, K are mass, damping, and stiffness characteristics at the drill tip; Fx, Fy are lateral forces; thrust force is Fz; and Tc is the torque.

The total displacement in any direction can be calculated by summing vibrations in all natural modes in a chosen direction, giving the overall effect:

$$q(t) = \sum_{i=1}^{k_m} \Delta q(t)$$

These deflections influence the chip thickness, changing the force distribution and effecting vibrations in all directions; therefore, the cutting force system must be coupled through the dynamic chip thickness distribution along the cutting edges.

The influence of pilot hole size, spindle speed, and torsional-axial chatter on lateral forces was observed experimentally and compared with simulated results. As the authors point out, this paper makes a fundamental contribution to the investigation of drilling by creating a complex numerical model that incorporates cutting force models and rigid body motion, as well as vibrations that occur in the process of drilling. In addition to these effects, the authors also include both grinding errors and tool misalignment effects. The data collected from these experiments, show that coupling of lateral and torsional vibrations does not occur; and torsional-axial chatter is seen as dominant, while lateral chatter is absent. The lateral chatter could become more important in longer, slender drills. Contrary to the experimental work, the model actually predicts a small value for the coupling of vibrations and actually predicts both torsional-axial and lateral chatter vibrations. Therefore, it can be seen that the model does not actually agree with the experimental results. The predicted values for stiffness and damping are not in agreement with the obtained experimental data, and issues like the rubbing of the cutting lips against the bottom of the hole, which is wavy in the first place, hole wall contact with drill flutes, the change in the contact boundaries as the drill penetrates deeper in the hole, and the friction effect all add to the complexity of accurate prediction of chatter vibrations.

The second part of the two-part paper [25] presents a general solution of drilling chatter stability in the frequency domain. In order to determine the regenerative chip thickness, the drill's flexibility in torsional, axial, and lateral directions are analyzed. All the vibrations mentioned have an influence on the regenerative chip thickness. Mathematically, the change in chip thickness due to regenerative displacements can be defined as:

$$dh_1 = \frac{dx\cos\Omega t - dy\sin\Omega t}{\tan\kappa t}$$

$$dh_2 = -\frac{dx\cos\Omega t - dy\sin\Omega t}{\tan\kappa t}$$

Where κt is the tip angle of the drill and Ωt is the tool rotation angle.

The effect of torsional vibrations on the chip thickness can be defined as:

$$dh_1 = dh_2 = -d\theta \times \frac{1}{2\pi} \times f_r,$$

Where f_r is feed per revolution.

The effect of the axial vibrations on the chip thickness can be defined as:

$$dh_1 = dh_2 = -dz$$

When all vibrations are taken into consideration, the effect of the whole system on the change in chip thickness is:

$$\begin{Bmatrix} dh_1 \\ dh_2 \end{Bmatrix} = \begin{Bmatrix} \frac{1}{\tan\kappa_t}(dx\cos\Omega t - dy\sin\Omega t) - dz - \frac{f_r}{2\pi}d\theta \\ -\frac{1}{\tan\kappa_t}(dx\cos\Omega t - dy\sin\Omega t) - dz - \frac{f_r}{2\pi}d\theta \end{Bmatrix}$$

The dynamic forces can therefore be expressed as:

$$\begin{Bmatrix} Fx \\ Fy \\ Fz \\ Tc \end{Bmatrix} = \begin{Bmatrix} k_{tc}b(dh_1 - dh_2)\sin\Omega t - k_{tc}k_{rc}(dh_1 - dh_2)\cos\Omega t \\ k_{tc}b(dh_1 - dh_2)\cos\Omega t - k_{tc}k_{rc}(dh_1 - dh_2)\sin\Omega t \\ k_{tc}k_{ac}b(dh_1 - dh_2) \\ \{k_{tc}b(dh_1 + dh_2)\}(1 - k_{rc})R_t \end{Bmatrix}$$

Where Fx, Fy, Fz are cutting forces acting on the tool tip; Tc is the cutting torque acting on the tool tip; Rt is the torque arm – from tangential and radial forces; ktc is the tangential cutting stiffness; krc, kac are radial and axial force cutting stiffness factors; and b is the radial depth of cut.

The dynamic cutting forces can be summarized as:

$$\{F(t)\} = bk_{tc}[B(t)]\{\Delta r\}$$

Where [B(t)] is the dynamic drilling coefficient matrix dependent on time and contains the negative regenerative displacements. In this paper, the frequency response method was used and compared with results from other models. The frequency response matrix at the tool tip has been defined as:

$$\Phi(i\omega) = \begin{bmatrix} \Phi_{xx}(i\omega) & 0 & 0 & 0 \\ 0 & \Phi_{yy}(i\omega) & 0 & 0 \\ 0 & 0 & \Phi_{zz}(i\omega) & \Phi_{z\theta}(i\omega) \\ 0 & 0 & \Phi_{\theta z}(i\omega) & \Phi_{\theta\theta}(i\omega) \end{bmatrix}$$

Where Φaa (iω) is the direct frequency response function and Φab (iω) is the cross-frequency response function.

After extensive computation, the authors express the eigenvalues as:

Lateral: $\Lambda_1 = \dfrac{-\gamma_1 + \sqrt{\gamma_1^2 - 4\gamma_0}}{2\gamma_0}$, $\Lambda_2 = \dfrac{-\gamma_1 + \sqrt{\gamma_1^2 - 4\gamma_0}}{2\gamma_0}$

Torsional-axial: $\Lambda_1 = \dfrac{-\gamma_3 + \sqrt{\gamma_3^2 - 4\gamma_2}}{2\gamma_2}$, $\Lambda_2 = \dfrac{-\gamma_3 + \sqrt{\gamma_3^2 - 4\gamma_2}}{2\gamma_2}$

Where γ_0, γ_1, γ_2, γ_3 are coefficients of the characteristic equation.

The results show that frequency domain solutions actually give a more conservative stability estimate than the time domain model. A comparison of the model with results from other researchers [3] show that the values for critical depth of cut are less conservative in this particular model, while the shape of stability lobes is different. The lateral chatter stability was compared to a rotating coordinate approach, where the difference is in using an average of time varying coefficients instead of time invariant cutting coefficients. This comparison shows good agreement. The frequency domain solutions are compared with the experimental data to establish the agreement and validity of the model used. Unfortunately, the stability lobes obtained are both too small and too narrow to be used for selection of actual cutting conditions. The experiments are not in agreement with the linear stability prediction laws and show that if the spindle speed is below 2000 rpm, drilling is always stable and stability rises with decrease in spindle speed. It was even stated that stability increases with chisel edge engagement. Although this paper [25] presents the most comprehensive model and covers the largest number of influential parameters, the model is not in agreement with the experimental results and needs further work.

Present In addition, current models of drilling motion do not include the gyroscopic moments and rotary inertia effects and their influence on natural frequencies of the drills [26]. Work conducted by Timoshenko et al. [27] shows that there is a slight decrease in the value of the natural frequency calculated with the effect of rotary inertia. The nonlinear equations of motion linearized and including gyroscopic moments and rotary inertia are expressed as follows:

This paper confirms Timoshenko's statement showing that the effects of the gyroscopic moment and rotary inertia are not significant if the rotational speed is lower than 60,000 rpm for longer drills and up to 120,000 rpm for short drills. We also explore the effect of drill geometry—coolant holes, web taper, and number of flutes, showing that increasing the web taper causes a slight decrease in bending frequency, that the increase in the number of flutes increases the stiffness of the drill bit, and also that the increase in the coolant hole size slightly increases the bending frequency of the drill. The authors also investigated the effect of material properties along the drill bit, showing that the natural frequency of a steel tool increases by 26% if steel is replaced by carbide and by 20% if the drill is a combination of steel and carbide.

$$m\ddot{v}_1 - 2m\omega_0\dot{v}_2 - m\omega_0^2 v_1 - J_1\ddot{v}_{1zz} + J_p\beta_0\ddot{v}_{2zz} + J_2\beta_0^2\ddot{v}_1 + J_2\omega_0^2\beta_0^2\dot{v}_1 + J_p\omega_0^2\beta_0 v_{2z} - J_1\omega_0^2 v_{1zz} - EI_1 v_{1zzzz} -$$
$$2E(I_1+I_2)\beta_0 v_{2zzz} - E(2I_1+4I_2)\beta_0^2 v_{1zz} + 2E(I_1+I_2)\beta_0^3 v_{2z} + EI_1\beta_0^4 v_1 + Fz\left[v_{1zz} - 2\beta_0 v_{2z} - \beta_0^2 v_1\right] = F_\Xi$$

$$m\ddot{v}_2 - 2m\omega_0\dot{v}_1 - m\omega_0^2 v_2 - J_2\ddot{v}_{2zz} + J_p\beta_0\ddot{v}_{1zz} + J_1\beta_0^2\ddot{v}_2 + J_1\omega_0^2\beta_0^2\dot{v}_2 + J_p\omega_0^2\beta_0 v_{1z} - J_2\omega_0^2 v_{2zz} + EI_2 v_{2zzzz} +$$
$$2E(I_1+I_2)\beta_0 v_{1zzz} - E(4I_1+2I_2)\beta_0^2 v_{2zz} - 2E(I_1+I_2)\beta_0^3 v_{1z} + EI_2\beta_0^4 v_2 + Fz\left[v_{2zz} - 2\beta_0 v_1 - \beta_0^2 v_2\right] = F_\Psi$$

2.4. Chatter Suppression

Chatter suppression is one of the main directions of research in the area of vibrations in machining. Chatter control in turning and milling is extensively covered in [28-37]. Spindle speed variation methods are successfully applied for chatter suppression in turning and milling operations. Reference [38] presents a spindle speed variation method for chatter suppression in drilling. It has been shown that based on the stability charts chatter can be suppressed by changing the spindle speed to a certain range. Using thrust force as the basis for chatter detection, these authors present both theoretical and experimental results. The predicted stable spindle speeds are considered to be:

$$n_s = \frac{30 f_c}{N}$$

Where N = 1, 2, 3, …n, and f_c is the chatter frequency.

Further, in [39], the relationship between the spindle speed and the phase shift are studied. It is shown that phase shift can be modified by changing spindle speed. Therefore, the cut width limit can be increased or decreased by a change in spindle speed, as well. Chatter occurs when the width of cut is greater than the limiting value, and chatter is suppressed if the spindle speed is changed so that the width of cut is smaller than the limiting value. Two strategies are investigated: the first one states that the regulation of the spindle speed to phase shift will be optimal at 2π, and the second one states that the regulation of the spindle speed to phase shift will be optimal at π/2. While experiments show that the π/2 optimal phase shift is impossible, the 2π phase shift has been confirmed.

The value of the width of cut at the stability limit is defined as:

$$b_{\lim} = \frac{-1}{k_s G_m(j\omega)(1 - e^{-j\varepsilon})}$$

Where $G_m(j\omega)$ is the frequency response function of the machine tool structure, ω(rad/s) is the chatter frequency, and $\varepsilon = \omega T$ is the phase shift between the inner and outer modulation.

The relationship between the tooth passing frequency and chatter frequency is defined as:

$$\frac{f_c}{f_t} = N + \frac{\varepsilon}{2\pi}$$

Where $f_t=zn_s$ (z – number of cutting edges, ns – spindle speed), N is the largest integer to satisfy the equation. and $0 < \varepsilon/2\pi \leq 1$.

When the phase shift was chosen to be 2π, the stable spindle speed was calculated to be:

$$n_{ss} = \frac{f_c}{(N+1)z}$$

In this case. it was shown that the chatter frequency curve and the stable spindle speed curve intersect close to the largest stability limit of the width of cut, b_{lim}. With a phase shift of π/2, the stable spindle speed is:

$$n_{ss} = \frac{f_c}{(N+1/4)z}.$$

It was shown that in this case, the chatter frequency curve and the stable spindle speed curve do not intersect, therefore proving that it is impossible to use this combination.

After defining the chatter frequencies and the optimal shifts, the authors continued their research in the area of chatter detection working on neural networks for reduction of chatter in drilling [7]. Many research challenges must be considered in chatter recognition and suppression in drilling operations. Modeling the process becomes harder with increasing complexity and modification of the chisel edge of the drill. Further, prediction of cutting forces in the chisel edge region is difficult due to the variety of cutting speeds and rake and oblique angles along the cutting edge. As stated, many researchers have simplified in order to solve these problems by analyzing one model or one problem at a time. However, the fact is that the axial, torsional, and lateral vibrations can be acting at the same time in drilling, and the effects of chisel edge, margin effect, pilot hole size, etc., are always present as well. The optimal model would need to be able to incorporate all these effects to be successfully used in industrial applications.

3. Investigation of Chatter in Micro Drilling

It is well known in machining operations that chatter affects the surface of the part, as well as the tool used. What studies of various chatter issues since the 1950s have in common is that the need for accurate measurements of the frequency response function (FRF) at the tip of the tool when it is attached to the tool holder and spindle. For macro testing, the impact test is sufficient for dynamic measurements, but it is impossible to apply the same system to micro tools due to their size and fragility. Impact hammer forces would be detrimental to the tools used in micro machining. Impact testing is not applicable if an accelerometer is to be used due to the mass loading necessary. Many times, tools have been modeled as cantilever beams, incorrectly representing the actual dynamics of the tool tip. Receptance coupling technique developed by Park et al. [40] suggests the use of coupling of analytical and experimental FRFs of the components with the aim of obtaining the response of the assembly as a whole.

The accuracy of the receptance coupling technique depends on accurate identification of the joint dynamics of the substructures at the assembly joint, and the FRFs of each substructure. Any noise in the FRF or inaccuracies in the joint dynamic parameters amplify errors in obtaining the FRF for the assembled tool-spindle system. This paper proposes an algorithm that allows analytical extraction of rotational dynamics at the joints from linear displacements and impact force tests. The authors analytically identify the rotational dynamic response of the spindle-holder by substituting the direct and cross FRF measurements taken at the free end of the assembly and joint of the blank cylinder in the receptance coupling expressions. Combining finite element (FE) analysis and experimental work lead to the acquisition of the overall dynamics of the system.

The FRF of the substructure A can be defined as:

$$\begin{Bmatrix} X_1 \\ X_{A,2} \end{Bmatrix} = \begin{bmatrix} H_{A,11} & H_{A,12} \\ H_{A,21} & H_{A,22} \end{bmatrix} \begin{Bmatrix} F_1 \\ F_2 \end{Bmatrix}$$

The FRF of the substructure B can be defined as:

$$\{X_{B,2}\} = [H_{B,22}]\{F_{B,2}\}$$

Where X, F are displacement and force vectors applied on the structure, and $H_{A,ij}$ vectors are FRFs between points i and j.

Boundary conditions used for coupling substructures A (free – free model of the end mill) and B (spindle) can be defined as:

$$F_2 = F_{A,2} + F_{B,2}$$
$$X_2 = X_{A,2} = X_{B,2}$$

The displacements X_1 and X_2 can be expressed in a matrix form as:

$$\begin{Bmatrix} X_1 \\ X_2 \end{Bmatrix} = \begin{bmatrix} (H_{A,11} - H_{A,12}H_2^{-1}H_{A,21}) & (H_{A,12} - H_{A,12}H_2^{-1}H_{A,22}) \\ (H_{A,21} - H_{A,22}H_2^{-1}H_{A,21}) & (H_{A,22} - H_{A,22}H_2^{-1}H_{A,22}) \end{bmatrix}$$

The direct and cross-receptance at the tool tip can be expressed as:

$$\frac{X_1}{F_1} = H_{A,11} - H_{A,12}(H_{A,22} + H_{B,22})^{-1} H_{A,21} = H_{11}$$
$$\frac{X_1}{F_2} = H_{A,12} - H_{A,12}(H_{A,22} + H_{B,22})^{-1} H_{A,22} = H_{12}$$

Since each FRF contains translational and rotational displacements, the FRF expressed earlier needs to be expanded as:

$$\begin{Bmatrix} x_1 \\ \theta_1 \end{Bmatrix} = \begin{bmatrix} h_{11,ff} & h_{11,fM} \\ h_{11,Mf} & h_{11,MM} \end{bmatrix} \begin{Bmatrix} f_1 \\ M_1 \end{Bmatrix} \rightarrow \{X_1\} = [H_{11}]\{F_1\}$$

$$\begin{Bmatrix} x_1 \\ \theta_1 \end{Bmatrix} = \begin{bmatrix} h_{12,ff} & h_{12,fM} \\ h_{12,Mf} & h_{12,MM} \end{bmatrix} \begin{Bmatrix} f_2 \\ M_2 \end{Bmatrix} \rightarrow \{X_1\} = [H_{12}]\{F_2\}$$

Where each h_{ij} is evaluated from the receptance coupling expression by including both translational (x) and rotational (q) displacements due to lateral force (f) and moment (M).

The direct and cross-receptances at the tool tip also need to be expanded as:

$$[H_{11}] = \begin{bmatrix} h_{A11,ff} & h_{A11,fM} \\ h_{A11,Mf} & h_{A11,MM} \end{bmatrix} - \frac{1}{h_{2,Mf}^2 - h_{2,MM}h_{2,ff}} \begin{bmatrix} h_{A12,ff} & h_{A12,fM} \\ h_{A12,Mf} & h_{A12,MM} \end{bmatrix} \begin{bmatrix} -h_{2,MM} & h_{2,fM} \\ h_{2Mf} & -h_{2,ff} \end{bmatrix} \begin{bmatrix} h_{A21,ff} & h_{A21,fM} \\ h_{A21,Mf} & h_{A21,MM} \end{bmatrix}$$

$$[H_{12}] = \begin{bmatrix} h_{A12,ff} & h_{A12,fM} \\ h_{A12,Mf} & h_{A12,MM} \end{bmatrix} - \frac{1}{h_{2,Mf}^2 - h_{2,MM}h_{2,ff}} \begin{bmatrix} h_{A12,ff} & h_{A12,fM} \\ h_{A12,Mf} & h_{A12,MM} \end{bmatrix} \begin{bmatrix} -h_{2,MM} & h_{2,fM} \\ h_{2Mf} & -h_{2,ff} \end{bmatrix} \begin{bmatrix} h_{A22,ff} & h_{A22,fM} \\ h_{A22,Mf} & h_{A22,MM} \end{bmatrix}$$

The elements that represent the rotational degrees of freedom are complex to determine, and they can be expressed by rewritten equations with common FRF terms. With extensive rewriting, rotational degrees of freedom can be expressed as:

$$h_{B22,Mf} = \beta - h_{A22,Mf}$$
$$h_{B22,MM} = \delta - h_{A22,MM}$$

Where β, δ are explained in detail in [40] through a symbolic non-linear analytical toolbox and will not be further analyzed here.

Comparison of these results with previously suggested approaches show that this design has improved accuracy in experiments carried out with various sizes of end mills and successfully indentified chatter free conditions in milling operations.

Developing capability for tool point frequency response prediction for micro end mills is discussed in reference [41]. In order to predict the response of the whole system as an assembly, receptance coupling substructure analysis was used. Response functions of the tool, tool holder, and spindle were coupled showing an integrated system. The authors obtain receptance of the tool and collet holder using Timoshenko beam models due to their size and fragility, while the spindle translational and rotational receptances were obtained experimentally.

The spindle holder base receptances were determined by defining the receptance for the standard tool holder:

$$L_{55}(\omega) = \frac{H_{55} - H_{55a}}{S}$$

Where H_{55} is the direct FRF at coordinate 5, H_{55a} is the cross FRF at coordinate 5, and S is considered to be 50.5mm.

The rotation to moment receptance is determined as:

$$P_{55} = \frac{L_{55}N_{55}}{H_{55}} = \frac{L_{55}^2}{H_{55}}$$

Where H_{55} was measured and N_{55} was assumed equal to L_{55} by reciprocity.

An assembly receptance matrix is defined as:

$$G_{55}(\omega) = \begin{bmatrix} H_{55} & L_{55} \\ N_{55} & P_{55} \end{bmatrix}$$

Further, Timoshenko beam models were used for substructures I and II, and the tool and tool holder were rigidly coupled using:

$$RS_{55} = R_{55} - R_{56}(R_{66} + R_{6a6a})^{-1} R_{65}$$
$$RS_{77} = R_{77} - R_{76a}(R_{66} + R_{6a6a})^{-1} R_{6a7}$$
$$RS_{57} = R_{56}(R_{66} + R_{6a6a})^{-1} R_{6a7}$$
$$RS_{75} = R_{76a}(R_{66} + R_{6a6a})^{-1} R_{65}$$

Where RS represents coupled substructures receptance matrices, and R represents substructure matrices.

The tool holder-spindle assembly receptances are defined through beam models:

$$G_{41} = \begin{bmatrix} H_{41} & L_{41} \\ N_{41} & P_{41} \end{bmatrix}$$
$$G_{44} = \begin{bmatrix} H_{44} & L_{44} \\ N_{44} & P_{44} \end{bmatrix}$$

Where H, L, P, and N are different assembly responses. Subassembly receptance matrices were developed through Timoshenko beam models for substructures I-VI and spindle holder dynamics. Substructures IV – VI were rigidly coupled to each other and then rigidly coupled to the spindle holder base receptances to give the direct receptance matrix for the subassembly.

In order to combine data and couple the tool to the holder, a flexible connection and a scalar stiffness matrix was used:

$$Kc = \begin{bmatrix} k_{xf} & k_{xm} \\ k_{\theta f} & k_{\theta m} \end{bmatrix}$$

Where k represents stiffness values from displacement (x) and rotation (θ) due to force (f) and moment (m).

Mascardelli et al. [42] apply receptance coupling to predict the FRF at the tip of the micro tool and consequently use the FRF for chatter analysis. These authors divide the machining system into two subsections in their research. The dynamics of the first subsystem, the cutter and the taper, are performed using FE analysis due to the fragile and slender nature of the tool, while the dynamics of the rest of the structure including the tool and the tool holder are defined experimentally. By identification of the rotational degree of freedom (RDOF) at the joint, these two subsystems are linked. in this way, stability lobes were determined, which provides information on the stability of the system for a given axial depth of cut and spindle speed. In FE analysis, two different models were used: a perfectly constrained cantilever beam and an imperfectly constrained beam. Micromilling experiments were performed in order to confirm the chatter-stability analysis based on the receptance coupling (RC) method. Information on high-frequency forced response tool dynamics was obtained by using acoustic emission sensors.

The response of the substructures is defined as:

$$\begin{bmatrix} X_1 \\ X_2 \end{bmatrix} = \begin{bmatrix} H_{11} & H_{12} \\ H_{21} & H_{22} \end{bmatrix} \begin{bmatrix} F_1 \\ F_2 \end{bmatrix}$$

$$[X_3] = [H_{33}] F_{33}$$

$$X = \{x, \theta\}$$

$$F = \{f, M\}$$

Where Hij is the transfer function of the substructures, subscripts indicate the location of the displacement and force; X is represented by translational displacement x as well as rotation θ; and F is represented by force f and moment M.

The matrices of the assembled structures are defined as:

$$\begin{bmatrix} X_1 \\ X_2 \\ X_3 \end{bmatrix} = \begin{bmatrix} G_{11} & G_{12} & G_{13} \\ G_{21} & G_{22} & G_{23} \\ G_{31} & G_{32} & G_{33} \end{bmatrix} \begin{bmatrix} F_1 \\ F_2 \\ F_3 \end{bmatrix}$$

Where Gij are the transfer functions of the assembly. The TF matrices are obtained either using EMA (experimental modal analysis) or FEA. The responses at the beginning and end of the substructure 1 can be predicted mathematically, taking into account both translational and rotational degrees of freedom. The rotational dynamics at the end of substructure 2 are identified using the technique covered in [40], giving the result:

$$b_{mf} = \frac{(-b_{ff}G_{11,ff}h_{22,mf} + b_{ff}h_{21,mf}G_{12,ff} + b_{ff}h_{11,ff}h_{22,mf} - b_{ff}h_{h21,mf}h_{12,ff} + h_{21,mf}h_{22,ff}h_{12,ff} - h_{21,ff}h_{12,ff}h_{22,mf}}{h_{11,ff}h_{22,ff} - G_{11,ff}h_{22,ff} - h_{21,ff}h_{12,ff} + h_{21,ff}G_{12,ff}}$$

$$b_{mm} = \frac{1}{h_{11,ff}h_{22,ff} - G_{11,ff}h_{22,ff} - h_{21,ff}h_{12,ff} + h_{21,ff}G_{12,ff}} b_{ff}h^2_{21,mf}G^2_{12,ff} + (2b_{ff}h_{11,ff}h_{22,mf}h_{21,mf} + h_{12,ff}h^2_{21,mf}h_{22,ff}$$
$$- 2b_{ff}G_{11,ff}h_{22,mf}h_{21,mf} - h_{12,mf}h^2_{21,ff}h_{22,mf} + h_{22,ff}h_{21,ff}h_{21,mf}h_{21,ff} - 2h_{12,ff}b_{ff}h^2_{21,mf} - h_{12,ff}h_{21,mf}h_{21,ff}h_{22,mf})G_{12,ff}$$
$$- h^2_{22,ff}h_{12,mf}h_{21,mf}G_{11,ff} + h^2_{22,ff}h_{12,mf}h_{21,mf}h_{11,ff} + h^2_{22,mf}b_{ff}h^2_{11,ff}h_{22,ff}h_{12,mf}h_{21,ff}h_{22,mf}h_{11,ff} + h_{22,ff}h_{12,mf}h_{21,ff}h_{22,mf}G_{11,ff}$$
$$+ h^2_{22mf}b_{ff}G^2_{11,ff} + h_{12,ff}h_{22,ff}h_{22,mf}h_{21,mf}h_{11,ff} - 2h^2_{22,mf}b_{ff}G_{11,ff}h_{11,ff} + h_{12,ff}h_{12,mf}h^2_{21,ff}h_{22,mf} - h_{12,ff}h_{22,ff}h_{22,mf}h_{21,mf}G_{11,ff}$$
$$- h_{12,ff}h_{22,ff}h_{12,mf}h_{21,mf}h_{21,ff} + 2h_{12,ff}b_{ff}G_{11,ff}h_{22,mf}h_{21,mf} + h^2_{12,ff}b_{ff}h^2_{21,mf} - 2h_{12,ff}b_{ff}h_{11,ff}h_{22,mf}h_{21,mf} + h_{12,ff}h^2_{22,mf}h_{21,ff}G_{11,ff}$$
$$+ h^2_{12,ff}h_{21,mf}h_{21,ff}h_{22,mf} - h^2_{12,ff}h^2_{21,mf}h_{22,ff} - h_{12,ff}h^2_{22,mf}h_{21,ff}h_{11,ff}$$

Where $h_{ff} = x/f$, $h_{fm} = x/M$, $h_{Mf} = \theta/f$, $h_{MM} = \theta/M$; Gij are transfer functions of the assembly; x is the translational displacement of beams; and θ is the rotational displacement of beams.

Research suggests a 3% difference between the experimental and receptance coupling results, which is considered to be acceptable for chatter prediction purposes. It was confirmed that adding the effect of the shank, taper, and tip bending modes can be ignored for chatter prediction purposes because the additional lobes do not have any effect until the speeds reach 120,000 rpm. At that depth, the allowed axial depth of cut exceeds 100 microns, which would break the tool. The lobes did reduce the depth of cut to 38 microns at speeds in excess of 180,000rpm, which are beyond the speed ranges of available spindles. Based on this analysis, it is concluded that only the primary bending mode of the tool must be evaluated. This paper covers only the case where the feed rate is greater than the critical chip thickness, which means that it is applicable only to soft materials. Therefore, the RC technique for chatter prediction is applicable only to shearing-dominant cutting mechanisms, which occur only in machining soft materials. The plowing would bring non-linearity effects into the system, making the use of RC less suitable.

Paper [43] presents a review of the state of the art in micromachining, with emphasis on micro milling . A comparison of micro and macro scales was done, pointing out how the knowledge from one scale can be applied to the other. There are several problematic issues when dealing with micromachining. First, the accurate measurement of forces is extremely challenging due to the size of tools and small values of the forces themselves. Noise, if present, greatly disturbs the signals and gives distorted results. The issue of applying the Merchant's cutting model that is used for macro cutting processes is problematic to say the least. This is due to the effects of edge radius, which result in high negative rake angle and elastoplastic effects. Presence of high negative rake angles also affects the magnitude of shearing and plowing, resulting in a substantial increase in the axial force and increased friction on the rake face of the tool. The shearing and plowing effects would result in the need for a relatively large volume of material to become fully plastic for a relatively small amount of material to be removed, which would cause significant increase in the specific energy.

Apart from these problems, the interaction between tool and the workpiece also presents challenges. In micro cutting, the depth of cut needs to be larger than the critical chip thickness in order for chips to be formed due to the elastic effect present when machining such small thicknesses. The relationship between the tool radius and minimum chip thickness depends on the cutting edge radius and the material of the workpiece. The minimum chip thickness was been studied extensively [44-46] and can be obtained either by experiments or FE analysis (figure 2). Regenerative chatter also presents an important phenomenon that needs to be taken in consideration; however, accurate dynamic measurement of chatter between the tool and the workpiece is extremely challenging [41]. Various authors suggest that chatter, tool wear, monitoring, and work handling are issues that have not been sufficiently investigated on a micro scale. Chatter presents a very significant problem due to the fact that vibrations can have a catastrophic impact on micro systems, leading to failure. Micromachining also needs to incorporate two other aspects of chatter: (1) the fact that contrary to macro machining cases, the feed rate does influence micromachining due to the elasticity of the workpiece, and (2) the regenerative aspect of chatter, resulting in dynamic change of the chip thickness. Another important issue in chatter in micro drilling is that homogeneity of the material machined cannot be assumed. In macrodrilling, the effect of grains, grain boundaries, and anisotropy of material can be neglected, while in microdrilling, due to the size of the tool, these considerations will have a significant effect on the process.

Other issues important in investigating chatter in micromachining include tool run-out caused by misalignment of the axis of symmetry between the tool and tool holder or spindle, available sensing methods for monitoring and controlling high speed machining processes, thermal expansion, and the application of adaptive control techniques.

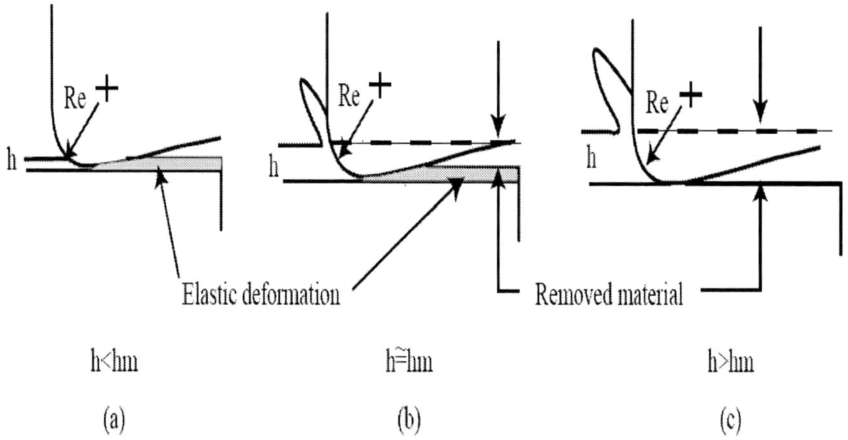

Figure 2. Schematic of the effect of the minimum chip thickness (Re), radius of cutting tool; h, undeformed chip thickness; hm, minimum chip thickness in micromachining.

4. Conclusions

The chapter summarizes the effect and suppression of chatter in drilling and micromachining operations. Research has been conducted into chatter recognition and suppression since the 1950s. Since then, substantial advances were made in the field of modeling and many of the resulting theoretical conclusions have been applied in practice, especially in milling and turning operations. Drilling has not been researched to the same extent and requires additional theoretical and experimental development. In addition, the need is identified to incorporate axial-torsional, lateral, and bending vibrations into one system, as well as to consider other influential parameters such as the chisel edge, margin effect, pilot hole size, tool grinding errors, misalignment, etc.

Microdrilling operations are yet to be addressed in the literature and require a significant amount of attention in order to model the process accurately and obtain explanations regarding its influential parameters. The idea of applying theory developed for macrodrilling operations to microdrilling has proved inappropriate, but analysis of the similarities and differences in micro- and macrodrilling provides important information. Future work will require the development of mathematical models for microdrilling operations and further work on receptance coupling techniques to obtain overall dynamic models of the tool-workpiece assembly.

REFERENCES

[1] Tobias, S.A., Machine Tool Vibration. 1965, New York: Wiley.

[2] Koenigsberger, F., Tlusty, J., Machine Tool Structures. Vol. 1. 1970, New York: Pergamon Press.

[3] Bayly, P.V., Metzler, S.A, Schaut, A.J., Keith, A.Y, Theory of Torsional Chatter in Twist Drills Model Stability and Composition to Test. Journal of Manufacturing Science and Engineering, 2001. 123: p. 552-561

[4] Minis, I., Yanusheshevsky, R., A new theoretical approach for the prediction of chatter in milling. Journal of Engineering for Industry, 1993. 115: p. 1-8.

[5] Altintas, Y., Budak, E., Analytical prediction of stability lobes in milling. Annals of the CIRP, 1995. 44(1): p. 357-362.

[6] Roukema, J.C., Altintas, Y., Generalized modeling of drilling vibrations. Part I: Time domain model of drilling kinematics, dynamics and hole formation. International Journal of Machine Tools & Manufacture, 2007. 47: p. 1455–1473

[7] Tarng, Y.S., Li T.C., Adaptive Pattern Recognition of Drilling Chatter. Journal of Materials Processing Technology. 1995. 48: p. 247-253.

[8] Kalmar, T., Stepan, G., Moon, T., Subcritical Hopf Bifurcation in the Delay Equation Model for Machine Tool Vibrations. Non-linear Dynamics, 2001. 26: p. 121-142.

[9] Ema, S., Fujii, H., Marui, E., Whirling Vibration in Drilling. Part 3: Vibration Analysis in Drilling Workpiece with a Pilot Hole. Journal of Engineering for Industry, 1988. 110: p. 315-321.

[10] Fujii, H., Marui, E., Ema, S., Whirling Vibrations in Drilling. Part 1: Cause of Vibration and Role of Chisel Edge. ASME Journal of Engineering Industry, 1986. 108(3): p. 157.

[11] Fujii, H., Marui, E., Ema, S., Whirling Vibration in Drilling. Part 2: Influence of Drill Geometries, Particularly of the Drill Flank on the Initiation of Vibration. ASME Journal of Engineering Industry, 1986. 108(3): p. 163.

[12] Bayly, P.V., Lamar, M.T., Calvert, S.G., Low-Frequency Regenerative Vibration and the Formation of Lobed Holes in Drilling. Journal of Manufacturing Science and Engineering, 2002. 124: p. 275-285.

[13] Altintas, Y., Weck, M., Chatter Stability of Metal Cutting and Grinding. CIRP Annals – Manufacturing Technology, 2004. 53(2): p. 619-642

[14] Tlusty, J., The dynamics of high-speed milling. Journal of Engineering for Industry, 1986. 108: p. 50-67.

[15] Stone, E., Askari, A., Non-linear model of chatter in drilling processes. Dynamical systems, 2002. 17(1): p. 65-85.
[16] Stone, E., Campbell, S.A., , Stability and Bifurcation Analysis of a Nonlinear DDE Model for Drilling. Journal of Nonlinear Science. 2004. 14: p. 27–57.
[17] Campbell, S.A., Stone, E. , Analysis of Chatter Instability in a Nonlinear Model for Drilling. Journal of Computational and Nonlinear Dynamics, 2006. 1: p. 294-306
[18] Dilley, D.N., Bayly, P.V., Schaut, A.J. , Effects of the chisel edge on the chatter frequency in drilling. Journal of Sound and Vibration, 2005. 281: p. 423-438.
[19] Dilley, D.N., Stephenson, D. A., Bayly, P.V., Schaut, A.J., Frequency Shift in Drilling Due To Margin Engagement. Journal of Manufacturing Science and Engineering, 2005. 127: p. 271-276.
[20] Arvajeh, T., Ismail, F., Machining stability in high-speed drilling—Part 1: Modeling vibration stability in bending. International Journal of Machine Tools & Manufacture, 2006. 46: p. 1563–1572.
[21] Ulsoy, A.G., A lumped parameter model for the transverse vibration of the drill bits. Control of Manufacturing Processes and Robotic Systems, 1983: p. 15–25.
[22] Ema, S., Fujii, H., Marui, E. , Chatter vibration in drilling. Journal of Engineering for Industry, 1988. 110: p. 309-314
[23] Rincon, D.M., Coupled force and vibration modeling of drills with complex cross sectional geometries, in Department of Mechanical Engineering and Applied Mechanics. 1993, University of Michigan: Michigan.
[24] Arvajeh, T., Ismail, F., Machining stability in high speed drilling—Part 2: Time domain simulation of a bending–torsional model and experimental validations. International Journal of Machine Tools & Manufacture. 2006. 46: p. 1573–1581.
[25] Roukema, J.C., Altintas, Y. , Generalized modeling of drilling vibrations. Part II: Chatter stability in frequency domain. International Journal of Machine Tools & Manufacture. 2007. 47: p. 1474-1485.
[26] Rincon, D.M., Usloy, A.J., Complex geometry, rotary inertia and gyroscopic moment effects on drill vibration. Journal of Sound and Vibration, 1994. 188: p. 701-715.
[27] Timoshenko, S., Young, D.H., Weaver , W., Vibration Problems in Engineering. 1974, New York: John Willey and Sons, Inc.
[28] Altintas, Y., Chan, P.K., In-process detection and suppression of chatter n milling. International Journal of Machine Tools & Manufacture, 1990. 32(3): p. 329-347.
[29] Inamura T., S., T., Stability analysis of cutting under varying spindle speed. Annals of the CIRP, 1974. 23(1): p. 119-120.
[30] Li, C.J., Ulsoy, A.G., Endres, W.J., The effect of spindle speed variation of chatter suppression in rotating tool machining. Materials Science Forum, 2006. 505-507: p. 859-864.
[31] Radulescu, R., Kapoor, S.G., DeVor, R.E., An investigatino of Variable Spindle Speed Face Milling for Tool-Work Structures With Complex Dynamics, Part2: Physical Explanations. Journal of Manufacturing Science and Engineering, 1997. 119: p. 273-279.
[32] Sastry, S., Kapoor, S.G., DeVor, R.E., Compensation of progressive radial run-out in face-milling by spindle speed variation. International Journal of Machine Tools & Manufacture, 2000. 40: p. 1121-1139.

[33] Smith, S., Tlusty, J, Update on high-speed milling dynamics. ASME Journal of Engineering for Industry, 1990. 112: p. 142-149.

[34] Smith, S., Delio, T., Sensor-based chatter detection and avoidance by spindle speed selection. ASME Journal of Dynamic Systems, Measurement, and Control, 1992. 114: p. 486-492.

[35] Takemura, T., Kitamura, T., Hoshi, T., Active suppression of chatter by programmed variation of spindle speed. Journal of the Japan Society of Precision Engineering, 1974. 42(11): p. 1049-1055.

[36] Tsao, T., McCarthy, M.W., Kapoor, S.G., A new approach to stability analysis of variable speed machining sistems. International Journal of Machine Tools & Manufacture, 1992. 33(6): p. 791-808.

[37] Smith, S., Tlusty, J., Efficient simulation programs for chatter in milling. Annals of the CIRP, 1993. 42(1): p. 463-466.

[38] Tarng, Y.S., Li, T.C., Detection and suppression of drilling chatter. Journal of Dynamical System Measurement and Control, 1994. 116: p. 729-734.

[39] Tarng, Y.S., Lee, E.C., A Critical Investigation Of The Phase Shift Between The Inner And Outer Modulation For The Control Of Machine Tool Chatter. International Journal Of Machine Tools And Manufacture, 1997. Vol 37(Is 12): p. 1661-1672.

[40] Park , S.S., Altintas Y., Movahhedy M., Receptance coupling for end mills. International Journal of Machine Tools & Manufacture, 2003. 43 p. 889–896.

[41] Cheng, C.-H., Arakere N., Schmitz T.L., G. Duncan S., An Approach For Micro End Mill Frequency Response Predictions. American Society of Mechanical Engineers, Manufacturing Engineering Division, MED, 2005. 16(2): p. 1139-1145.

[42] Mascardelli, B.A., Park, S.S., Freiheit, T., Substructure Coupling of Microend Mills to Aid in the Suppression of Chatter. Journal of Manufacturing Science and Engineering, 2008. 130(0110101-01101012).

[43] Chae, J., Park,S. S., Freiheit, T., Investigation of micro-cutting operations. International Journal of Machine Tools & Manufacture, 2006. 46 p. 313–332.

[44] Kim, B., Schmittdiel M.C., Degertekin F.L., Kurfess, T.R., Scanning grating interferometer for MEMS metrology. Journal of Manufacturing Science and Engineering, 2004. 126: p. 807-812.

[45] Vogler, M.P., Devor, R.E., Kapoor, S.G., Microstructure-level force prediction model for micro-milling of multi-phase materials. Journal of Manufacturing Science and Engineering, 2003. 125: p. 202-209.

[46] Weule, H., Huntrup, V., Tritschler, H., Micro-cutting of steel to meet new requirements in miniaturization. *Annals of CIRP,* 2001. 50: p. 61-64.

In: Metal Cutting: Research Advances
Editor: J. Paulo Davim pp. 213-238

ISBN: 978-1-60876-207-1

Chapter 8

ASSISTED MACHINING PROCESSES

A. Celaya*[*], *L.N. López de Lacalle and A. Lamikiz
Department of Mechanical Engineering –
University of the Basque Country
Faculty of Engineering of Bilbao;
Alameda de Urquijo s/n. 48013 Bilbao, Spain

ABSTRACT

The objective of the assisted machining techniques is to improve the cutting process by acting on the chip removal mechanism, i.e., the basic operation is a machining process, but an external action is added to improve and facilitate the chip formation. Currently, four techniques are most common due to their feasibility and industrial expectations: the Jet Assisted Machining (JAM) process, also known as High-Pressure Assisted Machining; Cryogenic Assisted Machining (CAM); Ultrasonic Assisted Machining (UAM); and Thermal Enhanced Machining (TEM), the latter two of which use plasma or laser as external action. In this chapter, we report the basis of these techniques and some good results showing the improvement in material removal rate and part quality.

1. INTRODUCTION

The term assisted machining refers to basic material removal processes (such as turning, drilling, milling, grinding, polishing, etc) in which an external aid is added to achieve radical improvement in the material removal rate or in the manufactured part quality. The assisted machining aids are applied by the same machine tools as conventional machining processes, but include accessory machine equipment and modifications to the cutting tool, tool holder, or machine components. The required accessories may be medium-pressure pumps, plasma welding equipment, ultrasonic transducers, electromagnetic vibrators, or other components.

[*] Corresponding author, Email: ainhoa.celaya@bi.ehu.es

Several technological and economical factors support the use of so-called assisted machining processes, including the following:

a. In very difficult-to-cut materials, tool life using conventional machining with defined cutting edge tools is very low, necessitating the use of grinding or electrodischarge machining. Grinding is a consolidated technique in which nearly every part usually requires a shaped wheel. It is also time consuming. However, for fragile or very hard materials like ceramics or sintered carbides, grinding is the best option. The use of turning, drilling, or milling allows more flexibility and reduces production time, but only if used along with an external aid.
b. In low machinability alloys, the use of assisted machining can improve metal removal rates two or three fold, or more. In this case, cost of conventional versus assisted machining techniques must be considered in order to justify the additional cost of assisted techniques. In fact, the cost of assistance (accessories, power consumption, and consumables) must be keep low with respect the main removal process (except in the case of laser assisted processes).
c. In parts where very high quality machining is required, assisted techniques can be a way to improve machining roughness and reduce the need for subsequent polishing.

Several techniques under development include the positive action of an external power to the basic machining process. A good example can be the Magnetic Aided Machining (MAM), for polishing, cleaning, deburring and burnishing metal parts. The force necessary for machining is produced by an adjustable magnetic field between the two magnetic poles within the working area. In the works [ELM 2007] [SIN 2002], the application of external electromotive force sources (e.g., magnetic field) when dry cutting shows that the magnetic field strengthens the cut surface with a thin secondary layer as the chip slides along the tool rake face. Also, a surface plastic deformation occurs when an external superimposed magnetic field reworks the material as it is sheared by a single-point cutting tool.

However, in this chapter we consider only the most basic, typical assisted defined cutting edge machining processes—drilling, turning, and milling.

2. High-Pressure Assisted Processes

A fluid injected into the tool-chip interface at high pressure significantly reduces cutting force, chip temperatures, tool-chip contact length, and heat transfer towards the tool. In addition, chip breaking and removal rates are improved [VIG 1996, EZU 2004, WER 1992, LOP 2000, NAN 2009]. The use of a pressurised jet to enhance cooling began in 1952 by Pigott and Colwell in conjunction with turning processes. In their experiments, tool life was increased seven or eight times, surface quality was improved, and the build-up edge was eliminated.

Vigneau [VIG 1996] applied a high-pressure water jet for turning titanium alloys and nickel-based superalloys. The waterjet pressures were within two ranges: high (from 5 to 20 MPa, with flows of 15-25 litres per minute) and very high (from 100 to 380 MPa, with flows of 2-3 litres per minute). Tool wear was seen to decrease significantly, from 5 MPa upward, in which circumstances very high pressures are needed to achieve reductions in wear, and the effects of water-jet-assisted machining became very uncertain. Thus, it is desirable to work with pressures ranging from 5 to 40 MPa, which can be achieved with portable industrial equipments that can be easily installed. The results recorded [VIG 1996] indicate that this technique might be applicable where carbide tools are used on titanium, but the results indicate that the technique is less favorable for nickel-based superalloys.

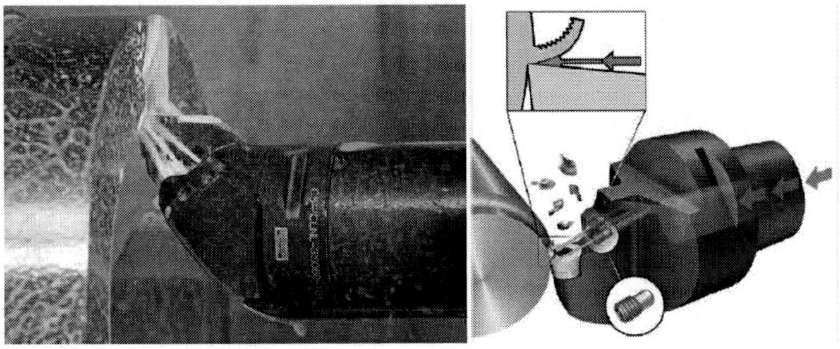

Figure 1. High-Pressure assisted turning system, Sandvik® Coro Turn HP.

Currently, some companies offer this technology to manufacturers of difficult-to-cut alloys. Their techniques require special toolholders and tools, and pumps with pressure ranging from 5 to 40 MPa, for example the ChipBlaster®, up to 13 MPa, and the Sandvik Coroturn HP®, with 7-8 MPa (see figure 1)

2.1. Test Results in Turning of Titanium Alloys

Various tests were carried out using a horizontal lathe. A special stainless-steel nozzle was used that conveyed a 2-KW waterjet (11MPa, 10 l/min) at a distance of 7-10 mm and at an angle of 10° on the rake face. When distances lay outside this range or when the angle of jet incidence was much different, the results were considerably inferior. In the case of Ti6Al4V (in annealed condition) chips tend to curl, so it is desirable that they be as fragmented as possible so that they leave the machine more easily. High-pressure lubrication is an aid to chip fragmentation. Figure 2 shows a titanium chip at a cutting speed of 140 m/min with the very typical serrated shape of this material, both with and without high-pressure lubrication. As may be observed, in the latter case the chip tends to break up.

The most satisfactory results (table 1) were in roughing operations, where in some cases the speed was doubled and the chip was severely fragmented. The pressure limitation of the equipment in use (11MPa maximum) determined the maximum cutting depth to 2 mm. A wear limit of 0.6 mm was adopted. Under conditions of high pressure in table 1, machining without a high-pressure jet would destroy the insert. With the jet, however, cutting forces diminish and temperature falls to such a degree that the cutting edge of the insert does not break or suffer plastic deformation, and fails only from continuous wear. Machining time was reduced by 50% and chip flow is doubled. After the success in these test, this technique were applied with a 20 MPa pump in an industrial case, applying 250 m/min in finishing.

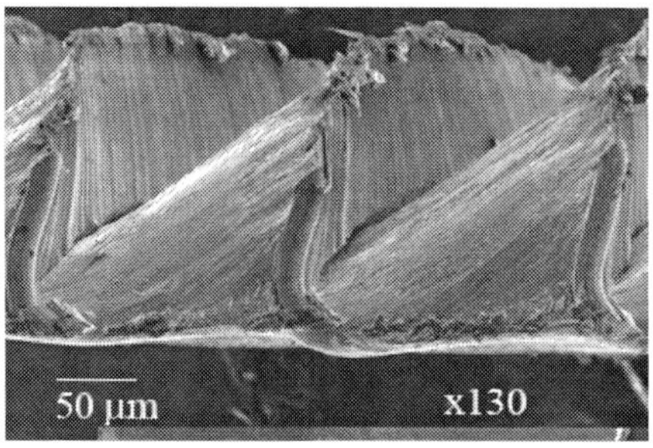

Figure 2. Left) Titanium chip at 140 m/min. Right) Titanium fragmented chip at the same speed with jet assistance.

Table 1. Conditions with conventional and with high pressure lubrication

	Conventional		High-pressure (11 MPa)	
Conditions	Roughing	Finishing	Roughing	Finishing
Vc (m/min)	70	80	140	120
a_p (mm)	2	0.5	2	0.5
f (mm./rev)	0.25	0.12	0.25	0.12
Tool life.(min)	9:30	44:08	3:10	25:50
Chip vol.(cm3)	332.5	211.84	221.67	186
flow(cm3/min)	35	4.80	72.7	7.20

2.2. High-Pressure Assisted Milling

In milling, the main requirement for the high pressure supply is the installation of a rotary joint to connect the external pump to the internal coaxial canal in the spindle, which must withstand more than 3 MPa. This problem has now been solved technically, and several successful applications are reported [CHI, 2008, KOV 1994] For example, Deublin Coolant Unions® offers a rotary joint for speeds up to 50,000 rpm and pressures up to 210 bar. This item is available with integral ball bearings and in bearingless configurations, suitable for coolant, MQL and air applications.

The mechanism of milling technology is such that every time an insert enters the material and removes a chip, the insert gets hot, but a moment later is rapidly cooled. This continuous thermal shock is even worse than the friction of the tool edge against the material. However, the injection of a high-pressure coolant directly on the edge-chip interface through holes made on the rake face decreases the tool edge temperature and the consequent thermal shock.

Another aid, the integral carbide endmill, has holes made along the tool flutes easily eject long chips from the cutting zone sliding on the internal flute faces. These carbide end mills are focused in the roughing of aluminium and titanium alloys.

The patent [KOV 1994] for this endmill claims as follows: "in an apparatus for machining of a workpiece, said apparatus including a drive motor and a rotary cutting tool, the improvement comprising: a high-pressure intensifier pump providing cutting fluid at a pressure greater than 30,000 psi; and conduit means for delivering and directing said high-pressure cutting fluid from said pump through said rotary cutting tool directly into an interface defined by said rotary cutting tool and a chip being cut from said workpiece; said conduit means including a main flow channel, a distribution manifold and a plurality of radially directed feed channels all formed in said rotary cutting tool and providing serial communication for the passage of cutting fluid." This language describes the basis of the high-pressure coolant-assisted milling, but in this case with a very high pressure (approximately 200 MPa); whereas the pressure in typical pumps installed in high performance machining centres is between 2 and 6 MPa.

2.3. High-Pressure Assisted Drilling

Internal lubrication is traditionally employed in the drilling of deep holes, with pressures from 0.17 up to 0.5 MPa. Such pressures, together with high lubricant flow, make it possible to remove chips from the cutting zone, but only slightly improved parameters can be used. Studies [LIN 1995] show that lubricant injection at pressures between 10.3 and 20.6 MPa inside the twist drill facilitates the successful machining of low-carbon steel and aluminium 380, to achieve a tool life of at least ten times longer. More importantly, when the cutting speed was doubled, the same number of holes were produced in half the time, with the same drill, .

A new twist drill with four holes for internal cooling (MNS by Mitsubishi®), (see figure 3) was designed especially for aluminium alloys. The drill uses 4 coolant holes, 2 positioned on the outer cutting edge and 2 on the inner cutting edge. The positioning of the holes was carefully studied using flow analysis to ensure that the chisel point and the cutting edge receive sufficient coolant to prevent weld build up during deep, single-step drilling.

Figure 3. MNS drill, with four holes for coolant, designed by Mitsubishi®.

The need for high pressure and high flow in gun drilling [SIH 2008] derives from the great amount of heat generated during the process. The problem with standard low-pressure coolant is that so much heat is produced during the process, the coolant boils before it can reach the chip tool interface at the gun drill tip. Then the hot steam forms a barrier that low-pressure coolant cannot penetrate. Consequently, effective cooling does not occur and little real lubrication is provided. Unfortunately, the vapour barrier that forms is not powerful enough to keep chips from falling back into the chip/tool interface and causing damage. Properly applied, high-pressure and high-volume coolant prevents this vapour barrier from forming by causing a localized pressure increase. Then, so much liquid is forced into the cutting zone that heat is removed and no vapour can form because of the pressurization.

Recent studies [SIH 2008] report results of gun drilling of carbon steels. The cutting parameters were as follows: guide hole length 12 mm and diameter 8mm; gun drilled hole depth 187 mm; spindle speed: 3,200 rpm (v_c 80 m/min); feed for the first 50 mm of the hole was 0.025 mm/r and after 50 mm was 0.03 mm/r; coolant pressure: 6 MPa (see figure 4).

Figure 4. Aspect of the gun drill used in [SIH 2008] after 105m, with coolant pressure of 6 MPa. Left) side view. Right) front view.

2.3.1. Test Results in Titanium Alloys

Our tests on Ti6Al4V alloy (in aged condition, hardness 42 HRC) were performed with a pump that supplied a jet of 10 litters per minute at 11 MPa, with the coolant injected into drilling tools of 6.4 mm diameter, with two or three teeth and internal-cooling channels of 0.5-1 mm diameter.

The first tested speed was *Vc* 120 m/min, i.e., three times the rate commonly used in industry, while the feed per tooth was 0.05 mm (*F 800* mm/min). Tool life was only thirty holes, a third of the minimum considered acceptable if tool changes are to be minimized. Accordingly, cutting conditions were introduced that ensured a minimum of 100 holes per drill. Drills were of hard metal coated with TiN, with two flutes at *Vc* 60 m/min. Feed per tooth was 0.05 mm. In these conditions, 252 holes of 7 mm depth (total length was 1,746 mm) were drilled, with flank wear greatly reduced to be lower than 0.26 mm as shown in figure 5.

Figure 5. Flank wear alter 252 holes at Vc 60m/min y and f 0.1 mm/rev.

2.3.2. Test Results in Stainless Steels

Stainless steels are Fe-C alloys with more than 11% of Cr. Within this family, austenitic[1] alloys are the most resistant to corrosion. Among the austenitic steels, the AISI 304 grade (EN1.4307) is often used for its low corrosion and high mechanical properties (ultimate tensile strength is approximately 590 MPa). This alloy contains 18% Cr and 8% Ni, with maximum carbon composition of 0.08%. It is used in chemical equipment, cooking equipment, cryogenic vessels, evaporators, feedwater tubing, food processing equipment, hospital surgical equipment, kitchen sinks, marine equipment, pressure vessels, valves, and shipping drums. The machinability of this material is very low; therefore, high pressure drilling was tested as an option to improve machinability.

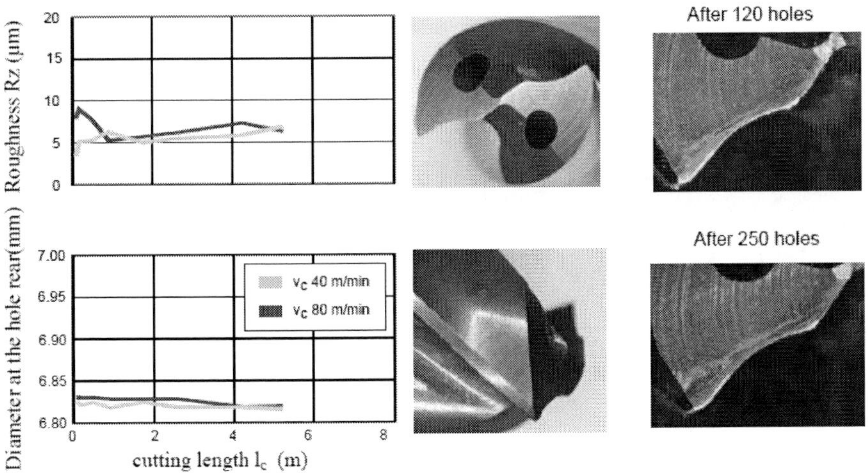

Figure 6. Performance of the in 304, at *Vc* 40-80 m/min and *f* 0.13 mm, coaxial internal coolant at 60 bars pressure.

High pressure up to 60 bars was applied, using emulsion coolant (4-5% of mineral oil in water). The drill in use presented a point angle 140°, carbide K20 with 10% Co, coated by a combined TiN–TiAlN with an external MoS2 layer. Figure 6 shows the performance of the two drills tested to *Vc* 40 m/min in AISI 304. The total hole length (each hole is 21 mm),

1 Austenitic refers to the application of a nonmagnetic solid solution of ferric carbide or carbon in iron, used in making corrosion-resistant steel. [After Sir William Chandler Roberts-Austen].

mean roughness, and entrance and bottom hole diameters are shown. The tool life criterion was the tool breakage. The drill made more than 1,600 holes, with good roughness (*Ra* 1μm) and hole dimension (diameter errors lower than 40 μm).

3. Cold and Cryogenic Machining

When a cold fluid is used as the main coolant injected into the cutting zone, three techniques are used: cold air machining, CO2-assisted machining, and liquid-nitrogen-assisted machining.

3.1. Cold Air

Cold air, (below 40°C , is produced from compressed air (5-8 bars) by using the vortex principle, and afterwards it is injected onto the cutting point. The vortex principle works in the following manner (see figure 7 left): the air is ejected tangentially into the vortex spin chamber at up to 1,000,000 rpms. The air revolves towards the hot end of the tube, where a portion of it escapes thorough a control valve. The remaining air, still spinning, is forced back to the tube. The inner airstream gives off kinetic energy to the outer stream and exits the vortex tube as cold air. The outer stream exists at the opposite end as hot air.

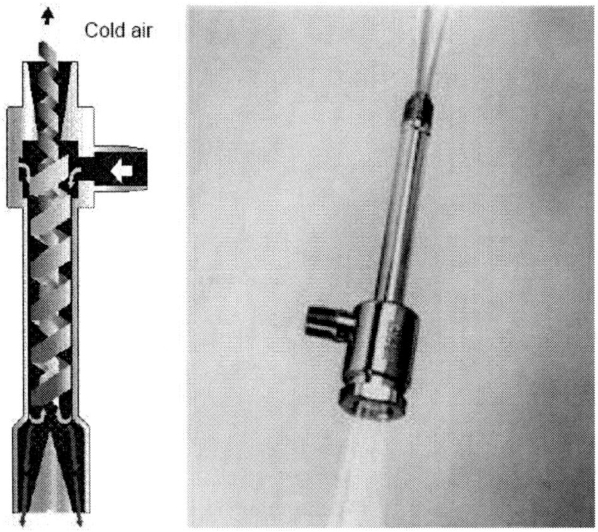

Figure 7. Vortex principle, showing vortex spin chamber and nozzle, by Exair®.

Turning experiments with treated steel at 32 HRC, using a Cermet tool TNMG 160408 with 300 m/min and a feed of 0.4 mm/rev, resulted in a tool life of 6 min with cold air; whereas only 2.5 min were achieved in dry conditions. In turning, drilling, and sawing of plastics, wood, and non-metal materials, the use of cold air increases tool life without damage to the materials themselves. Cold air is cheap and can be applied for each machining process along with other assistance techniques to reduce thermal damage.

3.2. CO_2 Assisted Machining

Carbon dioxide (CO_2) has been studied for the past 60 years as a machining fluid with regard to both lubricating and cooling qualities, in some cases with a solvent in the mist. In the earliest applications, CO_2 gas was injected into the atmosphere of the cutting zone, removing oxygen and reducing tool and material oxidation. It has been observed that when liquid CO_2 is sprayed, it retards crater wear in the case of carbide tools, while with HHS, a protective built up edge (BUE) forms on the HSS tools. The usual nozzle for spraying CO_2 is a Coanda-type nozzle [JAC, 2009]; the Coanda effect is the tendency of a fluid jet to stay attached to an adjacent curved surface that is very well shaped and polished.

CO_2 machining sprays can dissolve solid coolant particles or soy oil, penetrating the tool/chip interface easily [JAC 2008]. Also, CO_2 exhibits very strong hydrocarbon solubility, 600 percent higher solubility in oils as compared to that of compressed air.

A new cooling lubrication technology has been developed, called advanced minimum quantity cooling lubrication (AMQCL®, by Cool Clean Technologies, Inc.®). This method combines a source of propellant gas (i.e., compressed air), lubrication additives (i.e., soy oil), and solid and/or gaseous CO_2 (i.e., coolant). The combination of sublimating solid coolant and subcooled lubricants (mass) with near-sonic air flow (velocity between 50-400 m/s) creates significant penetration power (*Force = mass x velocity*), which allows the coolant and lubricant particles to penetrate deeply into a cutting zone.

Thus, for example, CO_2 cooling provides an economical method for turning carbide rods using PCD tooling, instead of grinding, as reported by Cool Clean Technologies.

3.3. Liquid Nitrogen as Coolant

Liquid nitrogen (LN2) is the most commonly used fluid in cryogenics. It is produced industrially by fractional distillation of liquid air, being so-called LN2. Nitrogen melts at −210.01 °C and boils at −198.79 °C. Nitrogen is a safe, noncombustible, and noncorrosive gas. Colourless, odourless, and tasteless, nitrogen is often used as an "inert" gas due to its nonreactive nature with many materials. The liquid nitrogen easily evaporates and returns back into the atmosphere, leaving no residue and with null environmental impact and eliminating disposal costs.

Cryogenic cooling is applied in four approaches, as compiled by Yilidiz and Nalbanta [YIL, 2008] as follows:

1. With cryogenic pre-cooling of the workpiece [BHA 1993][HON 1999], the aim is to cool workpiece or chip to change the properties of workpiece material from ductile to brittle. Bhattacharyya et al. [BHA 1993] tested two methods. In first method, they dipped the test rod in liquid nitrogen, and in second method the liquid nitrogen was poured continually onto the testpiece. The main conclusion of these authors is that chip breaking in machining AISI1008 can be improved only by combining chip material embrittlement through the use of LN2 with an adequate tool geometry and careful tool chipbreaker positioning.

2. With indirect cryogenic cooling or *cryogenic tool back cooling*, also known as *conductive remote cooling,* the aim is to improve heat conduction from the tool cutting by injecting liquid nitrogen to a chamber located at the back face of the tool or toolholder [EVA 1991, WAN 1996, WAN 1997, WAN 2000].
3. Cryogenic jet cooling by injection of LN2 into the cutting zone by (a) general flooding or (b) to the cutting tool tool–chip and tool–work interfaces by mini-nozzles. In this method, coolant holes are made in the tool rake face and/or the tool flank face [HON 2001][HON 1996]. With option (b) cooling power is not wasted on unnecessary areas and a lower flood is required; this localised cryogenic cooling reduces the tool face temperature, enhances its hardness, and so reduces its wear rate. This approach also embrittles the chip by cold temperature as in the case 1, and allows the bending of the chips with the help of a chipbreaker placed at exactly the right point on the rake face. On the other hand, the cryogenic machining approach eliminates the BUE problem, thereby achieving better surface quality.
4. Cryogenic treatment of cutting tools is very different from the three other methods. During this process, materials are subjected to deep cryogenic temperatures near -180°C, affecting the whole mass of the part [BAL 2008]. A conventional cryogenic treatment typically lasts two to three days. It is important to make a distinction between deep cryogenic treatments and subzero (sallow cryogenic) treatments, made at temperatures of around -80°C and sometimes used on certain steel parts after quenching for dimensional stabilisation. The cryogenic treatment process can be used to improve the wear and toughness properties of a wide range of materials, including steels (including stainless steels), aluminium, copper, carbide, and ceramics; even some polymers can be improved with this treatment. The improvement of mechanical properties in the case of steels can be ascribed to the complete transformation of the retained austenite into martensite, fine dispersed carbides precipitation, and the removal of residual stresses. This method therefore, is applied also in HSS cutting tools, with different results. However, the action of cryo treatments on sintered carbide is more difficult to study.
5. In our research, we had good results in the improvement of tool life (cryogenic treatment applied by Cryobest®) in several applications: treatment of drills for stainless steels (AISI 304 and 316) with and minimal quantities of lubricant, internally through the spindle, achieved in some cases 50% more cutting length at a cutting speed of 40 m/min and feed 0.05 mm/rev, in treatment of milling tools for routing of carbon fibre reinforced composites and milling of wood and gear hobbing.

Table 2. Improvement in machining times with ICEFLY® (approximate values)

Operation	Time with emulsion	Time with ICEFLY
Step 1: Rough turn tailstock end of roll	80	59
Step 2: Rough turn headstock end of roll	76	51
Step 3: Rough turn ring groove	21	12
Step 4: Finish turn headstock end of roll	46	20
Step 5: Finish turn ring groove	13	8
Step 6: Finish turn tailstock end of roll	48	20
Total time	284	170

The third approach above is the most frequently used for example, as , in [PAUL 2001] where AISI 1060 steel is turned. On the other hand, a commercial system on the market is Icefly®, which reports successful applications in titanium alloys and hardened steels (see table 2 for one example).

4. Vibration-Assisted Machining

Two common techniques for vibration-assisted machining are the Modulation-Assisted Machining (low frequency and large amplitude) and the Ultrasonic Assisted Machining (ultrasonic frequency and only a few microns amplitude). Both are described below.

4.1. Modulation-Assisted Machining

Modulation-Assisted Machining applies a controlled low-frequency modulation to the metal cutting process [M4S 2009][MAN 2007]. The frequency ranges up to 1,000Hz with amplitudes up to 500 micrometers. The controlled modulation may be applied on the cutting tool or workpiece in the direction of tool-feed, cutting velocity, or any combination offering major beneficial effects on the mechanics of machining. Chip breakage in machining processes (e.g., turning, boring, drilling) was found to be a consequence of the chip thickness reaching zero during each cycle of modulation under certain modulation conditions. Good results are reported in drilling [CHH 2002] of metal, fragile materials [BRE 2008], and composites using this method.

4.2. Ultrasonic Assisted Machining

The principle of ultrasonic assisted machining is adding high frequency (16-40KHz) and low vibration amplitude (3-30μm) to the tool or workpiece to improve the fabrication process. Ultrasonic assisted machining was first used in the 1960s [SKE 1969], but recently there has been a growing interest and tendency to employ this technique for both high precision machining applications and for difficult-to-cut material machining. This technology demonstrates a range of benefits in machining difficult-to-cut alloys and brittle materials. Nevertheless, ultrasonic assisted machining is still not widely used in industry.

In this method, an ultrasonic vibration is emitted by a transducer driven by a fixed-frequency generator. A booster attached to the transducer is used to intensify the amplitude of the vibration.

The ultrasonic assistance is applied to several different fabrication methods, with the expectation of several benefits on each of them, but most of the research on this method is focused on turning and drilling processes.

4.2.1. Ultrasonic-Assisted Turning

In ultrasonic-assisted turning (UAT), the vibration is applied to the cutting edge of the tool, in one of three possible directions: the cutting speed direction, the feed direction, and a radial direction. For the ultrasonic vibration effect to become noticeable, the vibration velocity must be an order of magnitude higher than the velocity of cutting in the direction of the vibration. In this way, during each vibration cycle there is a separation between the workpiece and the tool, which promotes a pumping action that decreases friction [ZHO 2002]. Hence, together the reduction in the cutting heat and the adherence of the work material to the cutting edge of the tool increase the accuracy of the workpieces obtained [MUR 1998].

Ultrasonic assisted turning has been widely studied and research shows the advantages of this technique in terms of cutting forces, surface finish, and tool wear in several materials. Conventional machining of nickel and titanium based alloys presents high tool temperatures and fast cutting edge wear, and studies report improvements in the machining of this type of alloys using ultrasonic assisted machining. In the machining of Inconel 718, high-frequency ultrasonic vibration of the cutting tool allows a reduction of cutting forces of 25-30% [AHM 2007, HSU 2008, NAT 2008], surface finish improvement of 15-85% [HSU 2008, NAT 2008], roundness improvements up to 40-50% [BAB 2003], as well as tool wear improvements of 25% [NAT 2008]. Differences in the material microstructure and hardness also are reported [AHM 2006]. In γ-TiAl turning, cutting forces reduction of 12%, and high variability in tool life values also have been reported [SHA 2001]. In other studies [BAB 2003, CEL 2009] improvements in roughness are found in the case of vibration assisted turning of mild steels. New tests on mild steel turning also were performed with vibration in feed direction. A commercial piezoelectric transducer Mastersonic MSG-2000 (17.5-24 kHz) was used to generate the ultrasonic vibration and an exponential booster was designed to amplify the vibration as shown in figure 8. The resonant frequency employed was 21.44 kHz, with an amplitude of 6 μm in the axial direction of the tool. The experimental results show an improvement in surface roughness up to 60% in the most favourable case as compared with conventional turning, as shown in table 3. However, the experimental roughnesses obtained without vibration are not in accordance with theoretical values due to low cutting speed and feed.

The topography of conventional and vibration-assisted turned surfaces is shown in figure 9. These results show a reduction in peaks when vibration is applied, although there are some areas with more peak amplitude. So, in order to get a reduction of surface roughness, it appears necessary to maintain a vibration of constant amplitude during the cutting process.

The ultrasonic vibration in the cutting of low alloy steels (DF2) indicates a reduction in cutting forces up to approximately 50%, and produces comparatively smaller chips and a better surface finish compared to conventional turning [NAT 2007]. Precision turning of stainless steel has also been tested by applying ultrasonic vibration to a diamond tool [MOR 1991] in order to reduce the tool wear. Another application of this technique is in the machining of brittle materials. For producing high quality surfaces on brittle materials, it is important to achieve a ductile deformation mode. Ultrasonic assisted machining can increase the critical depth of cut at which ductile-cutting can be obtained. For example, in fused silica machining, a higher depth is achieved in ductile cutting with a depth of cut up to 1.5μm, where conventional diamond cutting could reach only 0.5μm [ZHO 2002]. On the other hand, tool wear was also reduced by applying ultrasonic vibration, in comparison to conventional turning [ZHO 2006]. In short, there are several advantages to introducing ultrasonic vibration in the tool, including reduction of cutting forces, shorter chips, surface roughness improvement, tool wear reduction, roundness improvement, and, as a result, an increase of the machinability of difficult-to-cut materials.

Figure 8: Experimental set-up for ultrasonic assisted turning.

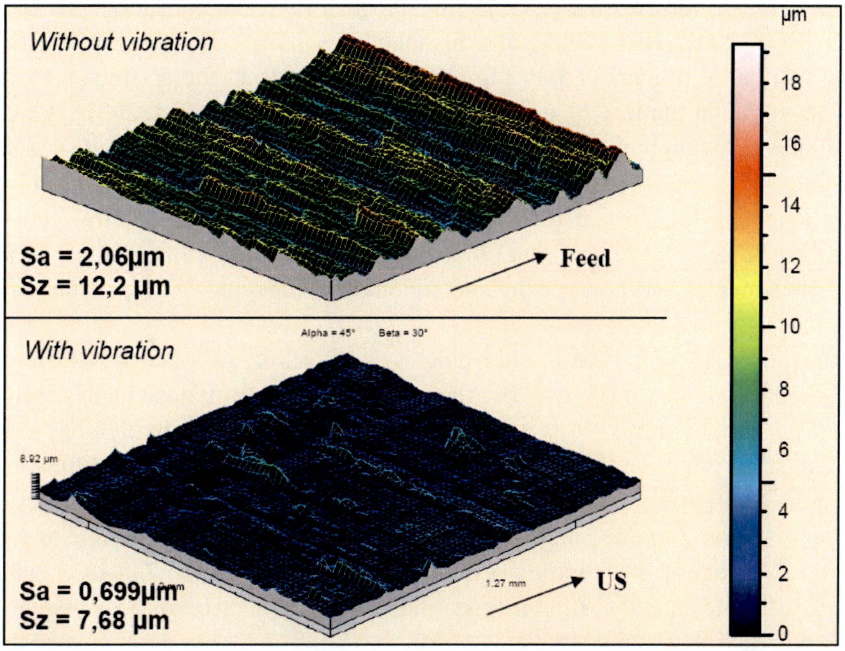

Figure 9. Surface topography without and with vibration assistance in mild steel turning, f=0,05mm/r, a_p=0,3mm and Vc=100m/min.

Table 3. Cutting conditions when vibration is applied in the feed direction and corresponding surface characteristics

Cutting parameter			Without vibration		With vibration		Roughness improvement (%)	
f (mm/rev)	a_p (mm)	Vc(m/min)	Ra(μm)	Rz(μm)	Ra(μm)	Rz(μm)	Ra(μm)	Rz(μm)
0.05	0.1	100	3.22	17.73	2.63	13.87	18	22
0.05	0.2	100	2.99	16.20	1.75	12.67	42	22
0.05	0.3	100	1.78	9.75	0.60	6.90	66	29
0.5	0.4	100	1.28	8.75	1.03	7.95	19	9
0.05	0.2	75	3.27	18.10	2.77	15.27	15	16
0.05	0.2	100	2.99	16.20	1.75	12.67	42	22
0.05	0.2	125	2.37	12.33	1.63	10.84	31	12
0.1	0.2	50	3.74	20.40	3.42	17.83	9	13
0.1	0.2	75	3.46	19.30	3.31	17.47	5	9
0.1	0.2	100	2.63	14.37	2.34	12.97	11	10
0.1	0.2	125	1.60	9.17	1.20	6.90	25	25

4.2.2. Ultrasonic-Assisted Drilling

Although ultrasonic-assisted drilling has been less studied than ultrasonic-assisted turning, several authors report successful results in drilling aluminium, steel- and nickel-based alloys with vibration assisted drilling. The principle of the ultrasonic assisted drilling is to add the high frequency vibration along the tool feed direction. This technique has been proven to be efficient for improving the machinability of several aeronautic materials, such as aluminium [SUZ 2005, BRE 2007] and nickel alloys [AZA 2008]. Chip breaking, burr generation, workpiece roughness, tool life or torque, and cutting forces are some parameters studied with vibration applied in drilling processes [ZHA 1994, LIA 2007]. In vibration-assisted drilling of aluminium alloys, force reductions of 30-50% are reported as well as chip fragmentation [NEU 2004]. Takeya and Kato [TAK 1991] have experimentally shown that it is possible to obtain a burr size reduction in aluminium drilling with ultrasonic vibration assistance. Cha [Cha 2005] reports that ultrasonic-assisted drilling of aluminium produces an average of 70% smaller burr height than conventional drilling with average reductions of 20% in thrust force.

Experimental results reported in ultrasonic-assisted drilling of Inconel 718 show that the chip size is reduced [LIA 2007], whereas the thrust force is also reduced and the tool life is prolonged. The authors of this study conclude that the amplitude of vibration has more effect on tool life than the vibration frequency, and therefore in vibration assisted drilling of Inconel 718 it is more appropriate to use an smaller amplitude with a higher frequency. With a combination of 4 μm vibration amplitude and 31.8 kHz, the authors were able to extend the tool life by 2.7 times. In vibration assisted drilling of Inconel 738-LC, comparative experiments of hole quality demonstrate improvement in workpiece average surface roughness and circularity of more than 60% [AZA 2007]. When ultrasonic-assisted drilling of titanium alloys is applied, the thrust force decreased by 10–20% on average, this force depends on the vibration amplitude as shown in figure 10 [PUJ 2009].

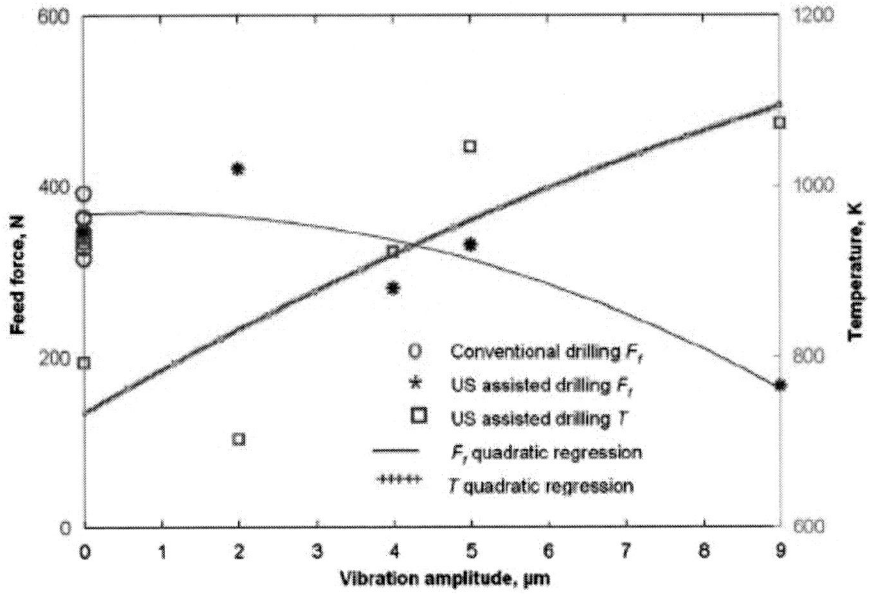

Figure 10. Relation between thrust force, temperature, and vibration amplitude when machining at 2,000rpm and f 200mm/min [PUJ 2009].

4.2.3. Ultrasonic Elliptical Cutting

Ultrasonic elliptical vibration cutting consists of applying a synchronised two directional vibration to the cutting edge in the plane, including the cutting direction and the chip flow direction. This technology was first presented in the 1990s [SHA 1994] and is used mainly in the field of the ultra-precision cutting. In elliptical vibration cutting, the tool motion has a velocity component in the chip flow direction in each cycle after penetrating into the workpiece. Shamoto and Moriwaki developed the resonant elliptical system for ultrasonic cutting shown in figure 11 [MOR 1995, SHA 1999]. The elliptical path of the cutting tool was generated with two orthogonally assembled piezoelectric actuators with some phase shift; the vibration amplitude is amplified by the step horn. The vibrator is attached at the two nodal points so as not to disturb the vibration. [SHA 1994] confirmed by observation with a scanning electron microscope that in this process, the frictional force between the tool rake and chip is reduced and also it is reversed, assisting the chip to flow out. In that way, the chip thickness and the cutting forces are reduced significantly, as compared with the conventional cutting and one dimension vibration cutting, by applying the elliptical vibration. In [MOR 1995], the authors also conclude that elliptical vibration cutting suppresses the formation of burrs and improves the geometrical shape accuracy.

Ultrasonic elliptical vibration also can be applied to ultraprecision diamond cutting of hardened tool steels [SHA 1999, SHA 2002]. While conventional cutting methods cannot be used in diamond cutting of steel because of the high wear of the diamond tool, elliptical vibration cutting presents as a suitable method for diamond cutting of steel due the lowest tool wear and chipping, and also because of the surface quality it renders. In [SHA 2002] the authors present a new control system to stabilize elliptical vibration. In previous designs, the modal frequencies of the two vibration directions were very close, causing "crosstalk," and as result, distorting the tool path. In the new design, however, a feedback control system is used to keep constant both the vibration amplitude and the phase difference, and the average resonant frequency is chassed by this control system. As a result, shape accuracy when turning a stainless steel mirror is improved from 2μm to 0.2μm when this control system is activated.

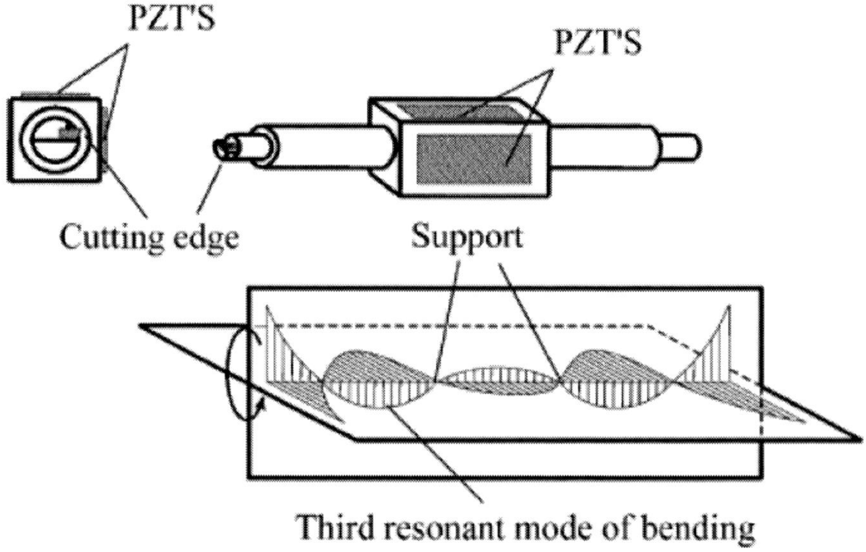

Figure 11. Ultrasonic elliptical vibrator developed by Shamoto and Moriwaki [MOR 1995, SHA 1999].

The theoretical study of ultrasonic elliptical cutting indicates that the pushing stress and bending stress of the deformation zone on the workpiece edge in burr formation is reduced because of the separating characteristic between the tool rake face and the chip [CHU 2005]. In aluminium (52S), the height of burrs generated in conventional cutting, of about 110μm, is reduced to 15μm with conventional vibration cutting and to 1μm with elliptical vibration cutting.

5. Thermal-Enhanced Machining

The local application of heat to the cutting zone during the machining of aeronautical alloys is addressed in several publications [WEI 1994, NOV 1997, KIT, 1990, LES 2001, KON 1990, LOP2004]. This technique is known as Thermal Enhanced Machining (TEM), in which the cutting process takes place at temperatures ranging from 400° to 1000°C. As a result, the shear strength of the material is significantly lower than at room temperature strength, resulting in considerably lower cutting forces.

The technique is economically feasible only when the machinability of the material being processed is very low and therefore the cutting parameters (depth of cut, feed per tooth, and cutting speed) must be very low. This is the case with heat-resistant alloys, nickel- and cobalt-base alloys, the ceramics. The former materials maintain their mechanical properties as well as an excellent corrosion resistance even over 600°C. Because of these properties, these alloys are used in the manufacturing of turbine components of both commercial and military airplane engines. Ceramics are refractory materials that retain their strength at very high temperatures, being very hard and brittle.

Laser [KON 1990, LESO 1994, ROZ 2000, THO 1999, CHA 2007, LEI 2001, TIA 2006, YAN 2008] and plasma [NOV 1997, KIT, 1990, LES 2001, LOP 2004] also can be used as an energy sources for TEM techniques, as discussed below.

5.1. Laser-Assisted Machining

The laser-assisted machining (LAM) technique is characterized by the use of a high-energy concentration on a small spot—less than 3 mm. However, the cost of the laser resonator and accessories is very high when compared to the machine tool on which it will be used. The application of laser is adequate for turning, because laser can be focused just ahead of the cutting tool in a very precise manner. However, in the patent for laser machining equipment [THO 1999], the inventors recommend "directing at least one laser beam onto said workpiece immediately upstream of said milling tool in said longitudinal direction during a milling operation so as to heat at least one zone of the workpiece parallel to a predetermined cutting profile; and…". Application to ceramic milling is also shown by Yang et al. [YAN 2008]. However, the difficulty in orientating the laser beam when milling complex shapes is a serious drawback, as it requires a rotary position device to be able to position the laser just beyond the milling tool when the toolpath differs from a straight line.

So far, laser-assisted machining of ceramics is the most successful. In this application, the laser heats ceramic material to more than 1,000 ° C, making it softer and more ductile. The soft, red-hot portion of the ceramic affected by laser is removed with Polycrystalline Cubic Boron Nitride (PCBN) tools. One of the first successful use of LAM to machine ceramic material was described by Konig et al. [KON 1990] who obtained continuous chips for hot-pressed silicon nitride when its temperature exceeded 1200 ° C. Some other good results in both Al2O3 [CHA 2007] and Si3N4 [LEI, 2001] [TIA 2006] are documented. In the latter ceramic type, a cutting speed of 25m/min, depth of cut of 1 mm, and feed of 0.02 mm/rev were used to achieve finishing of 0.3 Ra μm. These values are typical, with laser power ranging between 200 and 400 W. Figure 12 shows the surface of a ceramic cylinder heated by a laser so that the brittle cylinder can be machined with a lathe without cracking or splintering.

Laser-assisted micro-end milling (μ-LAM), now in research at the LAMSML (University of Wisconsin), is a hybrid machining method that uses a laser to locally heat a workpiece prior to material removal with a micro-end mill. This project investigates the effect of laser preheating using micro-end mills from 25-μm to 400-μm diameter.

5.2. Plasma-Assisted Processes

For plasma-assisted milling (PAM) processes, the cost of the generator is much lower, but the spot size is larger, 4–5 mm and the energy density on the heated spot is smaller. The larger spot becomes an advantage in the case of the milling process since . using a larger spot enables the heating of the spot just ahead of the endmill and involves the entire radial depth of the cut, up to 5 mm. If the whole chip section were heated using a smaller sized spot, the torch would require rotation at the same rotational speed as the tool, which would considerably increase operational problems of this technology.

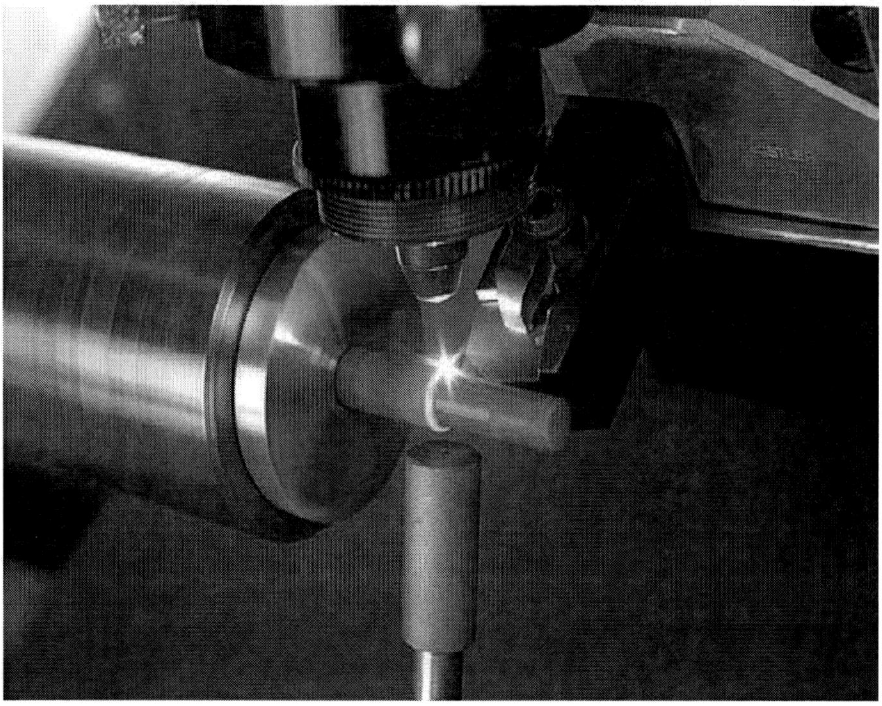

Figure 12. Laser-assisted turning of a ceramic cylinder, at Purdue University.

Figure 13 shows the plasma-machine system layout [LOP 2004]. The nozzle is focused at a distance of about 8-10 mm ahead of the milling tool in the direction of the feed; this distance is high enough to prevent the tool body from being directly affected by the plasma jet. The nozzle is placed at a height of 5-6 mm over the workpiece; thus, the electric arc responsible for the ionization of the channel (known as transferred arc) can be activated. The diameter of the heated spot is about 4-5 mm. The spot must be located exactly on the material to be removed and must avoid zones of the workpiece previously machined. The plasma generator is based in an electrical arc, but a High Velocity Oxy-Fuel (HVOF) also may be used.

The temperature of the ionized gas (Argon) is well over 15,000 K. The velocity of the plasma jet about 5 mm away from nozzle is 400 m/s; this is the speed at which the plasma jet impacts the surface of the material. The energy is transferred by convection to the workpiece [PFE 1990] and heats the work surface at temperatures ranging from 400 to 1,000°C. The heat distribution on the surface is supposed to follow a Gaussian distribution. Experimental measurements carried out using an infrared camera show that there are two clearly identified zones inside the spot heated by the plasma jet: a circle of maximum heating, where the temperature is between 500°C and 1,000°C, and the adjacent zone, where the heat is transferred by conduction to the workpiece.

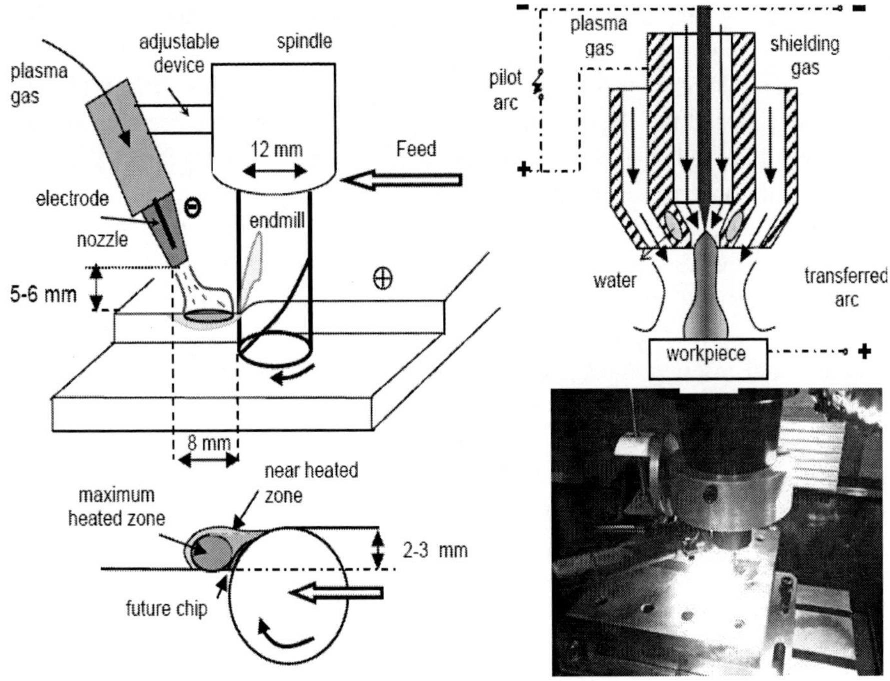

Figure 13. Left) Main components of plasma assisted milling system. Lower left) top view of the plasma spot and the milling tool. Right) Plasma nozzle. Lower right) General aspect during plasma milling.

5.2.1. Plasma-Assisted Milling Results in Nickel Alloys

Plasma-assisted milling (PAM) tests on Inconel 718 were carried out using the CC670 ceramic inserts, by Sandvik®. These Al2O3 ceramics, reinforced with CSi whiskers, are stable up to approximately 2,000°C. Thus, a higher productivity should be obtained from the application of the external heating provided by PAM.

A 50-mm diameter tool with 3 ceramic round inserts of 12-mm diameter was used. The inserts themselves are neutral and have no inherent rake and clearance angles. The axial and radial effective rake angles of the inserts, and the bolt on the toolholder, are –12° and –23°, respectively. The insert manufacturer recommends the use of 3.175 mm axial depth of cut (i.e., ¼ of the diameter). The radial depth of cut, a_e 4mm, was determined by the plasma spot size (≈4mm). The actual cutting speed was 942.5 m/min, which corresponds to the maximum spindle rotational speed (6,000 rpm). The feed per tooth was that recommended by the manufacturer, *fz* 0.054 mm/tooth, which means a linear feed of 972 mm/min. All these values were kept constant during all the tests, except the plasma intensity, which was varied in order to obtain different degrees of heating. Afterwards, the tool wear induced by each value of the plasma intensity was been measured. As heating increases, flank wear decreases. However, it is not possible to increase the plasma intensity to more than 110 A, since higher values are related to the formation of a weld bead. Therefore, the plasma intensity parameter was set at 110 A in order to minimize tool wear and avoid work surface melting.

The temperature increase of Inconel 718 when working at *F* 972 mm/min and *I* 110 A was measured on the surface between 850°C and 900°C just before machining, and just after an axial depth of cut of 3.175 mm was removed it was 400°C. Therefore, the temperature of the layer removed is between 400°C and 900°C, which is high enough to achieve the reduction of the cutting forces.

Figure 14 shows tool wear as a function of the machined length (*l*), and therefore as a function of time (*l/F*) and the chip volume ($l{*}a_p{*}a_e$). Thus, the time required to reach the same value of tool wear may be analyzed. In conventional milling, a flank wear V_B 0.4 is observed after 16.85 seconds, whereas if the material is heated with a plasma jet at 110 A, the same wear is reached after 33.7 seconds (twice the previous value). The same effect is observed for a wear of 0.67 mm, the tool life being 55 seconds in conventional milling and more than 100 seconds in PAM. Conventional milling effects a notch wear larger (in the zone of the insert in contact with the workpiece surface) than the flank wear (in the lower part of the insert). At 110A, both mechanisms exhibit a great reduction and deep notching nearly disappears.

PAM tests performed on a Kistler dynamometer with a plasma intensity of 110A,show a reduction of the cutting forces of up to 45% when compared with conventional milling.

These results pertain to the industrial interest in PAM, since tool life increases. Material integrity also was analyzed. Inconel 718 was analyzed in three different states: as supplied, after conventional machining, and finally after PAM. None of these materials presented carbide precipitation or the existence of new metallurgical phases.

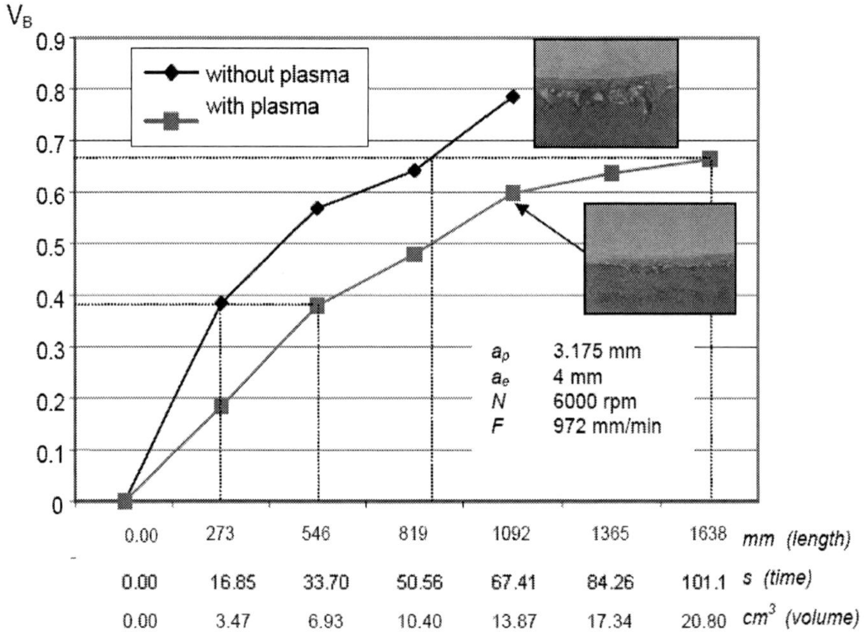

Figure 14. Flank wear as a function of machined length (and therefore machining time and chip volume). Dotted lines show tool life with and without PAM for the same value of flank wear.

After machining, the Inconel 718 work surface exhibited important strain hardening, which is inherent in materials with austenitic structure. This effect is even more obvious when round inserts with negative axial rake angle are used, since cutting then is very aggressive. However, in the case of conventional milling, the affected zone was 180 μm deep, with 435 $Hv_{0.2}$ (44 HRC) measured at 22 μm below the surface, and 405 $Hv_{0.2}$ (41.3 HRC) at 82 μm;

the hardness of the base material was 235 $Hv_{0.2}$ (20.5 HRC). In the case of PAM, the affected zone was very irregular, ranging from 80 to 100 μm deep, and the microhardness was between 340 and 399 $Hv_{0.2}$ (34.5 to 40.1 HRC) at 24 μm. Table 4 shows the microhardness measurements.

Table 4. Microhardness of PAM'ed workpieces, measured from the surface, after having removed the depth of cut a_p

Depth (mm) under machined surface	*Conventional. Microhardness $HV_{0.2}$*	*PAM. Microhardness $HV_{0.2}$*
Machined surface (≈0.022)	435 (44 HRC)	399.9 (40.1 HRC)
0.06	----	389.9 (39.3 HRC)
0.08	405 (41.3 HRC)	296 (28.9 HRC)
0.35 mm under surface	235 (20.5 HRC)	235 (20.5 HRC)

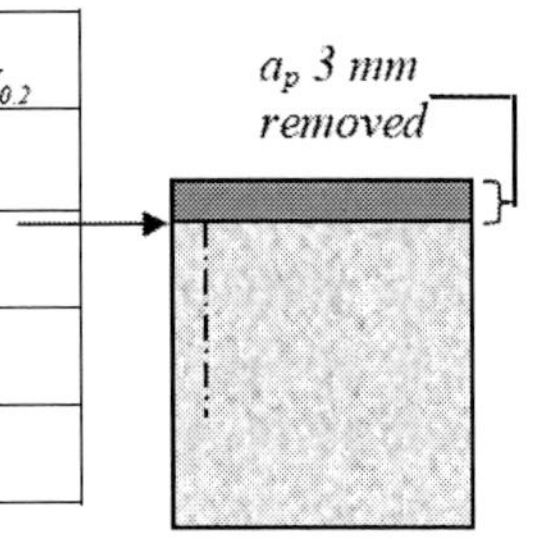

These results are coherent with the thermal softening suffered by Inconel after plasma heating was applied. Since the strength of the material to be machined is diminished, the layer affected by strain hardening is smaller both in depth and value. However, the strain hardening is still present even after applying PAM. This is due to the plowing effect associated with the small chip thickness at the low axial point of the inserts, which also results in high localized flank wear, and with the negative axial and radial rake angles of the ceramic cutting inserts.

The PAM processing of Inconel 718 does not affect the material integrity, and therefore this process could be recommended for industrial production on nickel alloys components, as well. Good result were also found for cobalt base alloys (Haynes 25 alloy), but experiments [LOP 2004] on titanium alloys showed damage of material structure, which makes this technique unsuitable for these alloys. Other fields where the technique may work might be with hardened steels to improve the material removal rate in roughing and semi-finishing conditions.

6. Conclusions

The current uses of the main assisted techniques for machining were discussed in this chapter. While the basis of these techniques differs, the aim is the same: to improve the productivity of the material removal processes without dramatically increasing the power consumption or requiring new expensive facilities. In the case of laser assisted techniques, the cost of additional equipment is high, but even there, the elimination of grinding may be a factor in overall reduction of production costs.

High-pressure coolant-assisted machining is now available, as are CO2 and LN2 assisted techniques. Laser-assisted machining is a good option for ceramic turning, and is nearly ready for application in industrial settings. Other techniques required more research not only in their basic of technologies but also in aspects of material integrity and user security.

In the future, new technologies will be proposed, and hybrid processes for machining will be devised. However, new processes will be recommended for industrial application only if the results show economic advantages.

Acknowledgments

We address our thanks to all cited companies for information and images and to Cryobest® for information on cryogenic treatments applied to tools. We also thank Mr. Gandarias for useful information about drilling of stainless steels.

References

[AHM 2006] Ahmed, N., Mitrofanov, A.V., Babitsky, V.I., Silberschmidt, V.V. (2006). Analysis of material response to ultrasonic vibration loading in turning Inconel 718. *Material Science and Engineering, 424*, 318-325

[AHM 2007] Ahmed, N., Mitrofanov, A.V., Babitsky, V.I., Silberschmidt, V.V. (2007). Analysis of forces in ultrasonically assisted turning. *Journal of Sound and vibration, 308,* 845-854

[AZA 2008] B. Azarhoushang, J. Akbari. (2008) Ultrasonic-assisted drilling of Inconel 738-LC, *International Journal of Machine Tools and Manufacture,* 47, 1027–1033.

[BAB 2003] Babitsky, V.I., Kalashnikov, A.N., Meadows, A., Wijesundara, A.A.H.P. (2003). Ultrasonically assisted turning of aviation materials, *Journal of Materials Processing Technology, 132,* 157-167.

[BAL 2008] P. Baldissera, C. Delprete. (2008). Deep Cryogenic Treatment: A Bibliographic Review, *The Open Mechanical Engineering Journal, 2/11*, 1-11.

[BHA 1993] D. Bhattacharyya, M.N. Allen and S.J. Mander, (1993) Cryogenic machining of Kevlar composites, *Materials and Manufacturing Processes, 8/.6*, 631–651

[BRE 2008] D.E. Brehl, T.A. Dow, (2008), Review of vibration-assisted machining, *Precision Engineering, 32,* 153–172.

[CEL 2009] A.Celaya, L.N. López de Lacalle, F.J. Campa, A.Lamikiz, Ultrasonic assisted turning of mild steels, *International Journal of Material and Product Technology*, In press.

[CHA 2005] Chang S.F., Bone G.M., (2005).Burr size reduction in drilling by ultrasonic assistance, *Robotics and Computer-Integrated Manufacturing, 21*, 442-450.

[CHA 2007] Chang Chih-Wei , Kuo Chun-Pao ,(2007) An investigation of laser-assisted machining of Al2O3 ceramics planning, *International journal of machine tools & manufacture, 47, 3-4,* 452-461.

[CHH 2002] Chhabra P. N., Ackroyd B., Compton W. D., Chandrasekar S, (2002). Low-frequency modulation-assisted drilling using linear drives, *Proceedings of the Institution of Mechanical Engineers. Part B. Journal of engineering manufacture, 216/ 3,* 321-330

[CHU 2005] Chunxiang, Ma., Shamoto, E., Moriwaki, T., Zhang, Y., Wang, L.,(2005). Suppression of Burrs in Turning with Ultrasonic Elliptical Vibration Cutting, *International Journal of Machine Tools& Manufacture, 45,* 1295-1300.

[ELM 2007] M. El Mansori, A. Mkaddem, (2007) Surface plastic deformation in dry cutting at magnetically assisted machining, *Surface and Coatings Technology, 202, 4-7,* 1118-1122.

[EVA 1991] C. Evans, (1991). Cryogenic diamond turning of stainless steel. *Annals of the CIRP, 40,* 571–575.

[EZU, 2004]E.O. Ezugwu, J. Bonney, (2004) Effect of high-pressure coolant supply when machining nickel-base, Inconel 718, alloy with coated carbide tools, *Journal of Materials Processing Technology, 153-154,* 1045-1050.

[HON 1996] Hong, S.Y., Cryogenic Machining, patent PCT/US1995/010282, 1996

[HON 1999] S.Y. Hong, Y. Ding and R.G. Ekkens, (1999) Improving low carbon steel chip breakability by cryogenic chip cooling, *International Journal of Machine Tools and Manufacture, 39,* 1065–1085.

[HON 2001] S.Y. Hong and Y. Ding, (2001) Cooling approaches and cutting temperatures in cryogenic machining of Ti–6Al–4V, *International Journal of Machine Tools and Manufacture, 41,* 1417–1437.

[HSU 2008] Hsu, C.Y., Lin, Y.Y., Lee, W.S., Lo, S.P., (2008) Machining characteristic of Inconel 718 using ultrasonic and high temperature-aided cutting. *Journal of Material Processing Technology, 198,* 359-365.

[JAC 2008] D.P. Jackson, United States Patent 7389941, Nozzle device and method for forming cryogenic composite fluid spray, 2008

[JAC 2009] D.P. Jackson , Machining with Advanced CO2 Machining Spray Technology, MoldMaking Technology, 2009 (available at www.moldmakingtechnology.com/articles)

[KIT 1990] Kitagawa,T. & K. Maekawa, K. (1990). Plasma Hot Machining for New Engineering Materials, *Wear, 139,* 251-267.

[KON 1990] Kónig, W., Cronjäger, L., Spur, G., Tónshoff, M. Vigneau & Zdebelick, W.J. (1990) Machining of New Materials, *Annals of the CIRP, 39/1,* 673-681.

[KOV 1994] Kovacevic, Radovan , United States Patent 5288186, Apparatus and method of high-pressure waterjet assisted cooling/lubrication in machining, 1994.

[LEI 2001] Lei, S., Shin, Y.C. and Incropera, F.P. (2001). Experimental Investigation of Thermo-Mechanical Characteristics in Laser Assisted Machining of Silicon Nitride Ceramics, *Transactions ASME, Journal of Manufacturing Science and Engineering, 123,* 639-646.

[LES 2001] Leshock,C.E., Kim, J.N., Shin,Y.C., (2001). Plasma enhanced machining of Inconel 718: modelling of workpiece temperature with plasma heating and experimental results, *International Journal of Machine Tools and Manufacture, 41,* 877-897.

[LESO 1994] Lesourd, B., LeMaitre, F. & Thomas, T, (1994) Laser assisted hot machining processes, *Proceedings of LANE-94, 1,* 389-404.

[LIA 2007] Y.S. Liao, Y.C. Chen, H.M. Lin, (2007) Feasibility study of the ultrasonic vibration assisted drilling of inconel superalloy, *International Journal of Machine Tools and Manufacture, 47,* 1988–1996.

[LIN 1995] Lindeke, R., Schoenig , F.C., Atherton, A.J., (1995). Improved hole making experiences with very high-pressure lubricating coolants delivered through twist drills., SME paper MR95-1 74

[LOP 2000] López de Lacalle, L.N., A.Gutiérrez , Llorente, J.I. , Sánchez, J.A. , J.Albóniga, (2000) Using High-pressure coolant in the drilling and turning of low machinability alloys, *International Journal of Advanced Manufacturing Technology, 6/ 2,* 85-91.

[LOP 2004] L. N. López de Lacalle, J. A. Sánchez, A. Lamikiz, and A. Celaya, (2004) Plasma Assisted Milling of Heat-Resistant Superalloys, *J. Manuf. Sci. Eng., 126/ 2,* 274-286.

[M4S 2009] M4 Sciences Corporation, available at http://www.m4sciences. com/news/

[MAN 2007] J. B. Mann, S. Chandrasekar, W.d. Compton, (2007) Tool holder assembly and method for modulation-assisted machining, PCT/US2007/000730.

[MOR 1991] Moriwaki, T., Shamoto, E. (1991) Ultraprecision diamong turning of stainless steel by applying ultrasonic vibration, *Annals of the CIRP, 40/1,* 559-562.

[MOR 1995] Moriwaki, T., Shamoto, E., (1995) Ultrasonic Elliptical Vibration Cutting, *Annals of the CIRP, 44/1*, 31-34.

[MUR 1998] Murakawa, M., Jin, M. (1998) Turning of beta-titanium alloys by means of ultrasonic vibration, *Trans. NAMRI/SME, 26.*

[NAN 2009] A.K. Nandya, M.C. Gowrishankara, S. Paul, (2009) Some studies on high-pressure cooling in turning of Ti–6Al–4V, *International Journal of Machine Tools and Manufacture, 49/2,* 182-198.

[NAT 2007] Nath, C., Rahman, M., Andrew, S.S.K. (2007) A study on ultrasonic vibration cutting of low alloy steel, *Journal of Materials Processing Technology, 192–193*, 159–165.

[NAT 2008] Nath, C., Rahman, M., (2008) Effect of machining parameters in ultrasonic vibration cutting, *International Journal of Machine Tools and Manufacture, 48/ 9,* 965-974.

[NEU 2004] Neugebauer, R., Stoll, A., (2004) Ultrasonic application in drilling, *Journal of Material Technology, 149*, 633-639.

[NOV 1997] Novak, J.W., Shin, Y.C. & Incropera, F.P., (1997) Assessment of Plasma Enhanced Machining for Improved machinability of Inconel 718, *Journal of Manufacturing Science and Engineering, 119*, 125-129.

[PAU 2001] S. Paul, , N. R. Dhar and A. B. Chattopadhyay, (2001) Beneficial effects of cryogenic cooling over dry and wet machining on tool wear and surface finish in turning AISI 1060 steel, *Journal of Materials Processing Technology, 116, 1, 3,* 44-48.

[PFE 1990] Pfender,E., Chen,W.L.T., Spores, R., (1990), A new look at the thermal and gas dynamic characteristics of a plasma jet, Thermal spray research and applications, *Proc. of the third National Thermal Spray Conference*, long Beach, CA, 1-10.

[PUJ 2009] J. Pujana, A. Rivero, A. Celaya, L.N. López de Lacalle, (2009) Analysis of ultrasonic-assisted drilling of Ti6Al4V, *International Journal of Machine Tools and Manufacture, 49/6,* 500-508.

[ROZ 2000] Rozzi, J.C., Pfefferkorn, F.E., Incropera, F.P. and Shin, Y.C., (2000), Experimental Evaluation of the Laser Assisted Machining of Silicon Nitride Ceramics, *Journal of Manufacturing Science and Engineering, 122*, 666-670.

[SHA 1994] Shamoto, E. and Moriwaki, T., (1994) Study on Elliptical Vibration Cutting, *Annals of the CIRP, 43/1,* pp. 35-38

[SHA 1999] Shamoto, E., Moriwaki, T., (1999) Ultraprecision Diamond Cutting of Hardened Steel by Applying Elliptical Vibration Cutting, *Annals of the CIRP, 48/1,* 441-444.

[SHA 2001]Sharman, A.R.C., Li, XP., Rahman, M., Bowen, P., Dewes, R.C., Aspinwall, D.K., (2001) Ultrasonic Assisted Turning of Gamma Titanium Aluminide. *Proceedings of the 13th International Symposium for Electromachining ISEM XIII, II,* 939-951

[SHA 2002] Shamoto, E., Suzuki, N., Moriwaki, T., Naoi, Y., (2002) Development of Ultrasonic Elliptical Vibration Controller for Elliptical Vibration Cutting, *CIRP Annals-Manufacturing Technology, 51/1,* 327-330.

[SIH 2008] I. Sihvo, J. Varis, (2008) The wear of single flute gun drill and tool life tests, *Mechanika. n.5, 73*, pp. 59-64.

[SIN 2002] Sehijpal Singh, H. S. Shan, P. Kumar, (2002) Wear behavior of materials in magnetically assisted abrasive flow machining, *Journal of Materials Processing Technology, 128, 1-3*, 155-161.

[SKE 1969] Skelton, R.C., (1969) Effect of Ultrasonic Vibration on the Turning Process, *Int.J.Mach. Tool Des. Res., 9,* 363-374.

[SUZ 2005] H. Suzuki, H. Yagishita, (2005) Burrless Drilling by Vibration Cutting Applying Ultrasonic Torsional Mode Vibration, *NAMRC 33,* New York.

[TAK 1991] Takeyama H., Kato S.,(1991) Burrless drilling by means of ultrasonic drilling, *Ann. CIRP, 40,* 83-86.

[THO 1999] T. Thomas, J.O. Vigneau, Joel Olivier, US Patent 5906459 - Laser-assisted milling process, 1999

[TIA 2006] Tian, Y. and Shin, Y.C., (2006) Laser-assisted Machining of Silicon Nitride with Complex Geometric Features via In-process Control, *Journal of the American Ceramic Society, 89*/ 11, 3397-3405.

[VIG 1996] Vigneau, J., 1996, Assistance a la Coupe des Alliages de Titane par Jet D'eau Haute Pression, CETIM, Dec., París.

[WAN 1996] Z.Y. Wang, K.P. Rajurkar and M. Murugappan, (1996) Cryogenic PCBN turning of ceramic (Si3N4), *Wear, 195, 1–2*, 1–6.

[WAN 1997] Z.Y. Wang and K.P. Rajurkar, (1997) Wear of CBN tool in turning of silicon nitride with cryogenic cooling, *International Journal of Machine Tools and Manufacture , 37,3*, 319–326.

[WAN 2000] Z.Y. Wang and K.P. Rajurkar, (2000) Cryogenic machining of hard-to-cut materials, *Wear, 239*, 168–175.

[WEI 1994] Weinert, K., (1994), Relation between Process Energy and Tool Wear when Turning Hardfacing Alloys, *Annals of the CIRP,. 43,. 1.*

[WER 1992] Wertheim, R., Rotberg, J., Ber, A., (1992) Influence of high-pressure flushing through the rake face of the cutting tool, *Annals of the CIRP, 41-1.*

[YAN 2008] B. Yang, , X. Shen, S. Lei,, (2008) Mechanisms of edge chipping in laser assisted milling of Silicon nitride ceramics, *International Journal of Machine Tools and Manufacture* doi:10.1016/j.ijmachtools.2008.09.006.

[YIL 2008] Yakup Yildiz, a, and Muammer Nalbanta , (2008) A review of cryogenic cooling in machining processes , *International Journal of Machine Tools and Manufacture, 48- 9,* 947-964.

[ZHA 1994] D. Zhang, X. Feng, L. Wang, D. Chen, (1994) Study on the drill skidding motion in ultrasonic vibration microdrilling, *International Journal of Machine Tools and Manufacture, 34-6,* 847–857.

[ZHO 2002] Zhou, M., Wang, X.J., Ngoi, B.K.A., Gan, J.G.K. (2002) Brittle-ductile transition in the diamond cutting of glasses with the aid of ultrasonic vibration, *Journal of Materials Processing Technology,121,* 243-251.

[ZHO 2006] Zhong, Z.W., Lin, G. (2006) Ultrasonic assisted turning of an aluminium-based metal matrix composite reinforced with SiC particles, *Int. J. Adv. Manufacturing Technology, 27,* 1077-1081.

[ZHO 2006] Zhou, M., Ngoi, B.K.A., Yusoff, M.N., Wang, X.J. (2006) Tool wear and surface finish in diamond cutting of optical glass, *Journal of Materials Processing Technology, Vol. 174,* 29-33.

INDEX

A

B

C

D

E

F

G

H

I

J

K

L

M

N

O

P

Q

R

S